W0042737

Progress in Gene Expression

Series Editor:

Michael Karin
Department of Pharmacology
School of Medicine
University of California, San Diego
La Jolla, CA 92093-0636
USA

Books in the Series:

Gene Expression: General and Cell-Type-Specific
M. Karin, editor
ISBN 0-8176-3605-6

Inducible Gene Expression, Volume 1: Environmental Stresses and Nutrients
P.A. Baeuerle, editor
ISBN 0-8176-3728-1

Inducible Gene Expression, Volume 2: Hormonal Signals
P.A. Baeuerle, editor
ISBN 0-8176-3734-6

Oncogenes as Transcriptional Regulators Volume 1

Retroviral Oncogenes

M. Yaniv
J. Ghysdael
Editors

Springer Basel AG

Editors:

Dr. M. Yaniv
Dept. of Biotechnology
Institut Pasteur
28, rue du Dr. Roux
F-75724 Paris Cedex 15
France

Dr. J. Ghysdael
CNRS UMR 146
Institut Curie, Section de Recherche
Centre Universitaire
F-91405 Orsay
France

A CIP catalogue record for this book is available from the Library of Congress, Washington D.C., USA

Deutsche Bibliothek Cataloging-in-Publication Data
Oncogenes as transcriptional regulators /
M. Yaniv and J. Ghysdael. - Basel ; Boston ; Berlin :
Birkhäuser.
(Progress in gene expression)
ISBN 978-3-0348-9816-4 ISBN 978-3-0348-8889-9 (eBook)
DOI 10.1007/978-3-0348-8889-9
NE: Yānîv, Moše
Vol. 1. Retroviral oncogenes. - 1997
ISBN 978-3-0348-9816-4

Originally published by Birkhäuser Verlag, Basel, Switzerland in 1997
Softcover reprint of the hardcover 1st edition 1997

Printed on acid-free paper produced from chlorine-free pulp

ISBN 978-3-0348-9816-4

9 8 7 6 5 4 3 2 1

Contents

Preface by the Series Editor
Michael Karin VII

Preface
Jacques Ghysdael and Moshe Yaniv IX

List of Contributors XIII

1 Myc Structure and Function
George C. Prendergast 1

2 The ETS Family of Transcriptional Regulators
Jacques Ghysdael and Anthony Boureux 29

3 *myb* Proto-Oncogene Product as a Transcriptional Regulator
Shunsuke Ishii 89

4 The v-erbA Oncogene
Anne Rascle, Olivier Gandrillon, Gérard Cabello and Jacques Samarut 117

5 Rel Proteins and Their Inhibitors: A Balancing Act
Mary Lee MacKichan and Alain Israël 165

6 Structure/Function and Oncogenic Conversion of Fos and Jun
Andrew J. Bannister and Tony Kouzarides 223

Index 249

Preface by the Series Editor

The control of gene expression is a central-most topic in molecular biology as it deals with the utilization and regulation of gene information. As we see huge efforts mounting all over the developed world to understand the structure and organization of several complex eukaryotic genomes in the form of Gene Projects and Genome Centers, we have to remember that without understanding the basic mechanisms that govern the use of genetic information, much of this effort will not be very productive. Fortunately, however, research during the past decade on the mechanisms that control gene expression in eukaryotes has been extremely successful in generating a wealth of information on the basic strategies of transcriptional control.

The progress in understanding the control of eukaryotic transcription can only be appreciated by realizing that twelve years ago we did not know the primary structure of a single sequence specific transcriptional activator, and those whose primary structures were available (e.g. homeodomain proteins) were not yet recognized to function in this capacity. Also, ten years ago transcription was thought to be carried out by an abstract assembly of transcription factors and RNA polymerases referred to as the "transcriptional machinery," while nowadays many of these basic components have been purified to homogeneity and are available as molecular clones. While the progress in this field has been incredible, it is far from reaching a plateau and it is likely that the next ten years will result in an even greater and faster increase in our understanding of gene regulation. However, we have reached a point at which some generalizations can be made, recurrent themes can be identified, and unifying hypotheses formulated. The purpose of this series is to summarize this overwhelming amount of information in a small number of volumes, each containing chapters written by well-recognized experts dealing with highly related topics. By studying the progress made in a select number of model systems, it is hoped that the reader will be able to apply this knowledge to his or her own favorite experimental systems.

It is our goal that the *Progress in Gene Expression* series serves as an important resource for graduate students and experienced researchers alike, in the fields of molecular biology, cell biology, biochemistry, biotechnology, cell physiology, endocrinology, and related fields. More exciting volumes are in the planning stages, and suggestions for future volumes are appreciated, and should be directed to the Series Editor.

Michael Karin La Jolla, CA, March 1997

Preface

The intensive study of molecular events leading to cellular transformation in tissue culture or in intact organisms culminated in the identification of 100 or more genes that can be defined as oncogenes or tumor suppressor genes. Functionally, these genes can be divided into several classes, each involved in a different step in transmission of signals from the exterior of the cell to the nucleus. The first oncogenes to be biochemically characterized included membrane receptors for growth factors, growth factors themselves, protein kinases or small GTP binding proteins involved in signal transduction. Later, the development of techniques to study proteins-DNA interaction in eucaryotes and the isolation and characterization of many promoter and enhancer sequences revealed that a number of the classical retroviral oncogenes were indeed transcription factors. In parallel, the rapid progress in the identification and cloning of chromosomal translocations in human and animal malignancies and the increased repertoire of known transcription factors families revealed that many other transcription factors can play a critical role in cancer. A more recent development concerns tumor suppressor genes. The realization that human tumors are frequently associated with a loss of function of one or several genes is also one of the landmarks of cancer research in the last 15 years. Again, as we will see below, some of these genes encode transcription factors.

It is becoming increasingly difficult to cover in a single monograph all oncogenes and tumor suppressor genes. Our task was facilitated somehow by the definition of the current series "progress in gene expression". We choose therefore to discuss only oncogenes and tumor suppressor genes that are *bona fidae* transcription factors. We did not try to be comprehensive – we rather decided to discuss extensively a limited number of representative gene families. We hope that these examples will be instructive enough in bringing a plausible explanation of why certain transcription factors behave like oncogenes or tumor suppressor genes. This is not an obvious issue. It is generally admitted that 5 to 10% of our genes are involved in transcription control, the majority being sequence-specific DNA binding proteins. The fact that only several dozens were isolated as retroviral transduced oncogenes can be explained perhaps by the rarity of viral transduction events. Still, some of these viral oncogenes were independently isolated more than once. The same question holds for genes involved in chromosomal translocations and in proviral insertions. These involve a rather limited number of transcription factor families.

Some rationale for this paradox can be brought forward if we follow carefully some of the chapters in this book. Many membrane-associated or cytoplasmic oncogenes products are involved in signal transduction. It is reasonable to postulate that transcription factors that respond directly to such signals may behave as oncogenes. This is certainly the case as discussed in this book. cFos and cJun transcription and protein phosphorylation is strongly stimulated by activation of the Ras-Raf-Mek-Erk (MapK) pathway. cFos and JunB transcription is also activated via Stat factors after cytokine stimulation. cMyc transcription is also strongly stimulated by mitogens. Members of the Rel family composing the NFκB transcription factor are translocated into the nucleus upon mitogenic stimulation of B or T cells. However, AP1 (Jun/Fos), Myc and NFκB are only a few examples among the transcription factors activated during a mitogenic response. Other transcription factors including Zn finger proteins, orphan receptors, etc., are activated by the same stimuli, yet they have so far not been identified as oncogenes. Thus, the response to the question we raised is not easy. Several other chapters in this book clearly demonstrate that the oncogenic activity of transcription factors may be frequently linked to their role in blocking a differentiation pathway. The classical examples of the v-erbA – v-erbB cooperation or the Friend erythroleukemia system clearly emphasizes the importance of the persistance of a mitogenic stimulus and a differentiation block. It is possible that many protein fusions that occur as a result of chromosomal translocations generate bifunctional proteins that not only interfere with the expression of normal target genes or activate new ones but may also perturb other circuits in the cell. It should be recalled that many transcription factors were shown to be involved in a cross-talk among different classes of transcription factors. The example of steroid receptors-AP1 (Jun/Fos) interaction will be discussed in chapter 4. Several chapters that describe such fusion proteins will discuss these issues.

As an example for a tumor suppressor gene that functions as a transcription factor, an obvious choice was p53. In this case, the identification of target genes like p21/Waf1 greatly enhanced our understanding of the biological role of this gene. This is also an excellent example of how the activity of a transcription factor can be modulated by different signals.

Recent studies render the classification of the E2F transcription factor difficult. E2F1 was initially shown to transform fibroblasts in culture, however gene knock-out studies suggest that it should be considered as a tumor suppressor gene. Since the function of E2F/DP heterodimers is crucial in cell cycle control and in response to Rb and modulators of its phosphorylation, we decided to include a chapter describing this factor in the present monograph.

Finally, many viruses interfere with the transcription program of the cell. We can cite RNA tumor viruses like HTLV1 with its Tax protein or

DNA tumor viruses like EBV. We decided to include a chapter on EBNA2, a viral protein that regulates the transcription of both viral and cellular genes.

In each chapter the authors attempted to include a description of the gene family comprising the oncogene in question. They discuss our knowledge about the domain structure of the proteins, the specificity of their DNA binding, modulation of DNA binding and activity, target genes and biological consequences of oncogenic mutations or simple overexpression. We are grateful to the authors for their efforts in presenting an up-to-date picture of the biological system and the genes they are studying. We hope that the comprehensive description achieved by the different chapters will help specialized readers by giving an in-depth picture of systems that are slightly outside their main field of everyday research, and challenge newcomers by giving the unique possibility to learn how diversion of the transcription program of a cell can result in its oncogenic conversion.

Jacques Ghysdael and Moshe Yaniv
January 1997

List of Contributors

Andrew J. Bannister, Wellcome/CRC Institute and Department of Pathology, University of Cambridge, Tennis Court Road, Cambridge CB2 1QR, UK

Anthony Boureux, CNRS UMR 146, Institut Curie, Section de Recherche, Centre Universitaire, F-91405 Orsay, France

Gérard Cabello, Institut National de la Recherche Agronomique, Unité d'Endocrinologie Cellulaire, Laboratoire de Différenciation Cellulaire et Croissance, Place Viala, F-34060 Montpellier Cedex 01, France

Olivier Gandrillon, Laboratoire de Biologie Moléculaire et Cellulaire de l'Ecole Normale Supérieure de Lyon, UMR 49 CNRS, Equipe associée 913 INRA, 46 Allée d'Italie, F-69364 Lyon Cedex, France

Jacques Ghysdael, CNRS UMR 146, Institut Curie, Section de Recherche, Centre Universitaire, F-91405 Orsay, France

Shunsuke Ishii, Laboratory of Molecular Genetics, Tsukuba Life Science Center, The Institute of Physical & Chemical Research (RIKEN), 3-1-1 Koyadai, Tsukuba, Ibaraki 305, Japan

Alain Israël, Unité de Biologie Moléculaire et Expression Génique, Institut Pasteur, 25, rue du Docteur Roux, F-85824 Paris Cedex 15, France

Tony Kouzarides, Wellcome/CRC Institute and Department of Pathology, University of Cambride, Tennis Court Road, Cambridge CB2 1QR, UK

Mary Lee MacKichan, G.W. Hooper Foundation, Box 0552, University of California, San Francisco, CA 94143, USA

George C. Prendergast, Wistar Institute, 3601 Spruce Street, Philadelphia, PA 19104, USA, and Department of Genetics, University of Pennsylvania Medical School, Philadelphia, PA, USA

Anne Rascle, Department of Developmental Biology, Beckman Center, B300, Stanford University School of Medicine, Stanford, CA 94305-5427, USA

Jacques Samarut, Laboratoire de Biologie Moléculaire et Cellulaire de l'Ecole Normale Supérieure de Lyon, UMR 49 CNRS, Equipe associée 913 INRA, 46 Allée d'Italie, F-69364 Lyon Cedex, France

1

Oncogenes as Transcriptional Regulators
Vol. 1: Retroviral Oncogenes
ed. by M. Yaniv and J. Ghysdael

Myc Structure and Function

GEORGE C. PRENDERGAST

Functional analysis of Myc continues to strengthen connections to transcriptional regulation. CTD interactions are growing in complexity, but, at a broad level of understanding, the new results reaffirm the role of this region in controlling DNA recognition. One should expect additional progress in defining b/HLH/LZ partners for both Myc and Max, as well as a dissection of higher order interactions between partner complexes and other transcription factors. The NTD has only begun to be examined mechanistically. Transactivation and repression domains have been roughly parsed, with repression turning out to play a critical role (perhaps the most important one) in cancer cells. Mechanisms should begin to emerge as several new NTD binding proteins are analyzed. These directions are exciting because they may also shed light on the persistent issues concerning nontranscriptional roles. Long a "citadel of incomprehensibility" (Lüscher et al, 1990), Myc is finally yielding answers to long-standing questions. For those interested in basic mechanisms and cancer, answers to current questions are likely to be both interesting and important.

Introduction

The Myc oncoprotein is a key regulator of cell cycle and apoptosis which plays an extensive role in human cancer (reviewed in Cole, 1986). Most work has centered on the product of the c-*myc* gene, which is broadly expressed, but many studies have examined other functionally related though less widely expressed *myc* genes activated in malignancy, for example, N-*myc*, initially identified in neuroblastoma, and L-*myc*, originally identified in lung carcinoma (reviewed in De Pinho et al, 1991). In tumor cells a large variety of regulatory mutations which lead to constitutive expression have been characterized, including, most commonly, chromosomal translocation and gene amplification (Kelly et al, 1986; Spencer et al, 1991; Marcu et al, 1992). Recent studies also have indicated that, like many other oncoproteins in cancer cells, Myc sustains structural mutations. Indeed, structural and regulatory mutations may be equally frequent, and in many cases may occur simultaneously (Bhatia et al, 1993; Yano et al, 1993). In the absence of mutations, Mcy may also be deregulated by constitutive activation of signal transduction pathways that induce *myc* transcription. Thus, a large number of neoplastic cells may contain deregulated Myc.

In normal cells, Myc is required to enter and transit the cell cycle (reviewed in Evan et al, 1993). Following mitogenic stimulation, Myc is rapidly induced and remains elevated throughout the cell cycle, suggesting it is needed for continuous cell growth. Cells containing reduced levels of Myc, due to hemizygous gene deletion, exhibit a delay in S phase entry and reduced rates of cell proliferation (Shichiri et al, 1993). Enforced expression is sufficient to drive quiescent cells into the cell cycle and sufficient to block cell differentiation. Conversely, inhibition blocks mitogenic signals and propels cells toward terminal differentiation. The critical role for Myc during the cell cycle is reinforced by the requirement of many oncoproteins for its activity, for example, Ras, Abl, Src, and SV40 T antigen (Sklar et al, 1991; Sawyers et al, 1992; Hermeking et al, 1994; Barone et al, 1995).

In addition to its ability to induce cell growth, under certain conditions Myc can induce apoptosis, a noninflammatory cell suicide program (reviewed in Evan et al, 1994; Packham et al, 1995). This feature of Myc is only manifested if its expression is deregulated and uncoupled from the orchestration of other cell cycle regulatory events. For example, following growth factor withdrawal, cells that contain normal Myc downregulate its expression and exit the cell division cycle, while cells that contain deregulated Myc maintain its expression and undergo apoptosis (Askew et al, 1991; Evan et al, 1992). Whether this phenomenon reflects a direct or indirect effect(s) of Myc activity is unknown. However, its molecular basis is of interest since the mechanism might point to a possible Achilles' heel in tumor cells which contain deregulated Myc.

Although originally discovered almost two decades ago, *myc* continues to attract considerable attention, in part because its exact function(s) has remained somewhat elusive. Progress has been hampered by the absence of the gene from the more genetically maleable organisms and by the fact that the Myc protein is rare and biochemically difficult to work with. Beginning from a line of work that essentially bootstrapped from structural clues in the Myc amino acid sequence, a large and compelling body of evidence has accumulated arguing that Myc acts largely in the guise of a transcription factor. Several recent reviews have surveyed this evidence as well as that indicating the links between Myc and apoptosis (Bernards, 1995; Packham et al, 1995; Henriksson et al, 1996). In this review, I will examine Myc structures and functions related to transactivation, but will also summarize results indicating a transcriptional repression function(s) and hints of nontranscriptional functions.

Myc Primary Structure

The human c-Myc primary structure will serve as the paradigm for Myc family proteins in this review. Myc is a short-lived phosphoprotein that exists in cells in two forms of apparent MWs of 64 kD and 67 kD. These two polypeptides are identical except for a 14 residue N-terminal extension in the 67 kD species, the result of translation initiation at a non-AUG site in the c-*myc* message (Hann et al, 1988). Myc is roughly tripartite, with the terminal domains necessary and the central region essentially dispensible for biological activity (Stone et al, 1987). Figure 1.1 shows a cartoon of the primary structure of Myc. This cartoon highlights important functional regions that will be discussed below.

Major landmarks of Myc primary structure are the basic/helix-loop-helix/"leucine zipper" (b/HLH/Z), which essentially comprises the C-terminal domain (CTD), and the evolutionarily conserved sequences called

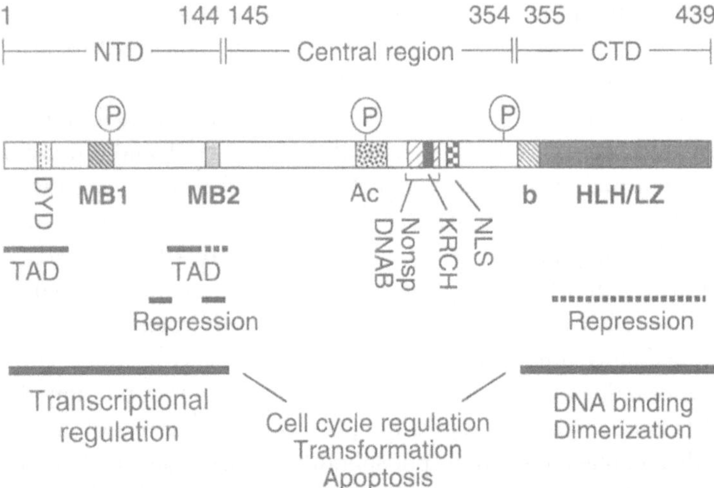

Figure 1.1 Primary structure of Myc. Functional domains and segments of evolutionarily conserved sequence that are discussed in the text are noted. NTD, N-terminal domain; CTD, C-terminal domain; TAD, transactivation domain. Circled P indicates sites of in vivo phosphorylation. Bars indicates regions required for the activities noted. Dashed bars indicate lesser importance.

Myc boxes 1 and 2 (MB1 and MB2), which are found in the N-terminal domain (NTD). MB1 is located at aa 41–65 and contains two sites of in vivo phosphorylation at T58 and S62. MB2 is located at aa 130–144 and constitutes the most hydrophobic region in the Myc polypeptide. The CTD (aa 355–439) mediates and controls specific DNA binding. Nucleotide contacts are made by the basic region (b), which specifically interacts with DNA after heterotypic dimerization through the HLH/LZ. The central region (aa 154–354) has been little studied. It contains a nuclear localization signal (NLS), a region capable of nonspecific DNA binding, and two clusters of phosphorylation sites. The presence of evolutionarily conserved sequences within the central region suggests that some negative regulatory functions may reside here. The NTD (aa 1–153) has been implicated in transcriptional regulation but has other roles as well. This domain contains at least two regions that can function independently as transcriptional activation domains (TADs), one of which includes MB2. MB1 fulfills mainly a regulatory role and contains at least two important sites for phosphorylation. MB1 is a hotspot for mutation in cancer, although mutations are seen throughout the NTD.

C-Terminus: Regulation of Physiological DNA Binding

The initial harbinger of Myc's transcriptional role was the discovery of a LZ motif at the Myc C-terminus (Landschultz et al, 1987). Subsequent identification of basic (Prendergast et al, 1989) and HLH (Murre et al, 1989) motifs located immediately upstream of the LZ completed the basic blueprint of the Myc CTD. The HLH and LZ are structures found in a wide variety of transcription factors controlling cell growth and differentiation. Initially defined as sequence motifs, structural studies essentially confirmed the hypothesis that they formed coiled-coil dimerization interfaces composed of amphipathic α-helices (O'Shea et al, 1989; Ferre-D'Amare et al, 1993). In most transcription factors, these domains are usually found alone or together in a variety of combinations. In Myc, the HLH and LZ are immediately adjacent to each other and act as a single composite dimerization domain. Basic regions constitute the DNA binding elements for HLH and LZ proteins, containing two or three clusters of basic amino acids that contact the DNA phosphodiester backbone and intervening residues that contribute to DNA recognition specificity. Basic regions must be dimerized to bind DNA.

Studies of the Myc b/HLH/LZ region initially focused on the DNA recognition specificity of the Myc basic region, leading to the definition of CACGTG as a binding site. Initial studies indicated that the LZ was necessary for Myc dimerization and biological activity (Dang et al, 1989; Smith et al, 1990), but later work showed that Myc could not homodimerize in vivo, indicating that it assumed a heterodimeric form (Dang et al, 1991). However, initial progress in defining the DNA binding specificity of the basic region was made using Myc homodimers, and by taking structural clues from Myc's relationship to the b/HLH family. b/HLH proteins recognize DNA sequences with the form CANNTG (the E box) and can be divided into two groups on the basis of their basic region primary structure. The amino acid sequence of the Myc-related b/HLH/LZ transcription factors, TFE-3 (Beckmann et al, 1990), AP-4 (Hu et al, 1990), and USF (Gregor et al, 1990) showed that their basic regions were nearly identical to Myc's. Since each recognized CACGTG it was very likely Myc would also recognize this element. This was subsequently shown to be the case, using approaches to dimerize the Myc basic region as a nonphysiological b/HLH/LZ homodimer (Blackwell et al, 1990; Fisher et al, 1991; Halazonetis et al, 1991; Kerkhoff et al, 1991) or as a Myc basic region-E2A HLH chimera (Prendergast et al, 1991). CACGTG represented a physiologically relevant site since in transformation assays the Myc-E2A chimera could dominantly inhibit Myc activity (Prendergast et al, 1991). Given that Myc likely assumed a heterodimeric form in vivo, this result implied that the structure of the basic region of a physiological Myc partner protein would be closely related to the Myc basic region.

The key advance in establishing Myc's role as a transcription factor was the cloning of Max, a physiological b/HLH/LZ partner protein (Blackwood et al, 1991; Prendergast et al, 1991). The HLH/LZ regions of each polypeptide are necessary and sufficient for dimerization. As anticipated, the Max basic region is essentially identical to Myc's and Myc/Max complex efficiently recognizes CACGTG in a manner requiring the basic regions of each protein. The Myc/Max complex represents the physiological form of Myc in vivo (Wenzel et al, 1991; Blackwood et al, 1992). Unlike Myc, Max readily forms homodimers under physiological conditions that also bind CACGTG (Prendergast et al, 1991; Berberich et al, 1992; Kato et al, 1992). DNA binding by Max homodimers is inhibited by casein kinase II-mediated phosphorylation at a site(s) proximal to the basic region (Berberich et al, 1992; Bousset et al, 1993). Phosphorylation also occurs in an acidic region downstream of the b/HLH/LZ (Prendergast et al, 1992; Bousset et al, 1993; Koskinen et al, 1994), although the physiological significance of this is unclear. Max overexpression can antagonize Myc-dependent cell transformation (Mukherjee et al, 1992; Prendergast et al, 1992; Amati et al, 1993a), suggesting that Max/Max dimers can compete with Myc/Max for DNA binding in vivo.

At least two features of DNA recognition by Myc/Max are notable. The first is that DNA binding by both Myc/Max and Max/Max is sensitive to CpG methylation in the CACGTG recognition site (Prendergast et al, 1991; Prendergast and Ziff, 1991). Whether this feature is relevant in vivo is unknown, but two proteins have been identified (MMBP-1 and MMBP-2) which can specifically bind methylated CACGTG under physiological conditions (Suetake et al, 1993). Thus, Myc's access to various target genes may be controlled by CpG methylation, which changes widely during development and carcinogenesis (Jones, 1986; Cedar, 1988; Bestor, 1990; Jones et al, 1990). The second feature is that Myc/Max can bind to CACGTG even when packaged in nucleosomes (Wechsler et al, 1994). Myc CTD homodimers lack this ability, indicating that Max acts not only to mediate physiological DNA binding but permits the complex to access sites in chromatin.

Other DNA sequences have been reported to be recognized by Myc/Max complexes, but their physiological significance is not yet clear. Using a PCR-based selection technique, a novel group of E box and non-E box sequences have been shown to be bound by Myc/Max (Blackwell et al, 1993). Also, in an interesting development, work from two different directions has implicated TCTTATGC as a potential binding site (Negishi et al, 1992; Hann et al, 1994). This sequence is shared by binding elements specifically recognized by the complex between Max and the larger, alternately initiated p67 Myc protein (Hann et al, 1994), or implicated as a binding site for Myc in a disputed role in DNA replication (Iguchi-Ariga et al, 1988; Ariga et al, 1989; Negishi et al, 1992). Although the flanking elements differ, the core sequence identified by both groups is the same. The reverse complement of this sequence (GCATAAGA) bears resemblance to the E box core consensus CANNTG and is no more divergent from CACGTG than the surprisingly diverse group of sequences identified by PCR selection (Blackwell et al, 1993). Although their physiological significance must be considered questionable at this point, these sequences deserve further study since they may represent relevant targets for Myc/Max regulation in vivo.

The structue of a Max b/HLH/LZ homodimer bound to CACGTG has been crystallographically determined (Ferre-D'Amare et al, 1993) (Figure 1.2). This sturcture provided the first glimpse of the HLH, a domain found in many transcription factors controlling cell growth and differentiation. The HLH forms a left-handed four-stranded coiled-coil. Each Max molecule contains two long helices. The first helix of the HLH is contiguous with the basic region, which reaches into the major groove of the DNA, and the second helix of the HLH is contiguous with the LZ, forming a long coiled-coil. The Myc/Max structue is expected to be

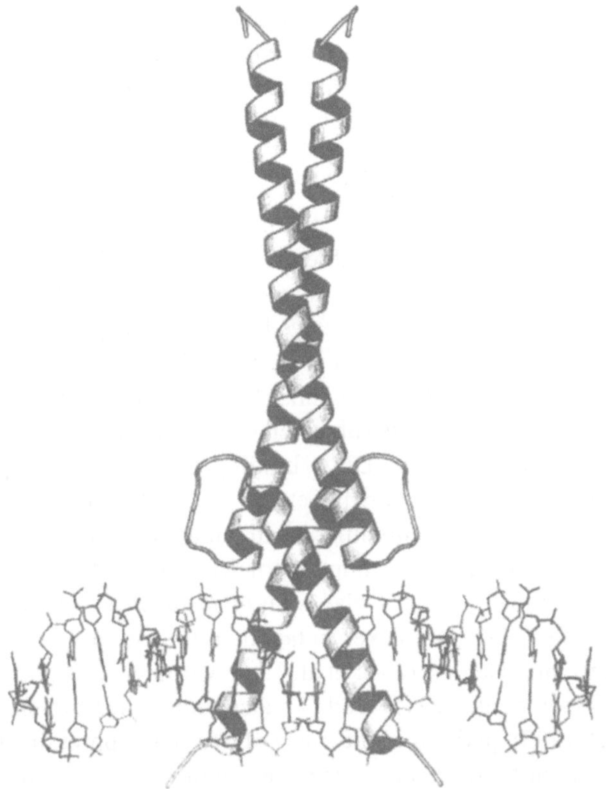

Figure 1.2 Structure of a Max b/HLH/LZ homodimer bound to CACGTG. The ribbon depicts the α carbon backbone of the b/HLH/LZ region.

similar, but may contain variations that give hints to the preference exhibited by each HLH/LZ for heterodimeration (Muhle-Goll et al, 1995), and how DNA bending is induced in different orientations by Myc/Max and Myc/Max dimers (Wechsler et al, 1992). In both Myc and Max, there is extensive evolutionary conservation of the amino acid sequences in the HLH but not in the LZ. The conservation in the HLH suggests that it may make proteins contact in addition to those involved in canonical heterodimerization.

The function of Max is to control the access of Myc to its physiological DNA sites. Max is essentially a b/HLH/LZ domain and lacks a TAD. Thus, its role is confined to DNA recognition and regulation of this process. Since Myc must oligomerize and bind DNA to exert its biological activity (Mukherjee et al, 1992; Prendergast et al, 1992; Amati et al, 1993 a; Amati et al, 1993 b), by controlling Myc's access to DNA Max acts as a master regulator of Myc activity. The importance of regulating this interaction has been underscored by the discovery of a set of b/HLH/LZ proteins called the Mad family (Mad1, Mad2/Mxi1, Mad3, and Mad4) which associate selectively with Max and control its access to Myc (Ayer et al, 1993; Ayer and Eisenman, 1993; Zervos et al, 1993; Hurlin et al, 1994). Thus, Max participates in a network of b/HLH/LZ proteins which influence Myc DNA binding.

Regulation of Max has important implications for cell growth, differentiation, and cancer. Max is a stable protein with a halflife of ≥24 hours (Blackwood et al, 1992). In cells induced to enter the cell cycle, the Max gene is induced a few-fold, with delayed early kinetics, in all cell types that have been examined (Prendergast et al, 1991; Shibuya et al, 1992; Martel et al, 1995), except BALB/c 3T3 cells (Berberich et al, 1992; Blackwood et al, 1992). The Max RNA is subjected to alternate splicing which generates proteins with different DNA binding and transforming activities (Mäkelä et al, 1992; Prochownik et al, 1993; Vastrik et al, 1995). As mentioned, Max/Max homodimers can antagonize Myc/Max activity. However, given the growing number of Max-interacting proteins, Max homodimers probably do not play a significant physiological role. Like Max/Max, Mad/Max complexes antagonize the activity of Myc/Max in cell transformation assays (Ayer et al, 1993; Cerni et al, 1995; Hurlin et al, 1995; Koskinen et al, 1995). Mad can also antagonize tumor cell growth (Chen et al, 1995), perhaps in the guise of a tumor suppressor (Eagle et al, 1995). A role for Mad proteins during development and cell differentation is also likely given their patterns of expression (Hurlin et al, 1994; Chin et al, 1995). While Max/Max probably acts passively by competing with Myc/Max for common DNA binding sites, Mad/Max both competes for DNA binding and actively inhibits transcription, by recruiting corepressor molecules such as mSin3 (Ayer et al, 1995; Schreiber-Agus et al, 1995). Thus, Mad/Max complexes can exert effects in the absence of Myc expression.

Myc is likely to be regulated by additional CTD interactions involving the b/HLH/LZ domain. Additional Myc partners are implied by the identification of at least one cell line that lacks Max but remains biologically responsive to Myc (Hopewell et al, 1995). One candidate for such a partner is the initiator (Inr)-binding transcription factor TF II-I, which has been reported to interact with Myc in a HLH/LZ-dependent manner (Roy et al, 1993 a). Canonical HLH/LZ complexes may be influenced by an additional level of protein interactions that affect Myc's DNA binding properties. For example, the transcription factors YY1 and AP2 each have been shown to associate with the Myc C-terminus and prevent DNA binding. The binding determinants in Myc for each protein may include parts of the central region as well as the CTD. YY1 association precludes Max binding to Myc

(Shrivastava et al, 1993). In contrast, AP2 does not block Max binding but interferes with DNA binding by Myc/Max (Gaubatz et al, 1995). Thus, the regulation of DNA binding by Myc is controlled at many levels, in addition to those controlled by Max itself. Nevertheless, despite indications of greater complexity, it is clear that the function of the CTD is to control access of Myc to its physiological DNA targets.

Central Region: Conserved Sequences But Unknown Functions

Relatively few studies have focused on the central region of Myc, because early structure-function analysis indicated that it was largely dispensible for cell transformation activity (Stone et al, 1987). A nuclear localization signal (aa 320–328) has been defined in this region (Stone et al, 1987; Dang et al, 1989). Additional functions of the central domain are implied by at least two segments of sequence that are conserved in Myc proteins throughout evolution. The fact that these segments are less important for transformation implies that they are only needed in certain cell types and/or that they may have a negative regulatory role. There are two clusters of casein kinase II phosphorylation sites in the central region, at aa 240–262 and aa 342–357 (Lüscher et al, 1989), the first but not the second group of which is important for cell transforming activity (Street et al, 1990).

The aa 240–262 phosphorylation sites are located in one of the two boxes of conserved sequence in the central region of Myc. This box, located at aa 250–271, is termed the acidic box (Ac) because of the stretch of acidic residues it contains. Ac has received some attention but less than the terminal domains because deletions encompassing it have different effects on transformation depending on the cell system used. For example, in the most widely used assay, the Ras cotransformation assay performed in primary rat embryo fibroblasts (REFs), Ac is dispensable for Myc activity (Stone et al, 1987). However, in other cell systems it is clearly important. Deletions encompassing Ac eliminate transformation of Rat1a cells, an established fibroblast line that is permissive for morphological transformation by Myc (Stone et al, 1987). Similarly, Ac deletions are reported to disable Myc immortalization activity in mouse embyro fibroblasts (Zoidl et al, 1993). Lastly, avian v-Myc requires Ac to transform hematopoeitic cells but not fibroblasts (Heaney et al, 1986). Cell type differences for an Ac requirement suggest that tissue-specific mechanisms operate. However, an alternate explanation is that there are differences in cell susceptibility to Myc transformation. If so, relatively small effects of Ac deletion on Myc function (for example, through folding) could be translated into significant biological effects. However, with this caveat, it is reasonable to infer that Ac function is important for full biological activity in certain cells. The function of Ac is probably not an "acid blob" transactivator, though, because Ac-containing regions lack TAD activity when assayed as chimeras that contain a heterologous DNA binding domain (Kato et al, 1990). One interpretation of the existing data is that Ac may be needed to relieve a negative function which operates in certain cells.

A second box of conserved sequence at aa 295–311, termed here the KRCH box, has not been studied but may have a negative regulatory role. The KRCH box falls into a region (aa 265–318) which, as above, is dispensable for Ras cotransformation of REFs, but is required for transformation of Rat1a cells (Stone et al, 1987). The aa 265–318 region can mediate nonspecific DNA binding activity by Myc (Dang et al, 1989). Whether the requirement in Rat1a is related to this activity is not known. Interestingly, a close look at the REF data indicates that deletion through the KRCH box actually leads to a slight increase in transforming activity. One interpretation of this result is that the KRCH box has a negative regulatory role, since its deletion slightly increases Myc activity. Supporting this possibility is that there is sequence relatedness between the 295–311 region of Myc and a functionally analogous part of adenovirus E1A, which shares other similarities with Myc (Ralston, 1991). Deletion of the related sequence in E1A (aa 232–240) also causes an increase in Ras cotransformation activity (Boyd et al, 1993). The E1A sequence has been implicated as a target for binding to the so-called C-terminal binding protein (CtBP), a 48 kD protein related to NAD-dependent acid dehydrogenases which has been cloned recently (Schaeper et al, 1995). Taken together, the structural and functional features of the KRCH box suggest it may be involved in negative regulation of Myc.

A third box of conserved sequence (aa 345–352) includes the cluster of casein kinase II sites mapped at aa 342–357 (Lüscher et al, 1989). Deletions through this region have little effect on cell transformation (Stone et al, 1987; Street et al, 1990), but the conservation suggests possible functional significance, perhaps negative regulatory.

N-Terminus: I. Transactivation

After the discovery of the b/HLH/LZ motifs in the Myc CTD, it was widely anticipated that the Myc NTD would function in transcriptional activation. Early support for this notion was provided by experiments in which the NTD was fused to the DNA binding domain of GAL4 and shown to transactivate transient reporter expression from a GAL4-dependent promoter (Kato et al, 1990). After discovery of a DNA recognition sequence and Max, these observations were validated by demonstrations that Myc/Max could transactivate reporter expression from artificial promoters containing multimerized CACGTG sequences and basal elements (Amati et al, 1992; Kretzner et al, 1992; Reddy et al, 1992; Gu et al, 1993). As would be predicted by its lack of a TAD, Max does not to contribute to transactivation by Myc/Max. Similar to their effect on cell transformation, Max/Max homodimers antagonize transactivation, as do Mad/Max heterodimers (Ayer et al, 1993; Hurlin et al, 1995).

Myc's transactivation activity has been definitively established in studies of a set of candidate target genes that contain CACGTG and are transcriptionally activated by Myc. These genes include the HMG-like protein prothymosin-α (pT) (Gaubatz et al, 1994), the purine biosynthesis enzyme ornithine decarboxylase (ODC) (Bello-Fernandez et al, 1993; Wagner et al, 1993; Tobias et al, 1995), the pyrimidine biosynthesis enzyme carbamoyl-phosphate synthase (CAD) (Miltenberger et al, 1995), the tumor suppressor p53 (Reisman et al, 1993), and an embryonic gene of undefined function, ECA39 (Benvenisty et al, 1993). Myc can also induce expression of cell cycle regulators, for example, cyclins A and E and the S phase-specific transcription factor E2F (Jansen-Durr et al, 1993), though whether these are direct targets is not clear. Although is not yet clear whether all the CACGTG-containing genes are physiological targets (see (Henriksson et al, 1996) for a more in depth discussion), one interesting feature found in each is that a Myc/Max binding site(s) needed for activation is located downstream of the RNA cap site. Since artificial promoters containing multimerized CACGTG sequences upstream of basal elements and cap site can respond to Myc/Max, the reason for the downstream locations in cellular target genes is unclear. One possibility is that this may reflect a CTD function, for example, in DNA bending (Fisher et al, 1992; Wechsler et al, 1992). In any case, the fact that Myc transactivation functions are invariably delivered 3′ of target gene promoters suggest some unique functional aspect.

Myc transactivation is probably necessary but is not sufficient for biological activity. The evidence supporting a requirement for transactivation is correlative. First, deletion analysis has indicated a perfect overlap between the NTD regions needed for transactivation and those needed for cell transformation and apoptosis (Stone et al, 1987; Kato et al, 1990; Evan et al, 1992). Second, the transactivation activity of Myc family proteins (assayed from artificial promoters) is directly proportional to their activity in cell transformation assays (Barrett et al, 1992). Third, induction of at least one Myc target gene, ODC, may be sufficient for cell transformation and/or apoptosis (Auvinen et al, 1992; Moshier et al, 1993;

Packham et al, 1994; Shantz et al, 1994). Finally, some NTD mutations that increase transformation activity have been reported to relieve inhibition of transactivation by two NTD-binding proteins (see below). However, transactivation is not sufficient for transformation activity. For example, MB2 mutations completely eliminate cell transformation but they do not affect transactivation (Li et al, 1994; Brough et al, 1995). In fact, mutation of MB2 augments activity relative to wild-type Myc, probably due to a reduction in the antagonism between NTD TADs (see below). Thus, transformation and transactivation are separable. One caveat to this interpretation is that MB2 mutants have been assayed to date only on artificial promoters, rather than on target gene promoters, so transactivation of certain target genes might actually prove to be MB2-dependent.

Deletion analysis to map NTD transactivation activity has been performed in the context of GAL4 chimeras (Kato et al, 1990). By this approach, two TADs have been identified whose activity seems to be antagonistic to each other. Features of the NTD sequence and transcriptional activities are shown in Figure 1.3 aa 1–41 and aa 103–144 are each strong transactivators (20-fold and 10-fold induction, respectively, above baseline), while the third, aa 41–103, has only weak transactivation activity (~ 2-fold induction).

The aa 1–41 TAD contains a box of sequence (aa 10–23), termed here the DYD box, that is partly conserved in evolution. It also includes stretches of glutamate (aa 26–28) and glutamine (aa 32–36) which may act as

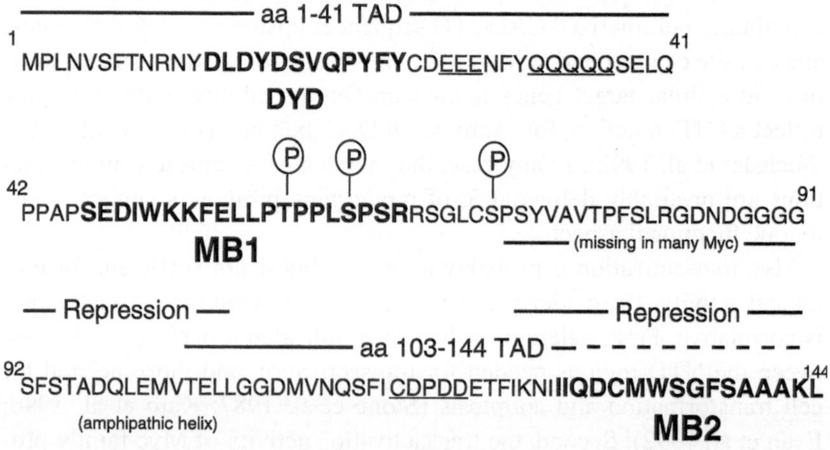

Figure 1.3 Features of the NTD sequence. Sequences that are highly conserved in evolution are shown in bold type. The aa 1–41 and aa 103–144 TADs are shown, as are the aa 92–106 and MB2 regions implicated in repression. The numbers refer to the human c-Myc amino acid sequence. A region missing in many Myc proteins (aa 72–90) and the relative position of a possible amphipathic helix discussed in the text are noted.

activators similar to that found in other transcription factors (e.g., Sp1). The aa 103–144 region includes functions involved in activation and/or repression; based on comparisons to the parts involved in repression, aa 107–130 would be inferred to comprise the TAD (see below). This region has a section of acidic character that may be relevant at aa 118–122. The aa 107–130 region is interesting because it is conserved in evolution for a particular Myc family protein (e.g., c-Myc) but divergent between members of the Myc family (e.g. c-Myc versus N-Myc). Variations here may explain the 10- to 20-fold differences in transactivation and transformation activities of the L-Myc and c-Myc NTDs (Birrer et al, 1988; Barrett et al, 1992). The conservation patterns suggest that, unlike the CTD, the functions of the NTD do not overlap completely, and that the aa 107–130 TAD on different Myc family proteins may interact with different coactivators.

The weakly transactivating aa 41–106 region includes the well-conserved MB1 sequences (aa 41–65) and a region implicated in repression (aa 92–106; see below). Most of the intervening segment (aa 72–90) is probably unimportant because it is not evolutionarily conserved (is actually missing in most Myc proteins). In addition to a role in repression, the aa 92–106 region may also be involved in activation, since it is reported to be structurally and functionally related to the HOB2 domain found in and necessary for transactivation by Jun, Fos, and c/EBP (Philipp et al, 1994). However, the virtual lack of activity of aa 41–106 argues that neither MB1 nor aa 92–106 have significant activation activity in the context of the NTD.

These regions interact positively and negatively with each other. aa 41–103 (MB1) is additive with aa 1–41 (DYD) but synergizes strongly with aa 103–144. In contrast, aa 1–41 (DYD) and aa 103–144 (MB2) antagonize one another (Kato et al, 1990). Taken together, the results argue that the MB1-containing aa 41–103 region has a minimal role in activation itself, but instead controls or integrates the adjacent TAD activities.

A role for MB1 as an integrator or modulator of NTD functions is supported by the presence there of three sites for in vivo phosphorylation, including T58, S62, and S71 (Henriksson et al, 1993; Lutterbach et al, 1994; Pulverer et al, 1994). S62 and T58 phosphorylation events functionally oppose each other. S62 modification acts positively and T58 modification acts negatively on Myc's transformation (Frykberg et al, 1987; Albert et al, 1994; Pulverer et al, 1994) and transactivation activities (Seth et al, 1991; Gupta et al, 1993; Henriksson et al, 1993; Born et al, 1994). Phosphorylation of S62 is cell-cycle regulated. Modification occurs during mid-G1 and G2 phases (Seth et al, 1993), when Myc activity is required for cell cycle progression (Evan et al, 1993). There is a strong block to S62 modification during M phase (Lutterbach et al, 1994). Whether T58 is similarly regulated is unknown. However, this is likely since S62 phosphorylation is a prerequisite for modification of T58 in vivo (Lutterbach et al, 1994). The hierarchical feature of MB1 events is consistent with the fact that the amino acid sequences near T58 and S62 constitute a consensus recognition site for glycogen synthase kinase-3 (GSK-3), which can modify T58 in vitro (Saksela et al, 1992; Lutterbach et al, 1994; Pulverer et al, 1994) and is activated by proximal downstream phosphorylation events (Fiol et al, 1990). Thus, T58 phosphorylation may represent feedback control of the activation event at S62. Which kinases modify T58 and S62 in vivo are not yet clear. In addition to GSK-3, MAPK and cell-cycle dependent kinases (cdks) can modify MB1 in vitro (Seth et al, 1991; Gupta et al, 1993; Hoang et al, 1995) but it is not clear they are relevant in vivo (Lutterbach et al, 1994). One unusual feature of T58 is that it can also be O-glycosylated in vivo (Chou et al, 1995). The significance of this modification is unknown. Finally, a third in vivo phosphorylation site has been identified immediately C-terminal to MB1 at S71 (Lutterbach et al, 1994). S71 is evolutionarily conserved but effects of its phosphorylation have not been explored yet. Taken together, the results suggest that modifications in MB1 control NTD activity(s).

Although the mechanisms underlying NTD transactivation have not been determined yet, possible roles for TFIID and potential regulatory proteins have been studied. One direction for identifying NTD interactions has drawn on the biological parallels between Myc and the adenovirus E1A protein (Ralston, 1991). Thus, links to the retinoblastoma (Rb) family of cell cycle regulators and the basal transcription factor TFIID were explored because of their interactions with E1A. Analogous to interactions with E1A, Rb and TFIID can in eract with overlapping regions of the NTD in vitro (Rustgi et al, 1991; Hateboer et al, 1993). Whether TFIID contacts occur physiologically is unknown, but transcriptional repression may involve them (see below). Rb has been reported to stimulate NTD transactivation activity in some cell types (Adnane et al, 1995), but this effect may be indirect since there is no evidence for formation of a Rb-Myc complex in cells. However, an in vivo complex has been observed between Myc and the Rb-related protein p107 (Beijersbergen et al, 1994; Gu et al, 1994). The p107-NTD interaction is mediated by the p107 "pocket" domain, which also mediates E1A binding (Zhu et al, 1993), and is correlated with inhibition of NTD transactivation activity. p107 has been suggested to deliver to the NTD an associated cdk2/cyclinA kinase, which can phosphorylate sites in MB1 (Hoan et al, 1995). One less appealing aspect of the Myc-p107 association is that it appears to have relatively small consequences for Myc transformation. A second NTD binding protein, termed Bin1 (myc Box-dependent Interacting protein-1) has recently been identified by a two hybrid approach (Sakamuro et al, 1996). Bin1 is related to RVS167, a negative regulator of the yeast cell cycle. Consistent with this relationship, Bin1 inhibits Myc transformation in a manner requiring MB1 integrity (Sakamuro et al, 1996). Bin1 will also inhibit Myc transactivation (M. Eilers, J. Feramisco, and G. C. Prendergast, unpublished data). Mutation of the negative-acting T58 site has no effect on binding by either p107 or Bin1, but relieves inhibition of transactivation by p107 (Hoang et al, 1995) and inhibition of both transactivation and transformation by Bin1 (Sakamuro et al, 1996). Certain other MB1 mutations derived from lymphoma *myc* genes may also relieve inhibition by p107 (Hoang et al, 1995). Thus, regulatory interactions between Myc and NTD-binding proteins may be interrupted by mutations that eliminate either phosphorylation or the downstream effects of phosphorylation events.

The NTD is the site of mutations in cancer cells which may act to relieve negative regulation of transactivation (and possibly other functions). MB1 is a hotspot for mutation, T58 in particular. T58 is universally mutated in retroviral *myc* genes. Formerly, it was believed that in cancer cells the only functionally significant alterations in Myc affected its regulation. NTD sequence variations were first noted over a decade ago (e.g., Papas et al, 1985; Showe et al, 1985), but were thought to be unimportant, despite some evidence to the contrary (Frykberg et al, 1987), because constitutive expression of wild-type Myc was shown to be sufficient for oncogenic activation (Land et al, 1983). However, it has become clear that Myc sustains structural mutations in cancer cells which also contribute to its deregulation. Recent surveys have revealed that as many as ~60% of the *myc* genes in lymphomas and several other tumor types are mutated. Mutations occur singly or in clusters in the NTD region, especially in and around MB1 (Bhatia et al, 1993; Yano et al, 1993; Bhatia et al, 1994). In many tumors, the mutations are homozygous (Bhatia et al, 1993), arguing for selection against negative functions in wild-type Myc. Consistent with this likelihood, some NTD mutations appear to (as mentioned above) relieve inhibition of transactivation by p107 and Bin1 (Hoang et al, 1995; Sakamuro et al, 1996). The appearance of NTD mutations in tumors suggests that they provide an escape from the inhibitory activity of a tumor suppressor. p107 is not believed to be a tumor suppressor, but Bin1 has some features consistent with this possibility (Sakamuro et al, 1996).

N-Terminus: II. Transcriptional Repression

In addition to activating gene expression, Myc can also repress it (see Suen et al, 1991; Yang et al, 1993 and references within). Recently, a repression function(s) has been mapped in the NTD that may explain this effect. In particular, two regions, aa 92–106 and MB2 (aa 130–144), have been implicated. The genetic target for repression by these elements is the initiator region (Inr) element (Roy et al, 1993 a; Li et al, 1994; Philipp et al, 1994). The mechanism of Inr-dependent repression is not yet known, but requires MB2, suggesting that interactions involving it are important (Li et al, 1994). Since mutations in MB2 eliminate Myc transformation, Inr-dependent repression is necessary for biological activity. Repression of some Inr elements require an intact CTD (Li et al, 1994), implying that b/HLH/LZ interactions are important. In addition to Max, one protein that may be germane is TFII-I, which directly interacts with the Inr DNA element and has been reported to interact with the Myc b/HLH/LZ (Roy et al, 1993 a). However, while all the targets examined to date require MB2 for repression, not all require an intact CTD (Philipp et al, 1994), suggesting there may be some variations in mechanism. There is little information concerning the basis for NTD action. There may be relevance for contacts with TFII-D (Hateboer et al, 1993), since Myc can repress Inr-dependent activation in vitro by precluding interactions between TFII-I and TFII-D (Roy et al, 1993 a).

A candidate for interaction with the MB2 and aa 92–106 regions is a ubiquitous NTD-binding nuclear factor whose association depends on MB2 integrity (Brough et al, 1995). Ectopic expression of a GAL4-NTD chimera can block transformation by either Myc or E1A, implicating the factor in the action of both oncoproteins. The dominant inhibitory effect of the chimera depends on the presence of MB2, implying that the factor is required. Based on the prominent role of MB2 in repression, this factor may prove to be a corepressor molecule. The high degree of hydrophobic character within MB2 makes it rather unlikely that it is involved directly in protein-protein interaction with a candidate corepressor. Instead, MB2 may act as a core organizing domain for another region, perhaps aa 92–106, which includes an amphipathic helix at its N-terminal end (G.C. Prendergast, unpublished data). That such a helix mediates contact with a corepressor could be predicted from the fact that a similar structure in Mad mediates interaction with the mSin3 corepressor (Schreiber-Agus et al, 1995).

The aa 92–106 region essentially fills the gap between MB1 and the aa 107–130 TAD (the non-MB2 part of the strong aa 103–144 transactivator; Kato et al, 1990). At its C-terminal end, the aa 92–106 region is quite sensitive to insertional mutagenesis. In REF transformation assays, Myc mutants containing a two-residue insertion mutant at aa 104–105 exhibit ~ 8-fold reduced activity (Stone et al, 1987). Thus, as is the case with MB2, deletion of aa 92–106 eliminates Myc biological activity, reinforcing the link between repression and cell transformation.

Recently, a lymphoma mutation has been reported within the aa 92–144 region (F115L) that augments both repression and transformation activities (Lee et al, 1996). Since it falls into the aa 107–130 region implicated as a TAD, this effect may reflect the indirect effect of a loss in transactivation activity. However, whatever the mechanism, the increased activity of this mutant provides additional evidence that repression, rather than transactivation, actually represents the critical NTD function for Myc's oncogenic activity. If so, Myc would resemble E1A in this regard.

Nontranscriptional Functions: Direct Effects on Apoptosis?

Myc-mediated apoptosis presents an interesting puzzle. Two significant issues are how Myc induces apoptosis and whether the mechanisms involved overlap with those involved in cell cycle activation (Evan et al, 1995). There are two models that have been widely entertained (Evan et al, 1995; Packham et al, 1995). In the first (the "conflict" model), apoptosis is proposed to be the indirect response of a cell to an inappropriate growth signal from Myc. In this model, Myc directly interacts only with growth regulators; apoptosis results due to activation of a safety mechanism that is triggered when positive and negative growth signals are received simultaneously (suggesting some loss of cellular regulation). The second model, the "dual signal" model, proposes that apoptosis is directly regulated by Myc. In this model, Myc interacts with both growth and death regulators; apoptosis occurs when Myc's death signals are not actively suppressed by other growth signals. Existing results support the "dual signal" model. The issue is germane to this review, because this model implies that Myc may directly interact with death regulatory proteins, perhaps ones which are not associated with transcriptional regulation.

Myc must oligomerize with Max and bind DNA to activate apoptosis as well as cell proliferation (Amati et al, 1993b). This implies that transcriptional activation of certain genes is required. However, in normal cells containing deregulated Myc, protein synthesis inhibitors do not prevent Myc-mediated apoptosis (Wagner et al, 1994). These observations can be rectified if Myc "primes" apoptosis by inducing expression of certain death genes (whose action is suppressed by other growth signals), but "triggers" apoptosis in primed cells by some other mechanism that does not involve gene activation. Support for the conflict model rested largely on demonstrations that growth inhibitory treatments caused apoptosis in cells containing deregulated Myc, which cannot withdraw from the cell cycle (Evan et al, 1992; Wagner et al, 1994). However, it has recently been shown that cells containing deregulated Myc can undergo growth arrest without dying, if treated with dibutyryl-cAMP (Packham et al, 1996). These results indicate that Myc's ability to drive cell growth and cell death can be separated, and therefore they argue against the conflict model and for the dual signal model.

How the trigger signal is configured is not yet known but may now begin to emerge. Given the frequency of NTD mutation, alterations that eliminate apoptotic trigger mechanisms might be found. Cell death mechanisms are emerging rapidly (Martin et al, 1995; Whyte et al, 1995) but there are as yet no connections to Myc. One line of work has explored the epistatic relationship between Myc and p53, the tumor suppressor protein. A prevailing viewpoint is that wild-type p53 monitors genomic integrity (Lane 1992) and induces apoptosis if its checkpoint operation fails (Perry et al, 1993; Lane et al, 1994; Lane et al, 1995). In fibroblasts, Myc requires p53 to kill fibroblasts (Hermeking et al, 1994; Wagner et al, 1994). However, this effect apparently reflects tissue-specific requirements, because p53 is not required for Myc to potentiate cytokine-induced death in M1 myeloid tumor cells (Selvakumaran et al, 1994) or to kill kidney epithelial cells (Sakamuro et al, 1995). Possible role(s) for the existing set of NTD-binding proteins have not yet been explored. Rb might be relevant, even if connections are indirect, because of the ability of Rb to suppress apotosis (Evan et al, 1995). Also, connections between p107 and cdk/cyclin A kinase are possible, since cyclin A induction is associated with Myc-mediated cell death (Hoang et al, 1994) and cdks have been reported to be required for apoptosis in some cells (Shi et al, 1994). In general, this area of research should be a major focus of future work, in part because of its possible applications to cancer.

Nontranscriptional Functions: Other Roles?

There are persistent hints that Myc may have some functions apart from transcriptional regulation. Immunocytochemisty performed by some (but not all) workers has localized Myc to "speckled" nuclear loci that are separated from transcriptionally active euchromatin (Persson et al, 1986; Spector et al, 1987). Also, Myc can rapidly regulate the expression of many genes at some posttranscriptional level in the nucleus (Prendergast et al, 1989; Inghirami et al, 1990; Gibson et al, 1992).

But the most provocative hints for nontranscriptional functions have come from studies of cell division in *Xenopus* oocytes. Following fertilization, maternal stores of Myc that are stored in the egg are rapidly localized to the nucleus, accumulating to exceptionally high levels during a period when there is no zygotic transcription (Gusse et al, 1989). Cells at this period of development are dividing rapidly, cycling between S and M phases, suggesting roles for Myc in DNA replication or chromosome structure regulation. Consistent with some different function, Myc in preblastula nuclei is not found bound to Max, and, in fact, extracts from preblastula cells dissociate preformed Myc/Max complexes (Lemaitre et al, 1995). Myc is believed to be required during the rapid cell divisions, because up to the midblastula transition, when zygotic transcription is turned on, the levels of Myc gradually drop to those seen in somatic cells. In mammalian somatic cells, the induction of S phase by Myc has been associated with upregulation of S phase cyclins (Jansen-Durr et al, 1993; Hanson et al, 1994), cell cycle kinases (Kim et al, 1994), and cyclin-activating enzymes (Steiner et al, 1995). However, these effects are due to transcriptional activation. Thus, there may be a fundamental difference in the way Myc is used during early development.

Connections to chromosome structure might be reconsidered in light of these results. Deregulated Myc induces high levels of sister chromatid exchange and chromosomal abnormality (Cerni et al, 1987). This effect is associated with immortalization, and therefore may be a feature of crisis, but it is not seen in fully transformed cells. Also, Myc overexpression leads to the generation in cells of abnormal chromatin structure and nuclear shape, as examined by electron microscopy (Henriksson et al, 1988) and immunocytochemistry (Koskinen et al, 1991). Finally, the recently identified NTD-binding protein Bin1 cor tains a region with structural similarity to several regulators of chromosome structure that act during M phase (G.C. Prendergast, unpublished data). Along with the report that Max-p67 Myc complexes recognize the same DNA sequence as identified as a Myc binding site in DNA replication studies (Negishi et al, 1992; Hann et al, 1994), the *Xenopus* work refuels long-standing disputes that hint at roles for Myc beyond that of a generic transcription factor.

References

Adane J, Robbins PD (1995): The retinoblastoma susceptibility gene product regulates Myc-mediated transcription. *Oncogene* 10: 381–387

Albert T, Urlbauer B, Kohlhuber F, Hammersen B, Eick D (1994): Ongoing mutations in the N-terminal domain of c-Myc affect transactivation in Burkitt's lymphoma cell lines. *Oncogene* 9: 759–763

Amati B, Dalton S, Brooks MW, Littlewood TD, Evan GI, Land H (1992): Transcriptional activation by the human c-Myc oncoprotein in yeast requires interaction with Max. *Nature* 359: 423–426

Amati B, Brooks MW, Levy N, Littlewood TD, Evan GI, Land H (1993a): Oncogenic activity of the c-Myc protein requires dimerization with Max. *Cell* 72: 233–245

Amati B, Littlewood TD, Evan GI, Land H (1993b): The c-Myc protein induces cell cycle progression and apoptosis through dimerization with Max. *EMBO J* 12: 5083–5087

Ariga H, Imamura Y, Iguchi-Ariga SMM (1989): DNA replication origin and transcriptional enhancer in c-*myc* gene share the c-*myc* protein binding sequences. *EMBO J* 8: 4273–4279

Askew DS, Ashmun RA, Simmons BC, Cleveland JL (1991): Constitutive c-*myc* expression in an IL-3-dependent myeloid cell line suppresses cell cycle arrest and accelerates apoptosis. *Oncogene* 6: 1915–1922

Auvinen M, Passinen A, Andersson LC, Holtta E (1992): Ornithine decarboxylase activity is critical for cell transformation. *Nature* 360: 355–358

Ayer DE, Eisenman RN (1993): A switch from myc-max to mad-max heterocomplexes accompanies monocyte/macrophage differentiation. *Genes Dev* 7: 2110–2119

Ayer DE, Kretzner L, Eisenman RN (1993): Mad: A heterodimeric partner for Max that antagonizes Myc transcriptional activity. *Cell* 72: 211–222

Ayer DE, Lawrence QA, Eisenman RN (1995): Mad-Max transcriptional repression is mediated by ternary complex formation with mammalian homologs of the yeast repressor Sin3. *Cell* 80: 767–776

Barone MV, Courtneidge SA (1995): Myc but not Fos rescue of PDGF signalling block caused by kinase-inactive Src. *Nature* 378: 509–512

Barrett J, Birrer MJ, Kato GJ, Dosaka AH, Dang CV (1992): Activation domains of L-Myc and c-Myc determine their transforming potencies in rat embryo cells. *Mol Cell Biol* 12: 3130–3137

Beckmann H, Su L-K, Kadesch T (1990): TFE3: A helix-loop-helix protein that activates transcription through the immunoglobulin enhancer mE3 motif. *Genes Dev* 4: 167–179

Beijersbergen RL, Hijmans EM, Zhu I , Bernards R (1994): Interaction of c-Myc with the pRb-related protein p107 results in inhibition of c-Myc-mediated transactivation. *EMBO J* 13: 4080–4086

Bello-Fernandes C, Packham G, Cleveland JL (1993): The ornithine decarboxylase gene is a transcriptional target of c-Myc. *Proc Natl Acad Sci USA* 90: 7804–7808

Benvenisty N, Leder A, Kuo A, Leder P (1993): An embryonically expressed gene is a target for c-Myc regulation via the c-Myc-binding sequence. *Genes Dev* 6: 2513–2523

Berberich SJ, Cole MD (1992): Casein kinase II inhibits the DNA binding activity of Max homodimers but not Myc/Max heterodimers. *Genes Dev* 6: 166–176

Berberich S, Hyde-deRuyscher N, Espenshade P, Cole M (1992): Max encodes a sequence-specific DNA binding protein and is not regulated by serum growth factors. *Oncogene* 7: 775–779

Bernards R (1995): Transcriptional regulation: flipping the Myc switch. *Curr Biol* 5: 859–861

Bestor TH (1990): DNA methylation: evolution of a bacterial immune function into a regulator of gene expression and genome structure in higher eukaryotes. *Phil Trans R Soc Lond B* 326: 179–187

Bhatia K, Huppi K, Spangler G, Siwarski D, Iyer R, Magrath I (1993): Point mutations in the c-Myc transactivation domain are common in Burkitt's lymphoma and mouse plasmacytomas. *Nat Genet* 5: 56–61

Bhatia K, Spangler G, Gaidano G, Hamdy N, Dalla-Favera R, Magrath I (1994): Mutations in the coding region of c-myc occur frequently in acquired immunodeficiency syndrome-associated lymphomas. *Blood* 84: 883–888

Birrer MJ, Segal S, DeGreve JS, Kaye F, Sausville EA, Minna JD (1988): L-myc cooperates with ras to transform primary rat embryo fibroblasts. *Mol Cell Biol* 8: 2668–2673

Blackwell TK, Weintraub H (1990): A new binding-site selection technique reveals differences and similarities between MyoD and E2A DNA-binding specificities. *Science* 250: 1104–1110

Blackwell TK, Huang J, Ma A, Kretzner L, Alt FW, Eisenman RN, Weintraub H (1993): Binding of myc proteins to canonical and noncanonical DNA sequences. *Mol Cell Biol* 13: 5216–5224

Blackwood E, Eisenman RN (1991): Max: A helix-loop-helix zipper protein that forms a sequence-specific DNA-binding complex with Myc. *Science* 251: 1211–1217

Blackwood E, Lüscher B, Eisenman RN (1992): Myc and Max associate *in vivo. Genes Dev* 6: 71–80

Born T, Frost J, Schönthal A, Prendergast GC, Feramisco J (1994): c-Myc and oncogenic *ras* induce the cdc2 promoter. *Mol Cell Biol* 14: 5741–5747

Bousset K, Henriksson M, Lüschner-Firzlaff JM, Litchfield DW, Lüscher B (1993): *Oncogene* 8: 3211–3220

Boyd JM, Subramanian T, Schaeper U, LaRegina M, Bayley S, Chinnadurai G (1993): A region in the C-terminus of adenovirus 2/5 E1a protein is required for association with a cellular phosphoprotein and important for the negative modulation of T24-ras mediated transformation, tumorigenesis, and metastasis. *EMBO J* 12: 469–478

Brough DE, Hofman TJ, Ellwood KB, Townley RA, Cole MD (1995): An essential domain of the c-Myc protein interacts with a nuclear factor that is also required for E1A-mediated transformation. *Mol Cell Biol* 15: 1536–1544

Cedar H (1988): DNA methylation and gene activity. *Cell* 53: 3–4

Cerni C, Mougneau E, Cuzin F (1987): Transfer of "immortalizing" oncogenes in rat fibroblasts induces both high rates of sister chromatid exchange and appearance of abnormal karotypes. *Exp Cell Res* 168: 439–446

Cerni C, Bousset K, Seelos C, Burkhardt H, Henriksson M, Luscher B (1995): Differential effects by Mad and Max on transformation by cellular and viral oncoproteins. *Oncogene* 11: 587–596

Chen J, Willingham T, Margraf LR, Schreiber-Agus N, De Pinho RA, Nisen PD (1995): Effects of the MYC oncogene antagonist, MAD, on proliferation, cell cycling and the malignant phenotype of human brain tumour cells. *Nat Med* 1: 638–643

Chin L, Schreiber-Agus N, Pellicer I, Chen K, Lee HW, Dudast M, Cordon-Cardo C, De Pinho RA (1995): Contrasting roles for Myc and Mad proteins in cellular growth and differentiation. *Proc Natl Acad Sci USA* 92: 8488–8492

Chou TY, Dang CV, Hart GW (1995): Glycosylation of the c-Myc transactivation domain. *Proc Natl Acad Sci USA* 92: 4417–4421

Cole MD (1986): The myc oncogene: Its role in transformation and differentiation. *Ann Rev Genet* 20: 361–384

Dang C, McGuire M, Buckmire M, Lee WMF (1989): Involvement of the "leucine zipper" region in the oligomerization and transforming activity of human c-myc protein. *Nature* 337: 664–666

Dang CV, Lee WMF (1989): Nuclear and nucleolar targeting sequences of c-*erb*A, c-*myb*, N-*myc*, p53, HSP70, and HIV *tat* proteins. *J Biol Chem* 264: 18019–18023

Dang CV, v. Dam H, Buckmire M, Lee WMF (1989): DNA-binding domain of human c-Myc produced in *Escherichia coli. Mol Cell Biol* 9: 2477–2486

Dang CV, Barrett J, Villa-Garcia M, Resar LMS, Kato GJ, Fearon ER (1991): Intracellular leucine zipper interactions suggest c-Myc hetero-oligomerization. *Mol Cell Biol* 11: 954–962

De Pinho RA, Schreiber-Agus N, Alt FW (1991): myc family oncogenes in the development of normal and neoplastic cells. *Adv Canc Res* 57: 1–46

Eagle LR, Yin X, Brothman AR, Williams BJ, Atkin NB, Prochownik EV (1995): Mutation of the MXI1 gene in prostate cancer. *Nat Genet* 9: 249–255

Evan G, Harrington E, Fanidi A, Land H, Amati B, Bennett M (1994): Integrated control of cell proliferation and cell death by the c-myc oncogene. *Philos Trans R Soc Lond B Biol Sci* 345: 269–275

Evan GI, Littlewood TD (1993): The role of c-myc in cell growth. *Curr Opin Genet Dev* 3: 44–49

Evan GI, Wyllie AH, Gilbert CS, Littlewood TD, Land H, Brooks M, Waters CM, Penn LZ, Hancock DC (1992): Induction of apoptosis in fibroblasts by c-myc protein. *Cell* 69: 119–128

Evan GI, Brown L, Whyte M, Harrington E (1995): Apoptosis and the cell cycle. *Curr Biol* 7: 825–834

Ferre-D'Amare A, Prendergast GC, Ziff EB, Burley SK (1993): Recognition by Max of its cognate DNA through a dimeric b/HLH/Z domain. *Nature* 363: 38–45

Fiol CJ, Wang A, Roeske RW, Roach PJ (1990): Ordered multisite protein phosphorylation: Analysis of glycogen synthase kinase 3 action using model peptide substrates. *J Biol Chem* 265: 6061–6065

Fisher DE, Parent LA, Sharp PA (1992): Myc/Max and other helix-loop-helix/leucine zipper proteins bend DNA toward the minor groove. *Proc Natl Acad Sci USA* 89: 11779–11783

Fisher F, Jayaraman P-S, Goding CR (1991): C-Myc and the yeast transcription factor PHO4 share a common CACGTG binding motif. *Oncogene* 6: 1099–1104

Frykberg L, Graf T, Vennström B (1987): The transforming activity of the chicken c-*myc* gene can be potentiated by mutations. *Oncogene* 1: 415–421

Gaubatz S, Meichle A, Eilers M (1994): An E-box element localized in the first intron mediates regulation of the prothymosin α gene by c-*myc*. *Mol Cell Biol* 14: 3853–3862

Gaubatz S, Imhof A, Dosch R, Werner O, Mitchell P, Buettner R, Eilers M (1995): Transcriptional activation by Myc is under negative control by the transcription factor AP-2. *EMBO J* 14: 1508–1519

Gibson AW, Ye R, Johnston RN, Browder LW (1992): A possible role for c-Myc oncoproteins in post-transcriptional regulation if ribosomal RNA. *Oncogene* 7: 2363–2367

Gregor PD, Sawadogo M, Roeder RG (1990): The adenovirus major late transcription factor USF is a member of the helix-loop-helix group of regulatory proteins and binds to DNA as a dimer. *Genes Dev* 4: 1730–1740

Gu W, Cechova K, Tassi V, Dalla FR (1993): Opposite regulation of gene transcription and cell proliferation by c-Myc and Max. *Proc Natl Acad Sci USA* 90: 2935–2939

Gu W, Bhatia K, Magrath IT, Dang CV, DallaFavera R (1994): Binding and suppresion of the myc transcriptional activation domain by p107. *Science* 264: 251–254

Gupta S, Seth A, Davis RJ (1993): Transactivation of gene expression by Myc is inhibited by mutation at the phosphorylation sites Thr-58 and Ser-62. *Proc Natl Acad Sci USA* 90: 3216–3220

Gusse M, Ghysdael J, Evan G, Soussi T, Mechali M (1989): Translocation of a store of maternal cytoplasmic c-myc protein into nuclei during early development. *Mol Cell Biol* 9: 5395–5403

Halazonetis TD, Kandil AN (1991): Determination of the c-Myc DNA binding site. *Proc Natl Acad Sci USA* 6162–6166

Hann SR, King MW, Bentley DL, Anderson CW, Eisenman RN (1988): A non-AUG translational initiation in c-*myc* exon 1 generates an N-terminally distinct protein whose synthesis is disrupted in Burkitt's lymphomas. *Cell* 52: 185–195

Hann SR, Dixit M, Sears RC, Sealy L (1994): The alternatively initiated c-Myc proteins differentially regulate transcription through a noncanonical DNA-binding site. *Genes Dev* 8: 2441–2452

Hanson KD, Shichiri M, Follansbee MR, Sedivy JM (1994): Effects of c-myc expression on cell cycle progression. *Mol Cell Biol* 14: 5748–5755

Hateboer G, Timmers H, Rustgi AK, Billaud M, Van'tVeer LJ, Bernards R (1993): TATA-binding protein and the retinoblastoma gene product bind to overlapping epitopes on c-Myc and adenovirus E1A protein. *Proc Natl Acad Sci USA* 90: 8489–8493

Heaney ML, Pierce J, Parsons JT (1986): Site-directed mutagenesis of the gag-myc gene of avian myelocytomatosis virus 29: biological activity and intracellular localization of structurally altered proteins. *J Virol* 60: 167–176

Henriksson M, Lüscher B (1996): Proteins of the Myc network: essential regulators of cell growth and differentiation. *Adv Canc Res* 68: 109–182

Henriksson M, Classon M, Ingvarsson S, Koskinen P, Sumegi J, Klein G, Thyberg J (1988): Elevated expression of c-myc and N-myc produces distinct changes in nuclear fine structure and chromatin organization. *Oncogene* 3: 587–591

Henriksson M, Bakardjiev A, Klein G, Lüscher B (1993): Phosphorylation sites mapping in the N-terminal domain of c-myc modulate its transforming potential. *Oncogene* 8: 3199–3209

Hermeking H, Eick D (1994): Mediation of c-Myc-induced apoptosis by p53. *Science* 265: 2091–2093

Hermeking H, Wolf DA, Kohlhuber F, Dickmanns A, Biollaud M, Fanning E, Eick D (1994): Role of c-myc in simian virus 40 large tumor antigen-induced DNA synthesis in quiescent 3T3-L1 mouse fibroblasts. *Proc Natl Acad Sci USA* 91: 10412–10416

Hoang AT, Cohen KJ, Barrett JF, Bergstrom DA, Dang CV (1994): Participation of cyclin A in Myc-induced apoptosis. *Proc Natl Acad Sci USA* 91: 6875–6879

Hoang AT, Lutterbach B, Lewis BC, Yano T, Chou T-Y, Barrett JF, Raffeld M, Hann SR, Dang CV (1995): A link between increase transforming activity of lymphoma-derived *MYC* mutant alleles, their defective regulation by p107, and altered phosphorylation of the c-Myc transactivation domain. *Mol Cell Biol* 15: 4031–4042

Hopewell R, Ziff EB (1995): The nerve growth factor-responsive PC12 cell line does not express the Myc dimerization partner Max. *Mol Cell Biol* 15: 3470–3478

Hu Y-F, Lüscher B, Admon A, Mermod N, Tjian R (1990): Transcription factor AP-4 contains multiple dimerization domains that regulate dimer specificity. *Genes Dev* 4: 1741–1752

Hurlin PJ, Ayer DE, Grandori C, Eisenman RN (1994): The Max transcription factor network: involvement of Mad in differentiation and an approach to identification of target genes. Cold Spring Harb. *Symp Quant Biol* 59: 109–116

Hurlin PJ, Queva C, Koskinen PJ, Steingrimsson E, Ayer DE, Dopeland NG, Jenkins NA, Eisenman RN (1995): Mad3 and Mad4: novel Max-interacting transcriptional repressors that suppress c-myc dependent transformation and are expressed during neural and epidermal differentiation. *EMBO J* 14: 5646–5659

Iguchi-Ariga SMM, Okazaki T, Itani T, Ogata M, Sato Y, Ariga H (1988): An initiation site of DNA replication with transcriptional enhancer activity present in the c-myc gene. *EMBO J* 7: 3135–3142

Inghirami G, Grignani F, Sternas L, Lombardi L, Knowles DM, Dalla-Favera R (1990): Down-regulation of LFA-1 adhesion receptors by c-Myc oncogene in human B lymphoblastoid cells. *Science* 250: 682–686

Jansen-Durr P, Meichle A, Steiner P, Pagano M, Finke K, Botz J, Wessbecher J, Draetta G, Eilers M (1993): Differential modulation of cyclin gene expression by MYC. *Proc Natl Acad Sci USA* 90: 3685–3689

Jones PA (1986): DNA methylation and cancer. *Cancer Res* 46: 461–466

Jones PA, Buckley JD (1990): The role of DNA methylation in cancer. *Adv Canc Res* 54: 1–23

Kato G, Lee WMF, Chen L, Dang C (1992): Max: Functional domains and interaction with c-Myc. *Genes Dev* 6: 81–92

Kato GJ, Barrett J, Villa-Garcia M, Dang CV (1990): An amino-terminal c-Myc domain required for neoplastic transformation activates transcription. *Mol Cell Biol* 10: 5914–5920

Kelly K, Siebenlist U (1986): The regulation and expression of c-myc in normal and malignant cells. *Ann Rev Immunol* 4: 317–338

Kerkhoff E, Bister K, Klempnauer K-H (1991): Sequence-specific DNA-binding by Myc proteins. *Proc Natl Acad Sci USA* 88: 4323–4327

Kim YH, Buchholz MA, Chrest FJ, Nordin AA (1994): Up-regulation of c-myc induces the gene expression of the murine homologues of p34cdc2 and cyclin-dependent kinase-2 in T lymphocytes. *J Immunol* 152: 4328–4335

Koskinen PJ, Sistonen L, Evan G, Morimoto M, Alitalo K (1991): Nuclear colocalization of cellular and viral *myc* proteins with HSP70 in *myc*-overexpressing cells. *J Virol* 65: 842–851

Koskinen PJ, Vastrik I, Makela TP, Eisenman RN, Alitalo K (1994): *Cell Growth Diff* 5: 313–320

Koskinen PJ, Ayer DE, Eisenman RN (1995): Repression of Myc-Ras cotransformation by Mad is mediated by multiple protein-protein interactions. *Cell Growth Diff* 6: 623–629

Kretzner L, Blackwood EM, Eisenman RN (1992): Myc and Max proteins possess distinct transcriptional activities. *Nature* 359: 426–429

Land H, Parada LF, Weinberg RA (1983): Tumorigenic conversion of primary embryo fibroblasts requires at least two cooperating oncogenes. *Nature* 304: 596–602

Landschultz WH, Johnson PF, McKnight SL (1987): The leucine zipper: a hypothetical structure common to a new class of DNA binding proteins. *Science* 240: 1759–1764

Lane DP (1992): p53, guardian of the genome. *Nature* 358: 15–16

Lane DP, Lu X, Hupp T, Hall PA (1994): The role of the p53 protein in the apoptotic response. *Philos Trans R Soc Lond B Biol Sci* 345: 277–280

Lane DP, Midgley CA, Hupp TR, Lu X, Vojtesek B, Picksley SM (1995): On the regulation of the p53 tumour suppressor, and its role in the cellular response to DNA damage. *Philos Trans R Soc Lond B Biol Sci* 347: 83–87

Lee LA, Dolde C, Barrett J, Wu CS, Dang CV (1996): A link between c-Myc-mediated transcriptional repression and neoplastic transformation. *J Clin Invest* 97: 1687–1695

Lemaitre JM, Bocquet S, Buckle R, Mechali M (1995): Selective and rapid nuclear translocation of a c-Myc-containing complex after fertilization of *Xenopus laevis* eggs. *Mol Cell Biol* 15: 5054–5062

Li L, Nerlov C, Prendergast G, MacGregor D, Ziff EB (1994): c-Myc activates and represses target gene through the E-box Myc binding site and the core promoter region respectively. *EMBO J* 13: 4070–4079

Lüscher B, Eisenman RN (1990): New light on Myc and Myb. Part I. Myc. *Genes Dev* 4: 2025–2035

Lüscher B, Kuenzel EA, Krebs EG, Eisenman RN (1989): Myc oncoproteins are phosphorylated by casein kinase II. *EMBO J* 8: 1111–1119

Lutterbach B, Hann SR (1994): Hierarchical phosphorylation at N-terminal transformation-sensitive sites in c-Myc protein is regulated by mitogens and in mitosis. *Mol Cell Biol* 14: 5510–5522

Mäkelä TP, Koskinen PJ, Västrik I, Alitalo K (1992): Alternative forms of Max as enhancers or suppressors of Myc-Ras cotransformation. *Science* 256: 373–377

Marcu KB, Bossone SA, Patel AJ (1992): Myc function and regulation. *Ann Rev Biochem* 61: 809–860

Martel C, Lallemand D, Cremisi C (1995): Specific c-myc and max regulation in epithelial cells. *Oncogene* 10: 2195–2205

Martin SJ, Green DR (1995): Protease activation during apoptosis: death by a thousand cuts. *Cell* 82: 349–352

Miltenberger RJ, Sukow KA, Farnham PJ (1995): An E-box-mediated increase in cad transcription at the G1/S-phase boundary is suppressed by inhibitory c-Myc mutants. *Mol Cell Biol* 15: 2527–2535

Moshier JA, Dosescu J, Skunca M, Luk GD (1993): Transformation of NIH/3T3 cells by ornithine decarboxylase overexpression. *Canc Res* 53: 2618–2622

Muhle-Goll C, Nilges M, Pastore A (1995): The leucine zippers of the HLH-LZ proteins Max and c-Myc preferentially form heterodimers. *Biochemistry* 34: 13554–13564

Mukherjee B, Morgenbesser SD, De PR (1992): Myc family oncoproteins function through a common pathway to transform normal cells in culture: cross-interference by Max and transacting dominant mutants. *Genes Dev* 6: 1480–1492

Murre C, McCaw PS, Baltimore D (1989): A new DNA-binding and dimerization motif in immunoglublin enhancer binding, daughterless, MyoD, and Myc proteins. *Cell* 56: 777–783

Negishi Y, Iguchi-Agriga SMM, Ariga H (1992): Protein complexes bearing *myc*-like antigenicity recognize two distinct DNA sequences. *Oncogene* 7: 543–548

O'Shea EK, Rutkowski RH, Kim PS (1989): Evidence that the leucine zipper is a coiled coil. *Science* 245: 538–541

Packham G, Cleveland JL (1994): Ornithine decarboxylase is a mediator of c-Myc-induced apoptosis. *Mol Cell Biol* 14: 5741–5747

Packham G, Cleveland J (1995): c-Myc and apoptosis. *Biochim Biophys Acta* 1242: 11–28

Packham G, Cleveland JL (1996) c-Myc induces apoptosis and cell cycle progression by separable, yet overlapping, pathways. *Oncogene* 13: 461–469

Papas TS, Lautenberger JA (1985): Sequence curiosity in v-*myc* oncogene. *Nature* 318: 237

Perry ME, Levine AJ (1993): Tumor-suppressor p53 and the cell cycle. *Curr Opin Genet Dev* 3: 50–54

Persson H, Gray HE, Godeau F, Braunhut S, Bellvé AR (1986): Multiple growth-associated nuclear proteins immunoprecipitated by antisera raised against human c-*myc* peptide antigens. *Mol Cell Biol* 6: 942–949

Philipp A, Schneider A, Väsrik I, Finke K, Xiong Y, Beach D, Alitalo K, Eilers M (1994): Repression of cyclin D1: a novel function of Myc. *Mol Cell Biol* 14: 4032–4043

Prendergast GC, Cole MD (1989): Posttranscriptional regulation of cellular gene expression by the c-myc oncogene. *Mol Cell Biol* 9: 124–134

Prendergast GC, Ziff EB (1989): DNA binding motif. *Nature* 341: 392

Prendergast GC, Ziff EB (1991): Methylation-sensitive sequence-specific DNA binding by the c-Myc basic region. *Science* 251: 186–189

Prendergast GC, Lawe D, Ziff EB (1991): Association of Myn, the murine homolog of Max, with c-Myc stimulates methylation-sensitive DNA binding and Ras cotransformation. *Cell* 65: 395–407

Prendergast GC, Hopewell R, Gorham B, Ziff EB (1992): Biphasic effect of Max on Myc transformation activity and dependence on N- and C-terminal Max functions. *Genes Dev* 6: 2429–2439

Prochownik EV, Van Antwerp ME (1993): Differential patterns of DNA binding by myc and max proteins. *Proc Natl Acad Sci USA* 90: 960–964

Pulverer BJ, Fisher C, Vousden K, Littlewood T, Evan G, Woodgett JR (1994): Site-specific modulation of c-Myc cotransformation by residues phosphorylated in vivo. *Oncogene* 9: 59–70

Ralston R (1991): Complementation of transforming domains in E1A/myc chimaeras. *Nature* 353: 866–869

Reddy CD, Dasgupta P, Saikumar P, Dudek H, Rauscher FJ, Reddy EP (1992): Mutational analysis of Max: role of basic, helix-loop-helix/leucine zipper domains in DNA binding, dimerization and regulation of Myc-mediated transcriptional activation. *Oncogene* 7: 2085–2092

Reisman D, Elkind NB, Roy B, Beamon J, Rotter V (1993): c-Myc transactivates the p53 promoter through a required downstream CACGTG motif. *Cell Growth Diff* 4: 57–65

Roy A, Carruthers C, Gutjahr T, Roeder RG (1993 a): Direct role for Myc in transcription initiation mediated by interactions with TFII-I. *Nature* 365: 359–361

Roy AL, Malik S, Meisterernst M, Roeder RG (1993 b): An alternative pathway for transcription initiation involving TFII-I. *Nature* 365: 355–359

Rustgi AK, Dyson N, Bernards R (1991): Amino-terminal domains of c-*myc* and N-*myc* proteins mediate binding to the retinoblastoma gene product. *Nature* 352: 541–544

Sakamuro D, Eviner V, Elliott K, Showe L, White E, Prendergast GC (1995): c-Myc induces apoptosis in epithelial cells by p53-dependent and p53-independent mechanisms. *Oncogene* 11: 2411–2418

Sakamuro D, Elliott K, Wechsler R, Prendergast GC (1996): Bin1, a novel Myc-interacting protein with features of a tumor suppressor. *Nature Genet* 14: 69–77

Saksela K, Mäkelä TP, Hughes K, Woodgett JR, Alitalo K (1992): Activation of protein kinase C increase phosphorylation of the L-myc trans-activator domain at a GSK-3 target site. *Oncogene* 7: 347–353

Sawyers CL, Callahan W, Witte ON (1992): Dominant negative MYC blocks transformation by ABL oncogenes. *Cell* 70: 901–1010

Schaeper U, Boyd JM, Verma S, Uhlmann E, Subramanian T, Chinnadurai G (1995): Molecular cloning and characterization of cellular phosphoprotein that interacts with a conserved C-terminal domain of adenovirus E1A involved in negative modulation of oncogenic transformation. *Proc Natl Acad Sci USA* 92: 10467–10471

Schreiber-Agus N, Chin L, Chen K, Torres R, Rao G, Guida P, Skoultchi AI, De Pinho RA (1995): An amino-terminal domain of Mxi1 mediates anti-Myc oncogenic activity and interacts with a homolog of the yeast transcriptional repressor SIN3. *Cell* 80: 777–786

Selvakumaran M, Lin HK, Sjin RT, Reed JC, Liebermann DA, Hoffman B (1994): The novel primary response gene MyD118 and the proto-oncogenes myb, myc, and bcl-2 modulate transforming growth factor beta 1-induced apoptosis of myeloid leukemia cells. *Mol Cell Biol* 14: 2352–2360

Seth A, Alvarez E, Gupta S, Davis RJ (1991): A phosphorylation site located in the N-terminal domain of c-Myc increases transactivation of gene expression. *J Biol Chem* 266: 23521–23524

Seth A, Gupta S, Davis RJ (1993): Cell cycle regulation of the c-Myc transcriptional activation domain. *Mol Cell Biol* 13: 4125–4136

Shantz LM, Pegg AE (1994): Overproduction of ornithine decarboxylase caused by relief of translational repression is associated with neoplastic transformation. *Canc Res* 54: 2313–2316

Shi L, Nishioka WK, Thng J, Bradbury EM, Litchfield DW, Greenberg AH (1994): Premature p34cdc2 activation required for apoptosis. *Science* 263: 1143–1145

Shibuya H, Yoneyama M, Ninomiya-Tsuji J, Matsumoto K, Taniguchi T (1992): IL-2 and EGF receptors stimulate the hematopoietic cell cycle via different signaling pathways: demonstration of a novel role for c-Myc. *Cell* 70: 57–67

Shichiri M, Hanson KD, Sedivy JM (1993): Effects of c-myc expression on proliferation, quiescence, and the G0 to G1 transition in nontransformed cells. *Cell Growth Diff* 4: 93–104

Showe LC, Ballantine M, Nishikura K, Erikson J, Kaji H, Croce CM (1985): Cloning and sequencing of a c-myc oncogene in a Burkitt's lymphoma cell line that is translocated to a germ line alpha switch region. *Mol Cell Biol* 5: 501–509

Shrivastava A, Saleque S, Kalpana GV, Artandi S, Goff SP, Calame K (1993): Inhibition of transcriptional regulator Yin-Yang-1 by association with c-Myc. *Science* 262: 1889–1892

Sklar MD, Thompson E, Welsh MJ, Liebert M, Harney J, Grossman HB, Smith M, Prochownik EV (1991): Depletion of c-*myc* with specific antisense sequences reverses the transformed phenotype in *ras* oncogene-transformed NIH 3T3 cells. *Mol Cell Biol* 11: 3699–3710

Smith MJ, Charron-Prochownik DC, Prochownik EV (1990): The leucine zipper of c-Myc is required for full inhibition of erythroleukemia differentiation. *Mol Cell Biol* 10: 5333–5339

Spector DL, Watt RA, Sullivan NF (1987): The v- and c-*myc* oncogene proteins colocalize in situ with small nuclear ribonucleoprotein particles. *Oncogene* 1: 5–12

Spencer CA, Groudine M (1991): Control of c-*myc* regulation in normal and neoplastic cells. *Adv Canc Res* 56: 1–48

Steiner P, Philipp A, Lukas J, Godden-Kent D, Pagano M, Mittnacht S, Bartek J, Eilers M (1995): Identification of a Myc-dependent step during the formation of active G1 cyclin-cdk complexes. *EMBO J* 14: 4814–4826

Stone J, de Lange T, Ramsay G, Jakobovits E, Bishop JM, Varmus H, Lee W (1987): Definition of regions in human c-*myc* that are involved in transformation and nuclear localization. *Mol Cell Biol* 7: 1697–1709

Street AJ, Blackwood E, Lüscher B, Eisenman RN (1990): Mutational analysis of the carboxyterminal casein kinase II phosphorylation site in human c-*myc*. *Curr Top Microbiol Immunol* 166: 251–258

Suen T-C, Hung M-C (1991): c-*myc* reverses *neu*-induced transformed morphology by transcriptional repression. *Mol Cell Biol* 11: 354–362

Suetake I, Tajima S, Asano A (1993): Identification of two novel mouse nuclear proteins that bind selectively to a methylated c-Myc recognizing sequence. *Nuc Acids Res* 21: 2125–2130

Tobias KE, Shor J, Kahana C (1995): c-Myc and Max transregulate the mouse ornithine decarboxylase promoter through interaction with two downstream CACGTG motifs. *Oncogene* 11: 1721–1727

Vastrik I, Makela TP, Koskinen PJ, Alitalo K (1995): Determination of sequences responsible for the differential regulation of Myc function by delta Max and Max. *Oncogene* 11: 553–560

Wagner AJ, Meyers C, Laimins LA, Hay N (1993): c-*myc* induces the expression and activity ornithine decarboxylase. *Cell Growth Diff* 4: 879–883

Wagner AJ, Kokonitis JM, Hay N (1994): Myc-mediated apoptosis requires wild-type p53 in a manner independent of cell cycle arrest and the ability of p53 to induce p21 waf1/cip1. *Genes Dev* 8: 2817–2830

Wechsler DS, Dang CV (1992): Opposite orientations of DNA bending by c-Myc and Max. *Proc Natl Acad Sci USA* 89: 7635–7639

Wechsler DS, Papoulas O, Dang CV, Kingston RE (1994): Differential binding of c-Myc and Max to nucleosomal DNA. *Mol Cell Biol* 14: 4097–4107

Wenzel A, Cziepluch C, Hamann U, Schürmann J, Schwab M (1991): The N-Myc oncoprotein is associated in vivo with the phosphoprotein Max(p20/22) in human neuroblastoma cells. *EMBO J* 10: 3703–3712

Whyte M, Evan G (1995): The last cut is the deepest. *Nature* 376: 17–18

Yang B-S, Gilbert JD, Freytag SO (1993): Overexpression of Myc suppresses CCAAT Transcription Factor/Nuclear Factor 1-dependent promoters in vivo. *Mol Cell Biol* 13: 3093–3102

Yano T, Sander CA, Clark HM, Dolezal MV, Jaffe ES, Raffeld M (1993): Clustered mutations in the second exon of the MYC gene in sporadic Burkitt's lymphoma. *Oncogene* 8: 2741–2748

Zervos AS, Gyuris J, Brent R (1993): Mxi1, a protein that specifically interacts with Max to bind Myc-Max recognition sites. *Cell* 72: 223–232

Zhu L, van den Heuvel S, Helin K, Fattaey A, Ewen M, Livingston D, Dyson N, Harlow E (1993): Inhibition of cell proliferation by p107, a relative of the retinoblastoma protein. *Genes Dev* 7: 1111–1125

Zoidl G, Brockmann D, Esche H (1993): Deletion of the β-turn/α-helix motif at the exon 2/3 boundary of human c-Myc leads to loss of its immortalizating function. *Gene* 131: 269–274

2

Oncogenes as Transcriptional Regulators
Vol. 1: Retroviral Oncogenes
ed. by M. Yaniv and J. Ghysdael

The ETS Family of Transcriptional Regulators

JACQUES GHYSDAEL AND ANTHONY BOUREUX

Introduction

The identification of the v-*ets* oncogene of avian leukemia virus E26 (Leprince et al, 1983; Nunn et al, 1983) has set the stage for the definition of a large protein family of over 30 members characterized by the presence of a conserved domain of about 85 amino acids, the ETS domain (Figure 2.1). For several years and despite the molecular cloning of the cDNAs of several *ETS* family members including c-*ets-1*, the cellular homolog of v-*ets* (Duterque-Coquillaud et al, 1988; Watson et al, 1988a; Leprince et al, 1988), c-*ets-2* (Boulukos et al, 1988; Watson et al, 1988a), *erg* (*e*ts *r*elated *g*ene) (Reddy et al, 1987) and *elk* (*e*ts *l*ike gene) (Rao et al, 1989), the analysis of the deduced amino acid sequence of the corresponding proteins failed to give any clue as to the nature of their biochemical function. The description of ETS-1 as a nuclear, chromatin-associated protein endowed with general DNA binding activity (Pognonec et al, 1989) and the fact that these properties all depended upon the integrity of the ETS domain (E domain) (Boulukos et al, 1989) were the first indications for a possible nuclear function for ETS proteins. Definitive experimental evidence supporting this view was provided by a series of independent observations which identified ETS-1 as a sequence-specific DNA binding protein and transcriptional activator of viral and cellular promoters (Bosselut et al, 1990; Gunther et al, 1990; Ho et al, 1990; Wasylyk et al, 1990) and the concomitant identification of the SV40 PU-box transcription factor as an ETS protein (Klemsz et al, 1990).

ETS proteins can be subdivided into subclasses, based upon the position and sequence similarities in their ETS domain and the presence of additional conserved subdomains essential to their functional specificity (Figure 2.1). For example, the ETS domain of PU-1/Spi-1 and the related Spi-B are the most divergent (36% identity) from the prototypic v-*ets* domain, a property which is reflected by the ability of PU-1 to specifically bind DNA sequences which deviates from consensus ETS binding sites (EBSs) and its ability to bind RNA (Hallier et al, 1996). Also, the

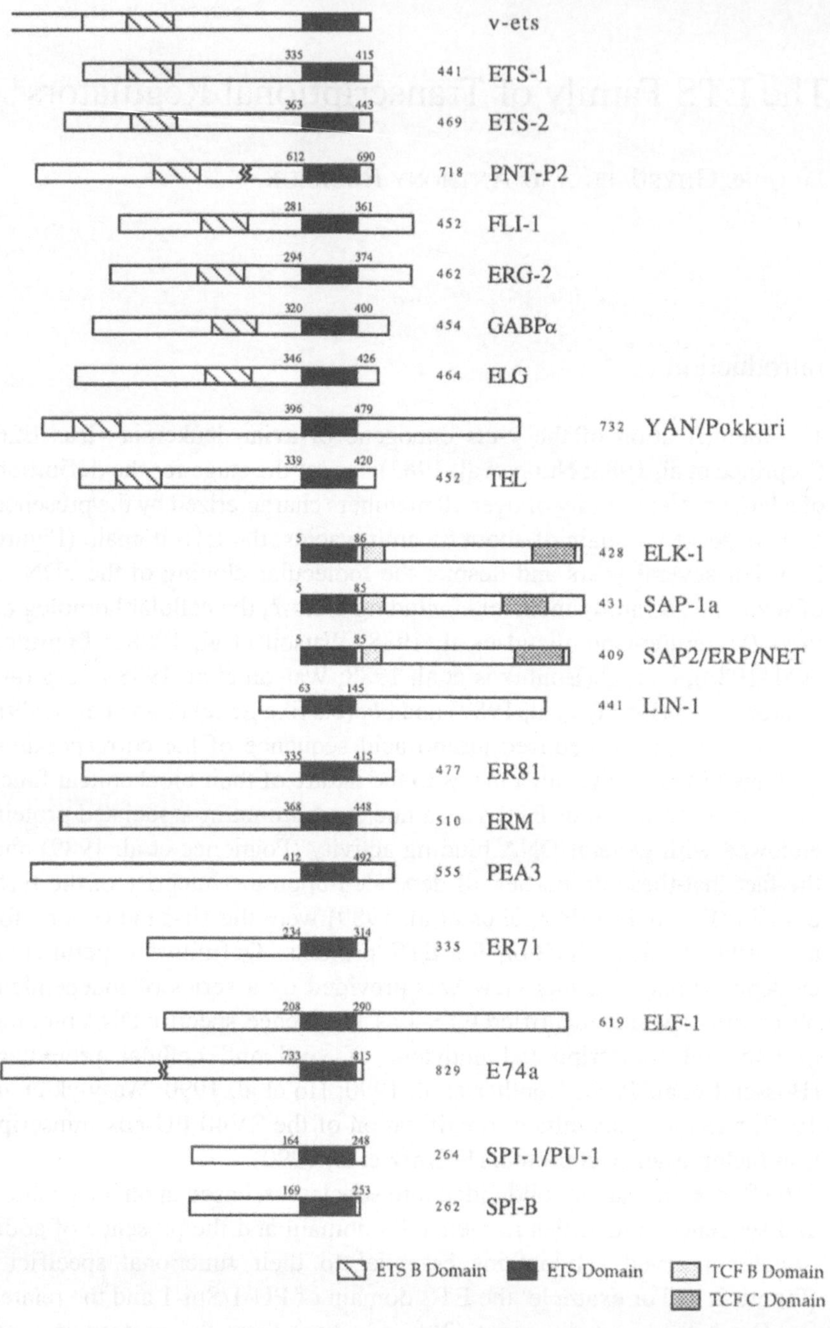
v-ets
335 415
441 ETS-1
363 443
469 ETS-2
612 690
718 PNT-P2
281 361
452 FLI-1
294 374
462 ERG-2
320 400
454 GABPα
346 426
464 ELG
396 479
732 YAN/Pokkuri
339 420
452 TEL
5 86
428 ELK-1
5 85
431 SAP-1a
5 85
409 SAP2/ERP/NET
63 145
441 LIN-1
335 415
477 ER81
368 448
510 ERM
412 492
555 PEA3
234 314
335 ER71
208 290
619 ELF-1
733 815
829 E74a
164 248
264 SPI-1/PU-1
169 253
262 SPI-B
ETS B Domain
ETS Domain
TCF B Domain
TCF C Domain

N-terminal positioning of the ETS domain of ELK-1, SAP-1a and SAP-2/ERP/NET is a characteristic feature of ETS proteins which display ternary complex factor activity (TCF) with SRF on the c-*fos* serum response element (SRE). These proteins are also characterized by the presence of other conserved domains (domains B and C) which play an essential role in their functional interaction with SRF and activation through the SRE (see section "Transcriptional Regulation"). Several ETS proteins including ETS-1, ETS-2, ERG-2, FLI-1, GABPα and TEL in vertebrates and PNT-P2, YAN/Pokkuri and ELG in *Drosophila* share a domain of about 65 amino acid residues (Figure 2.3). This domain has recently been shown to be sufficient for homotypic oligomerization of TEL (Jousset et al, 1997). Interestingly, the homologous domains of other vertebrate ETS proteins are not able to mediate homotypic oligomerization (Jousset et al, 1997), but are likely to encode specialized protein-protein interaction interfaces important to the specificity of otherwise highly similar proteins.

The Lessons of Genetics

Seven genes belonging to the ETS family have been identified so far in *Drosophila*, including *Pointed* (*PntP1* and *PntP2*; *D-ets2*) (Pribyl et al, 1988; Klambt, 1993), D-*elg* (Pribyl et al, 1991), *E74* (Burtis et al, 1990), *D-ets-3, D-ets-3, D-ets-6* (Chen et al, 1992) and *YAN/Pokkuri* (Lai and Rubin, 1992; Tei et al, 1992). The PNTP2 protein is similar in structure and function to vertebrate ETS-1 and ETS-2 (Klambt, 1993, see below). *D-Ets-3* and *D-Ets-6*, which encode proteins highly related to vertebrate FLI-1 and ERG-2 respectively, are uniformly expressed in early embryogenesis and become restricted to certain neurons in the ventral nervous system at later stages (Chen et al, 1992).

D-elg, the product of which shares extensive sequence identity with vertebrate GABPα, is a maternal effect gene in the egg and is later expres-

Figure 2.1 Schematic family portrait of ETS proteins. Representative members of the ETS family are depicted according to the alignment of their ETS domain. Position of ETS domain and the presence of specific protein subdomains has allowed to define several subclasses. In this figure, members of subclass I include ETS-1, ETS-2 and PNT-P2; subclass II: FLI-1 and ERG-2; subclass III: GABPα and ELG; subclass IV: YAN/Pokkuri; subclass V: TEL; subclass VI (TCF subfamily): ELK-1, SAP-1a, SAP2/ERP/NET; subclass VII: ER81, ERM and PEA3; subclass VIII: ELF-1 and E74; subclass IX: SPI-1/PU-1 and SPI-B. Other subclasses exist but are so far represented by a single member (ER71; LIN-1). The ETS domain is shown as a black box. The amino-terminal domain conserved in subclasses I- V is shown as a hatched box. The B and C domains of TCFs are shown as stippled boxes.

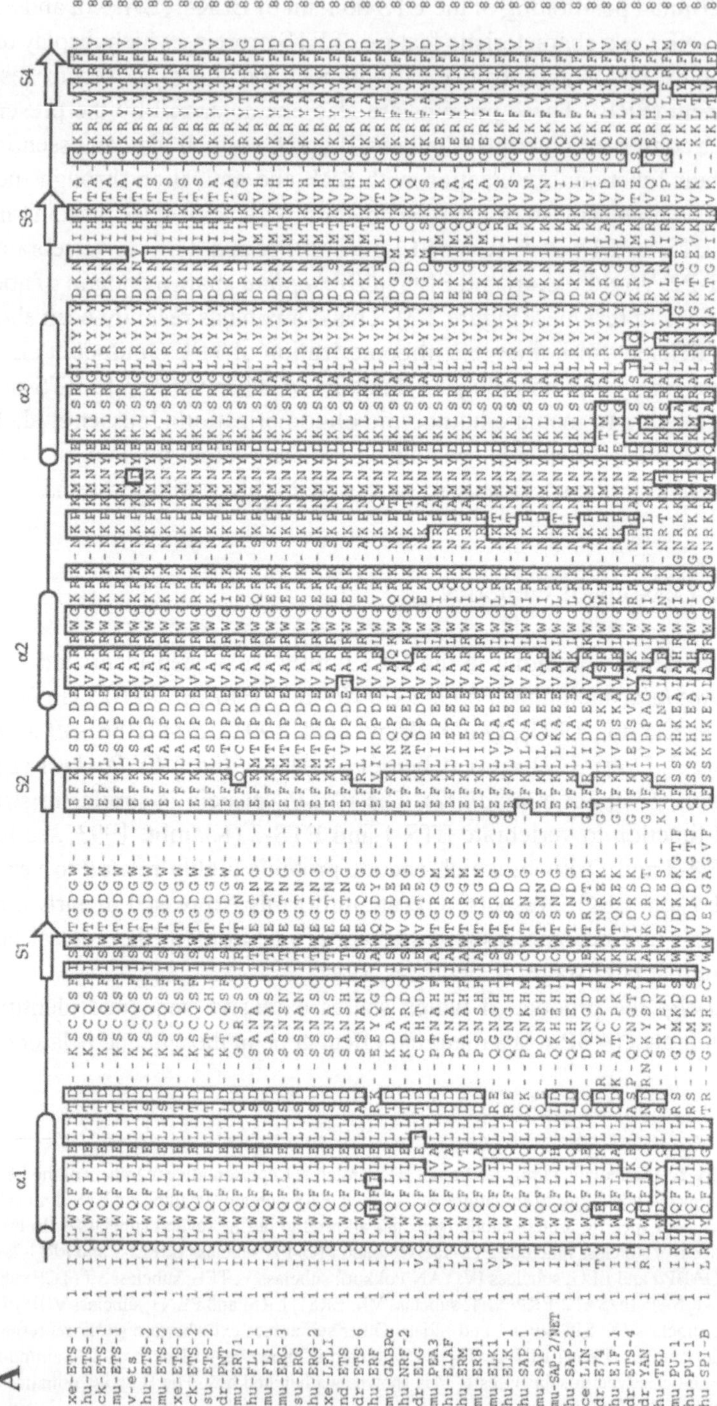
A
α1
S1
S2
α2
S3
α3
S4

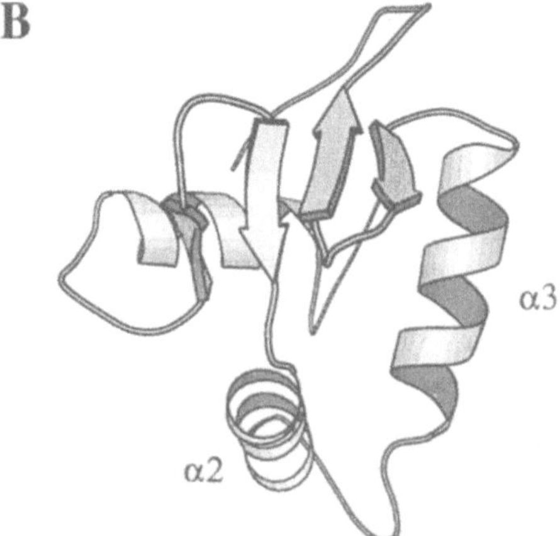

Figure 2.2 Multiple amino acid sequence alignment of the ETS domain of ETS protein and three-dimensional structure of the ETS domain of FLI-1.
Panel A: Primary amino acid sequence alignment of ETS domains. The optimal alignment was obtained by using the Clustal W program (Thompson et al, 1994). Boxed amino acids denote amino acid identity in at least 75 % of all sequences. The secondary structure is indicated on top by barrels for α-helices and arrows for β-sheets. The protein sequences are extracted from the Swissprot or EMBL/Genbank database sequence and include xe-ETS-1 (Stiegler et al, 1990), hu-ETS-1 (Watson et al, 1988 a; Reddy and Rao, 1988), ck-ETS-1 (Watson et al, 1988 b; Duterque-Coquillaud et al, 1988; Chen, 1988), mu-ETS-1 (Gunther et al, 1990), E26 v-ets (Nunn et al, 1983), hu-ETS-2 (Watson et al, 1990), mu-ETS-2 (Watson et al, 1988 a), xe-ETS-2 (Wolff et al, 1991), ck-ETS-2 (Boulukos et al, 1988), su-ETS-2 (Chen et al, 1988), dr-PNT (Klambt, 1993), mu-ER71 (Brown and McKnight, 1992), hu-FLI-1 (Delattre et al, 1992; Watson et al, 1992), mu-FLI (Ben-David et al, 1991; Zhang et al, 1993), mu-ERG-2 (Rivera et al, 1993), su-ERG (Qi et al, 1992), hu-ERG-2 (Rao et al, 1987), xe-LFLI (Meyer et al, 1993), nd-ETS (Lelievre-Chotteau et al, 1994), dr-ETS-6 (Chen et al, 1992), hu-ERF (Sgouras et al, 1995), mu-GABPα (LaMarco et al, 1991), hu-NRF-2 (Watanabe et al, 1993; Virbasius et al, 1993 b), dr-ELG (Pribyl et al, 1991; The et al, 1992), mu-PEA3 (Xin et al, 1992), hu-E1AF (Higashino et al, 1993), hu-ERM (Monte et al, 1994), mu-ER81 (Brown and McKnight, 1992), mu-ELK1 (Giovane et al, 1994), hu-ELK-1 (Rao et al, 1989; Janknecht and Nordheim, 1992), hu-SAP-1 (Dalton and Treisman, 1994; Dalton and Treisman, 1992), mu-SAP-1 (Giovane et al, 1994), mu-SAP-2/NET (Giovane et al, 1994; Lopez et al, 1994), hu-SAP-2 (Giovane et al, 1994), ce-LIN-1 (Beitel et al, 1995), dr-E74 (Burtis et al, 1990), hu-ELF-1 (Leiden et al, 1992; Wang et al, 1993), dr-ETS-4 (Chen et al, 1992), dr-YAN (Tei et al, 1992), hu-TEL (Golub et al, 1994), mu-PU1 (Moreau-Gachelin et al, 1989; Klemsz et al, 1990; Paul et al, 1991), hu-PU1 (Ray et al, 1990), hu-SpiB (Ray et al, 1992). (ce: *Caenorhabditis elegans*; ck: *Gallus gallus*; dr: *Drosophila melanogaster*; hu: *Homo sapiens*; mu: *Mus musculus*; nd: *Nereis diversicolor*; su: *Lytechinus variegatus*, xe: *Xenopus laevis*).
Panel B: Ribbon diagram of the winged HTH motif of the DNA binding of FLI-1 (Liang et al, 1994). The structure was extracted from PDB database (accession number 1FLI) and the figure was drawing using Molscript (Kraulis, 1991).

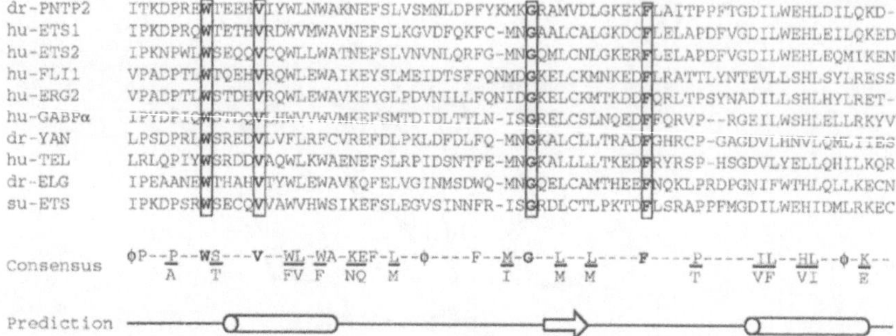

Figure 2.3 Multiple sequence alignment of the amino-terminal conserved domain of subclasses I–V ETS proteins. The CLUSTAL W program (Thompson et al, 1994) was used to align the amino acid sequences of human ETS-1, ETS-2, FLI-1, ERG-2, GABPα and TEL; *D. melanogaster* PNTP2, YAN/POK and ELG; sea urchin *S. purpuratus* ETS protein (NCBI, L19541). Invariant residues are boxed and positions displaying strong conservation or conservative substitution are indicated in the consensus lane. The secondary structure prediction was made using the SIMPA 95 program (Levin and Garnier, 1988) and is indicated on bottom by α-helix (barrel) and β-sheet (arrow) (Levin et al, 1993; Jousset et al, 1997).

sed in a relatively uniform pattern throughout early embryonic development (Pribyl et al, 1991). Complementation rescue experiments and sequence analyses show that the female sterile mutant tiny eggs (*tne*) is an allele of *D-elg* (Schulz et al, 1993), indicating an essential role for *D-elg* in various aspects of oogenesis, including follicle cell migration and chromosome decondensation in nurse cells. The sequence of a mutant *tne* allele showed the substitution for cysteine of an invariant tyrosine residue in the ETS domain, suggesting that D-ELG DNA binding activity is important to D-ELG function (Schulz et al, 1993).

E74 is a gene which was originally identified as the product of the E74 EF early puff induced by ecdysone in third instar larva (Burtis et al, 1990). Later studies showed *E74* to be transcribed in response to the six major ecdysone pulses during *Drosophila* development (Thummel et al, 1990). The gene is composed of two transcription units, E74A and E74B, each encoding a protein with unrelated amino-terminal sequences linked to a common carboxy-terminal domain (Burtis et al, 1990). The common carboxy-terminal domain includes the ETS domain and both E74A and E74B proteins bind classical EBSs in vitro (Urness and Thummel, 1990). The study of loss of function mutants of either *E74A* or *E74B* indicate that *E74* is dispensable for early embryonic development up to the pre-pupal stage, but is required for metamorphosis (Fletcher and Thummel, 1995). Furthermore, detailed analyses of several mutant alleles of *E74B* suggest a role for this protein in preventing the ecdysone-induced loss of larval muscles essential for the shaping of the pre-pupal body. The effect of *E74* mutations on gene expression as assayed by the analysis of the puffing pattern of polytene chromosomes indicates that E74A plays a role in proper induction of a subset of the late puffs (Fletcher and Thummel, 1995). This, together with the fact that the E74A protein is found to physically associate with both early and late puffs suggest a direct role for E74A in the induction of late puff formation and, possibly, in the repression of early puffs.

The analysis of the mutations affecting the development of presumptive R7 cells into differentiated photoreceptor neurons in response to the engagement of the sevenless receptor tyrosine kinase has played a major role in the identification of the components of the Ras/Raf/MAP kinase signaling pathway (Zipursky and Rubin, 1994). Two ETS family members, namely *PNTP2* and *YAN/Pokkuri* have been shown to play a key role in this process. *Pointed* was initially identified as a gene required for development of the *Drosophila* embryonic nervous system (Klambt, 1993). The gene is expressed in undifferentiated cells of the eye imaginal disk as two transcriptional units which encode proteins – the PNTP1 and PNTP2 proteins – which share a common carboxy-terminal domain, including the ETS domain, and unique amino-terminal moieties (Klambt, 1993). PNTP2 belongs to a subgroup of the ETS family characterized by an amino-terminal conserved domain of about 100 residues which has been referred to as the B domain (Boulukos et al, 1989) or the *pointed* domain (Klambt, 1993) (Figure 2.1 and 2.3). In both its *pointed* and ETS domains, PNTP2 is most homologous to vertebrate ETS-1 and ETS-2, a feature which is reflected by the ability of chicken ETS-1 to partially replace PNTP2 function in R7 photoreceptor cell development (O. Albagli, personal communication).

Genetic analyses show that *PNTP2* is required for R7 photoreceptor cell determination and acts downstream of Ras in the *sevenless* signaling pathway. Specifically, mutations that reduce *PNTP2* function were found to be suppressors of gain of function mutations in known components of the sevenless signaling pathway including mutations affecting the Rolled/ERK-A MAP kinase (Brunner et al, 1994; O'Neill et al, 1994). Conversely, the phenotype of hypomorphic alleles of several components of this pathway is enhanced when combined with a loss-of-function allele or a decreased dosage of *PNTP2* (O'Neill et al, 1994). Several lines of evidence suggest that the activity of the PNTP2 protein is under direct control of the Rolled/ERK-A activity. First, PNTP2 is phosphorylated in vitro by an activated version of Rolled/ERK-A (Brunner et al, 1994; O'Neill et al, 1994) on a single threonine residue (T151) located at the amino-terminal border of the *pointed* domain. Second, expression of activated forms of Ras or Rolled/ERK-A was found to increase the transcriptional activity of PNTP2 towards EBS-containing model reporter genes whereas the activity of a mutant PNTP2 in which T151 was replaced by alanine was unaffected (O'Neill et al, 1994). Third, targeted expression of a mutant PNTP2 carrying a (T151A) substitution in R7 precursor cells of transgenic flies was found to block normal R7 photoreceptor cell development in a dominant negative manner (Brunner et al, 1994). Interestingly, this residue is conserved in the corresponding domains of ETS-1 and ETS-2 and also plays an important role in the regulation of the activity of ETS-1 and ETS-2 in response to Ras signaling pathway in mammalian cells (see section "Transcriptional Regulation").

In contrast to *pointed, YAN/Pokkuri* is genetically defined as a cell-autonomous negative regulator of R7 photoreceptor cell development (Tei et al, 1992; Lai and Rubin, 1992). Loss of function mutant alleles of *yan* result in the generation of supernumerary R7 cells. Although this phenotype can occur in the absence of *sevenless* function, *yan* was found to interact genetically with Ras and downstream components of the Ras signaling pathway (Lai and Rubin, 1992; O'Neill et al, 1994). In flies carrying a mutant allele of *yan*, development of R7 photoreceptor cells still depends upon PNTP2 function since loss of function of both *yan* and *pointed P2* results in a *sevenless*-like phenotype (Brunner et al, 1994). Biochemically, YAN functions as a repressor of PNTP1 and PNTP2 transcriptional activity of model EBS-based reporter genes and is negatively regulated by the expression of activated forms of RAS and Rolled/ERK-A (O'Neill et al, 1994; Treier et al, 1995). YAN contains eight putative consensus Rolled/ERK-A phosphorylation sites and is a substrate for phosphorylation by this protein kinase in vitro (Brunner et al, 1994; O'Neill et al, 1994), suggesting that the repressor function of YAN is relieved by phosphorylation. In line with this notion, transcriptional repression by a YAN protein in which all putative Rolled/ERK-A phosphorylation sites are changed to alanine is found to be impaired in its response to RAS activation (Rebay and Rubin, 1995). Eye-specific expression of the YAN phosphorylation mutant – but not of wild type YAN – in transgenic flies results in severe impairment of ommatidial development and massive cell death of the cells of the eye imaginal disc (Rebay and Rubin, 1995). Further comparison of the properties of wild-type YAN and YAN phosphorylation mutants in transfected cells indicates that activation of the RAS signaling pathway and YAN phosphorylation correlates with the alteration of its subcellular localization from the nucleus to the cytoplasm, as well as with a decreased stability of the protein (Rebay and Rubin, 1995). Taken together, these observations form the basis of a model in which engagement of the *sevenless* receptor pathway results in the phosphorylation of PNTP2 and YAN by Rolled/ERK-A. These phosphorylation events result in the relief of YAN-mediated repression of R7 cells development and the concomitant activation of the transcriptional properties of PNTP2. Such a dual control might be essential to insure the precise and timely-controlled activation of the as yet unidentified target genes that specify R7 photoreceptor cell development. Expression of a constitutively activated form of JUN in ommatidial precursor cells is also sufficient to induce R7 photoreceptor cell development (Treier et al, 1995). Since PNTP2, PNTP1 and JUN co-operate to activate model reporter genes controlled by composite EBS/AP1 elements and since YAN can antagonize this interaction, it is likely that the activity of at least some of the target genes essential to photoreceptor cell development is controlled by composite EBS/AP1 elements (Treier et al, 1995).

In *Coenorhabditis elegans*, three of the six hypodermal blast cells (vulval precursor cells) adopt a non vulval fate, whereas the three others adopt a vulval fate. The latter developmental pathway is controlled both by the *Lin 3* (EGF)/*Let 23* (EGF-R) Ras-dependent signaling pathway and a *Lin 12*-dependent lateral signaling between vulval precursor cells (Eisenmann and Kim, 1994). In a way reminiscent of YAN function in *Drosophila* R7 cell fate determination, an ETS protein encoded by *lin-1* was found to be a negative regulator of vulval cell fate since, in *lin-1* null mutants, virtually all six vulval precursor cells adopt a vulval cell fate (Beitel et al, 1995). Genetic analyses have shown that LIN-1 acts downstream of the MPK-1 MAP kinase in the LIN 3/LET 23 signal transduction pathway (Lackner et al, 1994; Wu et al, 1994). However, in *lin-1* null mutants, vulval precursor cells were found to still respond to the LIN 3/LET 23 signaling pathway, indicating that *lin-1* either defines only one branch of a Ras-controled pathway or that *lin-1* identifies a parallel pathway which antagonizes events downstream of MPK-1 (Beitel et al, 1995). The ETS domain of LIN-1 is more closely related to that of the ETS proteins of the TCF subfamily and, like them, is located in the amino-terminal moiety of the molecule. LIN-1 is nevertheless distinct from the TCFs since it does not share the B and C domains of homology characteristic of this family of ETS proteins (Beitel et al, 1995). LIN-1 contains 18 putative consensus sites for protein kinases of the MAP kinase family, but it remains to be analyzed whether LIN-1 is a direct substrate of MPK-1 and whether LIN-1 phosphorylation at MPK-1 target sites relieves its activity as repressor of vulval precursor cells development.

The effect of hypomorphic alleles and of the complete loss of function of several ETS proteins has been analyzed in mouse. PU-1/Spi-1 is a transcription factor that is expressed exclusively in B lymphocytes, macrophages, mast cells, neutrophils and a subset of erythroid progenitors

(Klemsz et al, 1990; Galson et al, 1993; Henkel and Brown, 1994), but not T-cells (Ray et al, 1992), suggesting an important role of PU-1 in the development of at least some of these lineages. Consistent with this notion, a null mutation in PU-1 was found to induce a multi-lineage defect in the generation of B and T lymphoid progenitors, monocytes and granulocytes (Scott et al, 1994; McKercher et al, 1996). PU-1 null neonates lack mature macrophages, neutrophils, B-cells and T-cells, but develop low numbers of T-cells and neutrophils after a few days. This indicates that PU-1 function is not absolutely required for myeloid and lymphoid lineage commitment, but plays an essential role in the normal expansion and development of B-cells, neutrophils and monocytes/macrophages (Mc Kercher et al, 1996). The effect of PU-1 loss of function on granulocytic and monocytic progenitors is a cell-autonomous property as analyzed by ex vivo clonogenic assays (Scott et al, 1994). Additional studies show that loss of function of PU-1 in ES cells blocks their normal differentiation in macrophage in response to IL3 and M-CSF (Henkel et al, 1996) and that PU-1 is important for the proliferation of bone marrow-derived macrophages in tissue culture (Celada et al, 1996). PU-1 can regulate the activity of several myeloid and lymphoid-specific regulatory elements, including those of genes encoding specific cell surface adhesion proteins and myeloid-specific growth factor receptors (Table 1; see also section "Tissue-specific transcriptional activation ..."), suggesting that part of the observed phenotypes results from the lack or abnormal expression of several of these genes. In contast to its major effect on B lymphoid and myeloid cells development, loss of PU-1 function has no obvious effect on the erythroid and megakaryocytic lineages (McKercher et al, 1996).

In situ hybridization studies have shown the c-*ets-1* gene to be expressed in multiple tissues, mostly but not exclusively in cells of mesodermal origin during early chicken and mouse embryogenesis (Vandenbunder et al, 1989; Kola et al, 1993; Maroulakou et al, 1994). In contrast, in mouse neonates and adults, expression of c-*ets-1* is largely confined to lymphoid cells (Kola et al, 1993; Maroulakou et al, 1994). Of note, although highly related to c-*ets-1*, c-*ets-2* shows a clearly distinct pattern of expression both during embryonic development and in adult tissues (Maroulakou et al, 1994). ETS-1 is highly expressed in thymocytes and resting peripheral T-cells with higher levels in the single positive CD4+ subclass, suggesting a role of ETS-1 in T-cells maturation and survival (Chen, 1985; Ghysdael et al, 1986; Bhat et al, 1989; Pognonec et al, 1989). Engagement of the T-cell receptor results, within minutes, in the calcium dependent phosphorylation of ETS-1 on several serine residues followed, 1 to 2 hours later, by downregulation of c-*ets-1* gene expression and the induction of c-*ets-2* (Pognonec et al, 1988; Pognonec et al, 1989; Fujiwara et al, 1990;

Table 1. Gene promoters and enhancers regulated by ETS proteins

Cellular gene regulatory region	Known or alleged ETS protein involved	Known or alleged cooperating factor(s)	Reference
Class II MHC DRA promoter (human)	ETS1	nd	(Jabrane-Ferrat and Peterlin, 1994)
Interleukin 2 receptor β chain promoter (human)	ETS1; GABP	nd	(Lin et al, 1993)
Immunoglobulin μH chain enhancer	ETS-1; FLI-1; PU-1	E12	(Nelsen et al, 1993; Rivera et al, 1993)
CD4 promoter/enhancer (human)	ETS-1	nd	(Salmon et al, 1993)
Prolactin gene promoter/enhancer (rat)	ETS-1	PIT1/GHF1	(Conrad et al, 1994)
CD13 aminopeptidase myeloid-specific promoter/ enhancer (human)	ETS-1; ETS-2	Myb	(Shapiro, 1995)
T-cell receptor $\beta 3'$ enhancer (mouse)	ETS-1, ETS-2	CBF/PEBPα	(Prosser et al, 1992; Wotton et al, 1994)
Lck, type I promoter (human)	ETS-1; ETS-2	MYB	(Leung et al, 1993; McCracken et al, 1994)
Megakaryocyte Glycoprotein II B promoter (human)	ETS-1; FLI-1	GATA-1	(Lemarchandel et al, 1993)
T-cell receptor $\alpha 3'$ enhancer (mouse)	ETS-1	CBF/PEBP2α; ATFs; LEF-1	(Ho et al, 1990; Giese et al, 1995)
Endo A type II keratin enhancer (mouse)	ETS-1, ETS-2	nd	(Seth et al, 1994)
Interleukin 2 promoter/enhancer (human)	ETS-1	nd	(Romano-Spica et al, 1995)
Stromelysin promoter (rat)	ETS-1; ETS-2	nd	(Wasylyk et al, 1991)
Jun B promoter (mouse)	ETS-1; ETS-2	nd	(Coffer et al, 1994)
Collagenase promoter (human)	ETS-1; ETS-2	AP1	
GATA-1 promoter (chicken)	ETS-1; ETS-2; FLI-1	nd	(Seth et al, 1993)
Parathyroid Hormone-related Protein (PTHrP) P2 promoter (human)	ETS-1	Sp1	(Dittmer et al, 1994)
CD13/Aminopeptidase N promoter (human)	ETS-1; ETS-2	MYB	(Shapiro, 1995)
Interleukin 2 receptor α chain promoter (human)	ELF-1	nd	(John et al, 1995)

Table 1 (continued)

Cellular gene regulatory region	Known or alleged ETS protein involved	Known or alleged cooperating factor(s)	Reference
Interleukim2 enhancer (human)	ELF-1	nd	(Thompson et al, 1992)
GM-CSF promoter (human)	ELF-1	c-FOS; JUN-B	(Wang et al, 1994)
CD4 upstream enhancer (mouse)	ELF-1	nd	(Wurster et al, 1994)
Immunoglobulin H chain 3′ enhancer	ELF-1	c-FOS; JUN-B	(Grant et al, 1995)
Cytochrome c oxidase subunit V promoter (mouse)	GABP	nd	(Virbasius et al, 1993 b; Virbasius et al, 1993 a)
Cytochrome c oxidase subunit IV promoter (rat; mouse)	GABP	nd	(Virbasius and Scarpulla, 1991; Carter et al, 1992)
ATP synthase β subunit promoter (human)	GABP	nd	(Virbasius and Scarpulla, 1991)
Interleukin 2 receptor γc promoter (human)	GABP	nd	(Markiewicz et al, 1996)
rpL30 and rpL32 ribosomal protein promoter (mouse)	GABP	nd	(Genuario et al, 1993)
Male-specific P450 promoter	GABP	nd	(Yokomori et al, 1995)
6 phosphofructo-2-kinase/fructose 2,6 biphosphate F type promoter (rat)	GABP	nd	(Dupriez et al, 1993)
Fc γRIIIA promoter (mouse)	PU-1	TFE3/USF	(Feinman et al, 1994)
CD 11b (MAC-1α)	PU-1	nd	(Pahl et al, 1993)
Macrophage scavenger receptro promoter/enhancer (human)	PU-1	ETS-2; JUN-B; c-JUN	(Moulton et al, 1994; Wu et al, 1994)
Immunoglobulin V κ 19 promoter (mouse)	PU-1	OCT-2	(Schwarzenbach et al, 1995)
Immunoglobulin κ chain 3′ enhancer (mouse)	PU-1	NF-EM5	(Pongubala et al, 1992)

Table 1 (continued)

Cellular gene regulatory region	Known or alleged ETS protein involved	Known or alleged cooperating factor(s)	Reference
CSF1 receptor promoter (human)	PU-1	nd	(Zhang et al, 1994)
Spi-1/PU-1 promoter (mouse)	PU-1	nd	(Chen et al, 1995a)
Immunoglobulin $\lambda 2$–4 and $\lambda 3$-1 chains enhancers (mouse)	PU-1	NF-EM5	(Eisenbeis et al, 1993)
Immunoglobulin J chain enhancer (mouse)	PU-1	nd	(Shin and Koshland, 1993)
GM-CSF receptor α promoter (human)	PU-1	C/EBPα	(Hohaus et al, 1995)
FcγR1b promoter (human)	PU-1	nd	(Eichbaum et al, 1994)
CD18 (β2 leukocyte integrin) (MAC-1β)	PU-1	GABP	(Rosmarin et al, 1995)
Lysozyme (chicken)	PU-1	nd	(Ahne and Stratling, 1994)
Interleukin 1β (human)	PU-1	nd	(Kominato et al, 1995)
c-*fos* SRE (vertebrates)	ELK1; SAP1a, SAP2	SRF	(Hill et al, 1993; Price et al, 1995; Hipskind and Nordheim, 1991)
Terminal deoxynucleotidyl transferase (mouse)	nd	nd	(Ernst et al, 1993)
Perforin promoter/enhancer (mouse)	nd	nd	(Koizumi et al, 1993)
Cytosolic glutathione peroxidase 3′ enhancer (mouse)	nd	nd	(O'Prey et al, 1993)

Bhat et al, 1990). The serine residues targeted by these calcium-dependent phosphorylation events are all localized in a domain of ETS-1 adjacent to the ETS domain (Rabault and Ghysdael, 1994). The comparison of the properties of intact ETS-1 with an ETS-1 protein in which these residues were replaced by alanine show that the calcium-dependent phosphorylation of ETS-1 results in the inhibition of its specific DNA binding properties (Pognonec et al, 1989; Rabault and Ghysdael, 1994). This has led to a model which proposes that ETS-1 performs a specific function in quiescent T-cells and that inactivation of this function is required for T-cell activation. Further insight into the role of ETS-1 in the lymphoid compartment has been obtained from the analysis of the contribution of embryonic stem cells (ES cells) containing an homozygous deletion of part of the c-*ets-1* gene to the development and function of T- and B-cells in RAG-2 deficient chimeric mice. These studies have shown that ETS-1 is not required for thymocyte maturation as normal ratios of single positive CD4 and CD8 were observed in these chimeras, but is essential to maintain a normal number of thymocytes (Bories et al, 1995; Muthusamy et al, 1995). Ex vivo experiments with splenic T-cells obtained from chimeric animals show that inactivation of ETS-1 function renders quiescent T-cells more susceptible to apoptosis and also defective in their ability to proliferate following T-cell receptor engagement (Bories et al, 1995; Muthusamy et al, 1995). These mice also show a defect in B-cell maturation with the appearance of a novel population of B220low/IgM+ splenic B-cells, composed at least in part of IgM+ IgD- B lymphocytes and a high proportion of plasma cells. These experiments indicate an essential role of ETS-1 in the maintenance of both resting T- and B-cells and in their activation following antigen receptor stimulation.

FLI-1 shows a pattern of expression during mouse embryonic development which is reminiscent of that of ETS-1 (Melet et al, 1996). Mice containing a hypomorphic allele of FLI-1 were accidentally generated by homologous recombination at the *FLI-1* locus. These mice show a reduced cellularity in the thymus with no obvious effects on the distribution of CD4+ and CD8+ T-cells, a defect which can be corrected upon crossing these mutants with mice carrying a normal FLI-1 transgene (Melet et al, 1996). Since the FLI-1 mutation did not affect T-cells quantitatively and qualitatively in peripheral lymphoid organs, these experiments suggest a more specific role of FLI-1 in the survival and/or proliferation of early thymic progenitors (Melet et al, 1996).

DNA Binding by ETS Proteins: The ETS Domain Identifies a Novel Variation of the Winged Helix-Turn-Helix Motif

The 85 amino acids ETS domain is the common feature of all ETS proteins. Although the overall primary sequence identity in this domain ranges from 97% (ETS-1 versus ETS-2) to 35% (ETS-1 versus PU-1), 19 amino acids are invariant among all ETS proteins identified so far (Figure 2.2, see p. 32f).

The ETS domain is both necessary and sufficient for specific binding to DNA and in most cases analyzed, sufficient for nuclear accumulation of ETS proteins (Dalton and Treisman, 1992; Gégonne et al, 1992; Lim et al, 1992; Nye et al, 1992; Wang et al, 1992; Wasylyk et al, 1992; Janknecht and Nordheim, 1992). The in vitro sequence requirement for ETS proteins DNA binding has been determined for a dozen of ETS proteins by in vitro selection of ETS binding sites from random oligonucleotide libraries (Figure 2.4). ETS proteins bind to similar sequences, about 10 nucleotides

Proteins	Consensus Binding Sites									
	-3	-2	-1	**1**	**2**	**3**	**4**	5	6	7
ETS-1	$A/_{G}$	$C/_{G}$	$C/_{A}$	**G**	**G**	**A**	**$A/_{T}$**	G/A	T/C	
ETS-2	C/A	C	A/C	**G**	**G**	**A**	**A/T**	G/A	T/C	
FLI-1	A	C	$C/_{A}$	**G**	**G**	**A**	**A**	$G/_{A}$	$T/_{C}$	A/G
ERG-2		C/G	$C/_{A}$	**G**	**G**	**A**	**A/T**	$G/_{A}$	$T/_{C}$	
ER71	$G/_{C}$	C/G	$C/_{A}$	**G**	**G**	**A**	**A/T**	$G/_{A}$	$T/_{C}$	
ER81	$G/_{A}$	$G/_{C}$	$C/_{A}$	**G**	**G**	**A**	**A/T**	$G/_{A}$	$T/_{C}$	
GABPα	G/A	$C/_{G}$	$C/_{A}$	**G**	**G**	**A**	**$A/_{T}$**	$G/_{A}$	$T/_{C}$	
E74	C/T	C	$C/_{A}$	**G**	**G**	**A**	**A**	$G/_{A}$	T	
ELK-1	$A/_{G}$	C/A	C/A	**G**	**G**	**A**	**A/T**	$^{G}/_{A}$	$^{T}/_{C}$	
SPI-1	A	C	G	**G**	**G**	**A**	**A**	G/C	T	$A/_{G}$
SPI-B	A/T	$G/_{C}$	$A/_{C}$	**G**	**G**	**A**	**A**	$G/_{C}$	T	A/T

Figure 2.4 In vitro Ets binding consensus sites. The consensus sequences for ETS binding sites (EBSs) as identified by DNA-binding site selection. The central conserved GGAA/T is indicated in bold. These sequences were compiled as indicated: ETS-1 (Fisher et al, 1991b; Woods et al, 1992; Nye et al, 1992); ETS-2 (Klemsz et al, 1993; Woods et al, 1992); FLI-1 (Klemsz et al, 1993; Mao et al, 1994); ERG-2 (Murakami et al, 1993); SPI-1, SPI-B (Ray-Gallet et al, 1995); ER71, ER81, GABPα (Brown and McKnight, 1992); E74 (Urness and Thummel, 1990), ELK-1 (Treisman et al, 1992).

long, centered over a GGAA/T core sequence. Two exceptions to this rule are notable. First, binding of *Drosophila* E74 and vertebrate ELF-1 to DNA is strongly biased toward EBSs centered over a GGAA core (Urness and Thummel, 1990; Wang et al, 1992), a property which is determined by the substitution of a lysine residue in all other ETS proteins for a threonine in E74 and ELF-1 in helix $\alpha 3$ (the recognition helix, see below) (Bosselut et al, 1993). This substitution has been proposed to alter the water-mediated network of hydrogen bonds between helix $\alpha 3$ and residues on the anti-sense strand at the GGAA/T core (Pio et al, 1996). Second, the EBS bound by PU-1/Spi-1 and Spi-B which display the most divergent ETS domain of the family, show a strong bias toward purinic residues in the sequences flanking the central core (Zhang et al, 1993; Galson et al, 1993; Ray-Gallet et al, 1995); furthermore, the sequence requirement for central core recognition by PU-1 is less stringent than for other ETS proteins since several PU-1 response elements have been found to contain a variant AGAA core (Shin and Koshland, 1993; Pahl et al, 1993).

The three-dimensional structure of the ETS domains of human FLI-1 (Liang et al, 1994), mouse ETS-1 (Donaldson et al, 1994, Donaldson et al, 1996) and human ETS-1 (Werner et al, 1995; 1996) have been determined by NMR methods and that of PU-1 was elucidated by x-ray crystallography (Kodandapani et al, 1996). These structures converge to identify a variant type of the winged helix-turn-helix (wHTH) DNA binding motif originally identified in GH5, HNF3γ and HSF (Harrison et al, 1994; Ramakrishnan et al, 1993; Clark et al, 1993a). Specifically, the ETS domain is described to fold into three α-helices ($\alpha 1$, $\alpha 2$, $\alpha 3$; Figure 2.2, see p. 32f) which interact through multiple conserved hydrophobic contacts to a four-stranded anti-parallel β sheet scaffold. Together with an eight amino acids turn, helices $\alpha 2$ and $\alpha 3$ form the HTH motif with $\alpha 3$ being the recognition helix. Helix $\alpha 1$ which lies over the β sheet scaffold crosses the carboxy-terminal end of $\alpha 3$ and, together with specific residues in $\alpha 3$, the turn between $\alpha 2$ and $\alpha 3$ and the loop between $\beta 3$ and $\beta 4$ (forming the "wing" of the wHTH motif) contribute to specific contacts to DNA.

In line with the data obtained by chemical and nuclease protection assays (Nye et al, 1992) and the sequences of the EBSs obtained by random oligonucleotide selection for a number of ETS proteins, ETS-1 and PU-1 make strong DNA contacts over a length of 10 nucleotides through both major and minor groove interactions on one face of the double helix (Kodandapani et al, 1996). Major groove contacts by PU-1 to the first three bases of the central GGAA core involve the two arginine residues of helix $\alpha 3$. In line with the absolute conservation of the GGA core motif in all EBSs identified so far, these residues are invariant in all ETS proteins

(Figure 2.2, see p. 32f) and their mutation in ETS-1, PU-1 or FLI-1 abolishes DNA binding (Bosselut et al, 1993; Liang et al, 1994; Mavrothalassitis et al, 1994; Kodandapani et al, 1996). In the PU-1 DNA complex, contacts with the phosphate backbone on either side of the core motif are made by two lysine residues conserved in all ETS proteins and localized in the turn/loop between $\alpha 2$ and $\alpha 3$ and the loop forming the "wing" of the wHTH motif. The first loop is localized between β strands 3 and 4 and the second is formed by the extended turn/loop between the $\alpha 2$ and $\alpha 3$ helices (Kodandapani et al, 1996; Pio et al, 1996). Mutational analyses in a number of ETS proteins and the superposition of the structure of the PU-1 ETS domain with those obtained by NMR for DNA-bound FLI-1 and free ETS-1 show a close similarity in the structure of these proteins (Pio et al, 1996). Unexpectedly, the structure resolved for the DNA-bound ETS-1 by NMR spectroscopy was initially reported to diverge significantly from this model (Werner et al, 1995), a situation which has been recently reconsidered (Werner et al, 1996).

Regulation of DNA Binding

For a subset of the members of the ETS family, including ETS-1, ETS-2, ELK-1, SAP-1 and SAP-2/ERP/NET, DNA binding by the isolated ETS domain was found to be considerably more efficient than in the context of the full-length protein (Hagman and Grosschedl, 1992; Lim et al, 1992; Wasylyk et al, 1992; Dalton and Treisman, 1992; Janknecht et al, 1994; Giovane et al, 1994; Lopez et al, 1994). For ETS-1, this difference in binding activity is the consequence of both increased binding affinity and enhanced accessibility toward EBS sequences (Lim et al, 1992; Petersen et al, 1995) and does not result from a difference in structure between the ETS domain in the repressed and unrepressed state (Nye et al, 1992; Donaldson et al, 1996; Petersen et al, 1995). Inhibition of DNA binding in full-length ETS-1 is an intrinsic property and does not result from interactions with other proteins (Lim et al, 1992; Petersen et al, 1995). Deletion analyses shows that regions both amino-terminal and carboxy-terminal to the ETS domain are essential for repression (Lim et al, 1992; Hagman and Grosschedl, 1992; Wasylyk et al, 1992). Recently, the conformational changes that accompany the binding of ETS-1 to DNA have been studied by combining CD and NMR spectroscopy with an analysis of the susceptibility of ETS-1 to proteolytic degradation (Petersen et al, 1995). These studies show that DNA binding results in the specific unfolding of a 8-residues α helix (H1) located 19 residues upstream of the ETS domain. Helix H1, together with an adjacent helix, the first helix of the ETS domain ($\alpha 1$) and the $\alpha 4$ helix formed by the carboxy-terminal residues of ETS-1 are proposed to form a four-helix bundle in unbound (repressed) ETS-1 (Petersen et al, 1995; Skalicky et al, 1996). Binding of ETS-1 to DNA results in unfolding of helix H1 followed by disruption of the four helix bundle and transient relief from intramolecular repression. Since binding of ETS-1 to several natural EBSs often occurs synergistically with other sequence-specific DNA binding proteins (Gégonne et al, 1993; Wotton et al, 1993; Giese et al, 1995), these interactions could favor ETS-1 binding to DNA through stabilization of the derepressed state.

Intramolecular inhibition of DNA binding also operates for the ETS family members of the TCF subfamily. Analysis of deletion mutants of ELK-1, SAP-1a and SAP-2/ERP/NET shows that the conserved carboxy-terminal domain (C domain; see section "Transcriptional Regulation") is involved in repression of DNA binding (Dalton and Treisman, 1992; Janknecht et al, 1994; Lopez et al, 1994; Giovane et al, 1994; Price et al, 1995). Repression of DNA binding is also likely to result from an intramolecular mechanism since it is observed using purified preparations of these proteins (Lopez et al, 1994). Similarly to the model proposed for ETS-1, interaction with SRF could stabilize the binding of ELK-1, SAP-1a and SAP-2/ERP/NET to the suboptimal EBS of the c-*fos* SRE (Treisman, 1992; Treisman et al, 1992) by favoring these proteins to adopt a derepressed conformation. The situation is likely to be more complex however, since the different ETS proteins with TCF activity differ in their autonomous and assisted DNA binding activity (Price et al, 1995; Dalton and Treisman, 1992) and since other regions besides the conserved C domain appear to be involved in inhibition of DNA binding (Janknecht et al, 1994; Lopez et al, 1994; Giovane et al, 1994; Price et al, 1995).

Protein-protein interactions also clearly play an essential role in the regulation of the DNA binding activity of other ETS proteins. GABP is a multi-subunit DNA binding protein which was originally identified as a factor binding to a cis-regulatory element essential to VP16-mediated activation of herpes simplex virus (HSV) immediate early genes. Mouse GABP is composed of an α subunit (GABPα) which belongs to the ETS family and either of two isoforms of a β subunit (GABPβ1; GABPβ2) (LaMarco et al, 1991; Thompson et al, 1991). Association between each subunits occurs in the absence of DNA through an interaction between a domain of GABPα which includes the ETS domain plus 37 adjacent amino acid residues on the one hand, and, on the other, four imperfect 33 amino acid repeats in the amino-terminal domain of GABPβ resembling the repeats of the Notch protein of *D. melanogaster* (LaMarco et al, 1991;

Thompson et al, 1991). The $\beta 1$ and $\beta 2$ subunits differ in their carboxy-terminal end, with that of GABP$\beta 1$ encoding an homotypic dimerization domain. GABP exists therefore in solution either as a (GABPα) (GABP$\beta 2$) heterodimer or as a $(\text{GABP}\alpha)_2$ $(\text{GABP}\beta 1)_2$ tetramer. The situation is more complex both for mouse GABP and for its human E4TF1/NRF-2 homolog, due to the existence of variants β subunits resulting from alternative splicing and the existence of related genes (Watanabe et al, 1993; de la Brousse et al, 1994; Gugneja et al, 1995). GABPβ fails to bind DNA on its own and GABPα displays only weak binding to classical EBSs. Their association however, increases DNA binding through stabilization of the DNA bound complex as the result, at least in part, of the participation of the β subunit in contacting DNA (Thompson et al, 1991; Gugneja et al, 1995). Importantly, formation of the α_2 β_2 tetramer specifically enhances the binding of GABP to tandem EBS sites (Thompson et al, 1991; Virbasius et al, 1993b; Gugneja et al, 1995). Since the interaction of GABPβ is highly specific for GABPα and does not affect other ETS proteins (Brown and McKnight, 1992), the interaction of GABPα and GABPβ subunits are also critical determinants of target gene selection.

An additional level of control of DNA binding activity in several ETS proteins clearly operates via post-transcriptional modifications by phosphorylation. As described below (see section "Transcriptional Regulation"), transcriptional activation by PU-1 of several B-cell-specific enhancers depends upon its concomitant binding with a B-cell-specific factor, NF-EM5/Pip on a composite DNA element. Complex formation between PU-1 and NF-EM5/Pip depends upon phosphorylation of a single serine residue (S148) in the central PEST region of PU-1 (Pongubala et al, 1993). In line with its primary amino acid context, which conforms to casein kinase II consensus phosphorylation sites, this residue can be phosphorylated by purified casein kinase II in vitro (Pongubala et al, 1993). It is unclear however whether casein kinase II is the kinase responsible for PU-1 phosphorylation in vivo and whether PU-1 phosphorylation of S148 can be regulated by extra- and intracellular signaling events.

Transcriptional regulation at the *c-fos* SRE in response to the Ras signaling pathway is critically dependent upon the phosphorylation-dependent activation of the transcriptional activation domain of ELK-1 and related proteins, associated as ternary complexes with SRF on the SRE (see section "Transcriptional Regulation"). Although EGF stimulation of A431 cells does not affect SRE occupancy in vivo as assayed by footprinting analyses (Herrera et al, 1989) and in vitro, using extracts of different cells treated with a variety of mitogens (Marais et al, 1993; Zinck et al, 1993), other experiments have shown that ternary complex formation is increased in certain experimental conditions (Gille et al, 1992; 1995 a). This suggests that phosphorylation of at least certain members of the TCF subfamily may facilitate the exchange of different TCFs at SRF-bound SRE in response to specific signaling events.

As described above, ETS-1 plays an essential role in both T- and B-cell survival, proliferation and differentiation. Engagement of the antigen receptor of T- and B-cells induces the rapid and transient calcium-dependent phosphorylation of ETS-1 (Pognonec et al, 1988; Fisher et al, 1991 a; Rabault and Ghysdael, 1994) on four serine residues located immediately upstream of the DNA binding amino-terminal regulatory domain. Stoechiometric phosphorylation of these residues results in the inhibition of ETS-1 binding to EBSs and is associated with a reduced mobility of ETS-1 in denaturing gels, suggesting that phosphorylation negatively regulate DNA binding through conformational changes. This regulation is highly specific for ETS-1 since the increase in intracellular calcium that follows T- and B-cell activation fails to affect the DNA binding activity of the related ETS-2 and FLI-1 proteins (B. Rabault and J. Ghysdael, unpublished data).

Transcriptional Regulation

Consistent with their activity as transcriptional activators, one or several activation domains have been identified in ETS proteins by the study of deletion mutants and of Gal4 or LexA fusion proteins (ETS-1 and ETS-2: Punyammalee et al, 1991; Schneikert et al, 1992; Gégonne et al, 1992; Albagli et al, 1994; FLI-1 and ERG-2: Siddique et al, 1993; Rao et al, 1993; ELF-1: Wang et al, 1993; ELK-1, SAP-1a and SAP-2: Marais et al, 1993; Janknecht et al, 1993 b; Price et al, 1995; Spi-1/PU-1: Hagemeier et al, 1993; Shin and Koshland, 1993; Pio et al, 1995). In contrast to the conservation of the ETS domain, the activation domains of ETS proteins are highly divergent, suggesting that these domains could contribute to transcriptional specificity. For example, the physical interaction between ETS-1 and PEBP2α which is believed to play a key role in the cooperative formation of a complex on a composite DNA element in the TCRα enhancer has been shown to depend upon the integrity of the ETS-1 activation domain (Giese et al, 1995). A variable degree of conservation is however observed among members of specific ETS subfamilies, suggesting the maintenance of a common function. The activation domains of ELK-1, SAP-1a and SAP-2/ERP/NET share an overall amino acid identity of 70% and the conservation of aromatic residues and proline-directed serine phosphorylation sites, two features essential to their activity as signal-regulated transcriptional activation domains (Price et al, 1995) (see below).

Except for reported in vitro interactions between the amino-terminal activation domain of PU-1 and TBP, the interaction of ELF-1 with TFIIB and the interaction and synergy of ELK-1 and SAP1a with CBP (Hagemeier et al, 1993; John et al, 1995; Janknecht and Nordheim, 1996), the molecular mechanisms involved in transcriptional regulation by ETS proteins remain unknown. As detailed below and in Table 1, EBSs function in most cases as promoter or enhancer elements to regulate the activity of a variety of genes. In several instances, functionally important EBSs have been identified close from the transcriptional start site in TATA-less promoters, suggesting that ETS proteins could be components of initiation complexes regulating basal transcription (Virbasius and Scarpulla, 1991; Salmon et al, 1993; Feinman et al, 1994). For example, mutation of the tandemly arranged GABP binding site in the cytochrome oxidase subunit IV promoter was found to affect both transcriptional activity and initiation site selection (Carter and Avadhani, 1994). Also, mutation of a PU-1 binding site adjacent to the initiator sequence spanning the start sites of the FcγR1b promoter was found to suppress promoter activity, an effect which could be reversed by insertion of a TATA box at −30 bp in the promoter (Eichbaum et al, 1994).

Functional EBS have been identified in the regulatory regions of a number of genes (Table 1), including genes encoding ubiquitously expressed proteins as well as highly tissue-specific genes, especially in several hematopoietic lineages. Some ETS proteins like PU-1 are highly tissue-restricted and are therefore expected to play a role in cell-specific gene expression. Most ETS family members show however a broader expression pattern and yet have been shown to participate, through specific protein-protein interactions, in the regulation of the activity of tissue-specific genes. A second feature of ETS regulated transcription is the frequent involvement of ETS proteins in the regulation of gene in response to a wide variety of extracellular signals, an aspect which is described in more detail later in this section.

Although most studies reported so far have described ETS proteins as transcriptional activators, recent experimental evidence implicates particular members of this family in transcriptional repression. As discussed in section "The Lessons of Genetics", genetic evidence shows that *Drosophila* YAN/Pokkuri is a signal-dependent repressor and antagonist of Pointed function in R7 photoreceptor cell development. The mechanisms involved at the biochemical level remain however to be clarified. Similarly, ETS-1 has been proposed to be able to function as a repressor of the TCRβ enhancer in transient transfection assays (Prosser et al, 1992) and of AP1-regulated transcription in vitro (Goldberg et al, 1994). More recently, vertebrate *ERF* has recently been shown to encode a sequence-specific transcriptional repressor and antagonist of ETS-activated transcription, an activity which depends upon the presence of a strong and transportable repression domain localized at its carboxyterminus (Sgouras et al, 1995). ERF is hyperphosphorylated following activation of the Ras/Raf/ERK pathway and is found to be associated with ERK2 in vivo and to be a substrate for ERK2 in vitro. Activation of ERKs inhibits ERF repression activity in vivo but does not affect an ERF protein carrying serine-to-alanine substitutions in the ERK2 target sites (Sgouras et al, 1995; G. Mavrothalassitis, personal communication). These observations suggest that ERF may function as a repressor in G_0 of genes normally activated by ETS proteins during the entry of cells in G_1. A transportable repression domain has also recently been identified in SAP2/ERP/NET which affects both basal and Ras-activated transcription of this TCF (Maira et al, 1996). This domain is unique to SAP2/ERP/NET and is presumed to fold as a helix-loop-helix and to form a specific protein-protein interaction interface. At least part of the repressive activity results from inhibition of the DNA binding activity of SAP2/ERP/NET (Maira et al, 1996).

Tissue-specific transcriptional activation by ETS proteins frequently requires protein-protein interactions with other transcriptional regulators on composite DNA elements

The expression of PU-1/Spi-1 is largely restricted to hematopoietic cells of the B lymphoid and myeloid lineages (Klemsz et al, 1990; Ray et al, 1992; Chen et al, 1995b) and its inactivation by homologous recombination in mouse results in defective development of myeloid and lymphoid progenitors (Scott et al, 1994). In line with these properties, PU-1 has been shown to play a major role in the tissue-specific expression of itself (Chen et al, 1995b) and of many myeloid and lymphoid specific genes (Table 1).

The immunoglobulin light chain enhancers Eκ3, Eλ2–4 and Eλ3-1 contain an essential element which binds a heterodimeric complex composed of PU-1 and NF-EM5/Pip, a lymphoid-specific member of the interferon regulatory factor family (Pongubala et al, 1992; Eisenbeis et al, 1993; Eisenbeis et al, 1995). Tissue-specific activity of these enhancers in B-cells depends upon the integrity of both the PU-1 and NF-EM5/Pip binding sites (Pongubala et al, 1992; Eisenbeis et al, 1995). NF-EM5/Pip is unable to bind this element autonomously but is recruited by DNA bound PU-1 to form a stable ternary complex. Complex formation by these proteins is cooperative and requires the phosphorylation of a serine residue (S148) localized in the central PEST region of PU-1 (Pongubala et al, 1993). Complex formation is essential for activity of this element since PU-1 alone was found unable to activate transcription and since a S148A substitution mutant of PU-1 was inactive in the presence of a co-expressed NF-EM5/Pip protein (Pongubala et al, 1993; Eisenbeis et al, 1995). This indicates that the PU-1-mediated recruitment of NF-EM5/Pip to immunoglobulin light chain enhancers reflects the requirement of an activation domains in NF-EM5/Pip which critically contributes to the transcriptional activity of the complex.

PU-1 is involved in the tissue-specific regulation of several other B lymphoid specific genes (Table 1). An interesting situation has been described for the intragenic enhancer of the heavy chain immunoglobulin gene in which PU-1 has been shown to cooperate with other ETS family members. The 700 bp intragenic Eμ enhancer is a critical element in B-cell specificity of immunoglobulin heavy chain gene expression. At least two enhancer elements have been shown to be essential for B-cell specificity. One is the octamer motif also found in the enhancers of other immunoglobulin genes and the other is formed by the association of the μB and μA (or π) elements (Nelsen et al, 1993; Rivera et al, 1993). The μB element binds PU-1 and all μB mutations deleterious for enhancer activity in B-cells were found to inhibit PU-1 binding in vitro. Multimers of the μB element are however not active as enhancer in B-cells, suggesting the requirement for an additional element. This element, designated as μA or π, is bound by other members of the ETS family, including ETS-1, ERG-3 and FLI-1 and its mutation also suppresses enhancer activity in B-cells, indicating that occupancy of both μB and μA/π is required for enhancer activity. In line with this notion, co-expression of ETS-1 and PU-1 in non lymphoid cells was found to complement the absence of lymphoid-specific function in these cells and to restore activity of a core Eμ enhancer (Nelsen et al, 1993). Finally, activation of a reporter containing a multimer of the μA/π and adjacent μE2 element by ERG-3 or FLI-1 was found to occur in synergy with E12, suggesting that optimal activity of the μA/π element also requires the binding of bHLH proteins to the adjacent μE2 element (Rivera et al, 1993).

Although widely expressed during vertebrate embryonic development ETS-1 is largely restricted to cells of the T- and B-lymphoid lineages in adults, suggesting an important role for ETS1 in the lymphoid compartment (see section "The lessons of Genetics"). The activity of the TCRα and TCRβ enhancers has been shown to depend upon the integrity of several ETS binding sites (Ho et al, 1990; Prosser et al, 1992). In both enhancers, binding of ETS-1 is facilitated by the concomitant binding of core binding factor (CBF) to an adjacent site, a factor composed of a DNA binding α subunit (PEBP2α; AML1) and of an unrelated β subunit (Wotton et al, 1993; Giese et al, 1995). The Runt homology domain of PEBP2α is sufficient to recruit ETS-1 on the composite EBS/core binding factor element and a direct interaction between ETS-1 and the Runt domain PEBP2α protein has been identified by GST pull-down assays. This interaction is believed to antagonize the intramolecular repressing activity of ETS-1 toward DNA binding (Giese et al, 1995). Besides the ETS-1 and PEBP2α binding sites, activity of the TCRα enhancer in T-cells also critically depends upon the integrity of both an ATF/CREB binding site, located 65 bp upstream of the EBS, and of a binding site for the T-cell specific HMG domain protein LEF-1, located between the ATF/CREB site and the EBS (Giese et al, 1995). Substitution of the LEF-1 binding site for that of the distantly related HMG domain protein SRY was found to only marginally affect enhancer activity, suggesting a major role of LEF-1 induced DNA bending in the activity of the TCRα enhancer. Since LEF-1 and ATF/CREB proteins cooperate to stabilize DNA binding by PEBP2α and ETS-1 in DNase I footprinting experiments and since ETS-1 is shown to interact via its ETS domain with ATF-2, the activity of the TCRα enhancer was proposed to depend upon the LEF-1 coordinated assembly of a specific high-order multiprotein complex (Giese et al, 1995). In line with this model, a strong synergy is observed between these proteins to reconstitute TCRα enhancer activity in non T-cells (Giese et al, 1995). Of note, expression of the TCRα gene is maintained in T-cells derived from ETS-1$^{-/-}$ ES cells in mouse chimeras (Bories et al, 1995; Muthusamy et al, 1995), suggesting that other members of the ETS family expressed in T-cells can substitute for ETS-1 in regulating TCRα enhancer activity during T-cell development.

The ETS domain of ETS-1, besides its role in nuclear localization and DNA binding, also functions as a protein-protein interaction interface. MafB, a bZIP protein of the AP1 family has recently been identified as a partner for ETS-1, an interaction which involves the direct binding of the MafB basic and zipper region to the ETS domain of ETS-1 (Sieweke et al, 1996). Overexpression of Maf-B was found to inhibit the ability of ETS-1 to activate transcription of a model EBS-based reporter gene and of the promoter of the transferin receptor in transient transfection assays; the latter activity is believed to be important to the ability of overexpressed MafB to inhibit erythroid differentiation (Sieweke et al, 1996).

Analysis of the regulation of the IL2Rα promoter also provides evidence for the direct interaction of an ETS protein, in this case ELF-1, with the HMG box-containing protein HMG-I (Y) (John et al, 1995). The IL2Rα gene is induced at the transcriptional level following activation of quiescent T-cells. Induced transcriptional activation of the human IL2Rα enhancer/promoter region depends upon two regulatory domains: an enhancer domain consisting of adjacent NF-κB and CArG box motifs (PRR I element) and a more proximal regulatory element containing an EBS for the lymphoid specific ELF-1 as well as several HMG-I (Y) binding sites (PRR II element) (John et al, 1995). Mutation of either the ELF-1 or HMG-I (Y) DNA binding motifs inhibits the activity of the IL2Rα promoter in T-cells and co-expression of ELF-1 and HMG-I (Y) was found to be sufficient to transactivate through the PRR II element in non T-cells. Gel mobility shift assays show that T-cell extracts form a complex containing both ELF-1 and HMG-I (Y) on the PRR II element and GST pull-down experiments suggest a direct physical contact between the ETS domain of ELF-1 and HMG-I (Y) as well as NF-κB p50 and c-REL (John et al, 1995). These observations are interpreted in a model in which HMG proteins facilitate – through DNA bending – the assembly of a multiprotein complex involving multiple protein-DNA and protein-protein interactions. This complex is proposed to be essential to the cell-specific activity of the IL2Rα promoter and to its ability to respond to activation signals (see below). Of note, the genes encoding both the IL2Rβ chain and the γc chain – which together with the α chain form the high affinity IL2 receptor – are also dependent upon specific EBS for their tissue-specific expression and inducibility (Lin et al, 1993; Markiewicz et al, 1996).

ETS proteins are nuclear mediators of essential signal transduction pathways

The control of cell survival, proliferation and differentiation by cytokines and growth factors and the cellular esponses to stress conditions depend upon the propagation in all subcellular compartments of a series of phosphorylation-dephosphorylation events. Several ETS proteins have been shown to play an essential role as nuclear mediators of regulated gene expression controlled by several of these signaling pathways.

The best understood signaling pathways affecting the transcriptional activity of ETS proteins are those which link growth factors receptors with tyrosine kinase activity and cytokine receptors to the activation of protein kinases of the MAP kinase family through the activation of Ras and Ras-related GTP binding proteins. Several EBS-containing promoter elements have been identified that confer response to serum, polypeptidic growth factors, constitutively activated components of the Ras/Raf/MEK/MAP kinases pathway and phorbols esters. These include the TPA and oncogene-responsive unit initially identified in the α domain of the Polyoma virus enhancer, composed of adjacent ETS and AP1 binding sites (Wasylyk et al, 1989; Gutman and Wasylyk, 1990); the serum response element (SRE) of the c-*fos* promoter, composed of adjacent ETS and SRF binding sites (Shaw et al, 1989); the palindromic ETS binding site initially identified in the rat stromelysin promoter (Wasylyk et al, 1991; Yang et al, 1996); the composite ETS and PEBP2α binding site of the βE2 element in the context of the TCRβ enhancer (Prosser et al, 1992; Wotton et al, 1994); the PMA-responsive element of the Mo-MuLV LTR in which the two branches of a palindromic ETS binding site are separated by a PEBP2α/CBF binding site (Speck et al, 1990; Gunther and Graves, 1994); the composite EBS/GC-rich element of the HTLV1 LTR in which two ETS binding site are separated by a GC box element (Gégonne et al, 1993; Clark et al, 1993b; Coffer et al, 1994). In most of these situations, both the EBS and the binding site for the other factor were found to be required for maximal stimulation, an effect which reflects the cooperation of the respective factors to activate these elements (Wasylyk et al, 1990; Hill et al, 1991; Gégonne et al, 1993; Wu et al, 1994; Grant et al, 1995).

Since both the EBS and the associated binding sites of these signal-responsive elements are binding sites for families of transcriptional regulators composed of many members, the particular family members actually involved in basal and regulated expression have been difficult to identify. The ETS binding sites in the *junB* promoter (Coffer et al, 1994) and the ETS/AP1 composite element in the urokinase-type plasminogen activator (uPA) promoter (Yang et al, 1996) were found to be weakly activated by either ETS-1 or ETS-2 in co-transfection experiments but to be efficiently activated by either proteins when co-transfected with an activated form of Ras. This property appears to be rather specific to ETS-1 and ETS-2 since other ETS proteins including FLI-1, ELF-1 and PEA3, although able to increase the basal activity of the uPA promoter, failed to cooperate with an activated form of Ras (Yang et al, 1996).

Similarly, stimulation of the activity of the macrophage-specific scavenger receptor promoter by phorbol ester, a situation which is believed to mimic the activation of this gene in monocytes in response to CSF-1, depends upon the combined activity of a complex containing c-JUN, JUN-B and ETS-2. This complex is assembled onto ETS/AP1 elements located in both the promoter and enhancer region of the gene. An important part of the specificity for these particular members of the ETS and JUN families appears to come from the fact that their expression is rapidly and transiently induced following treatment of monocytes with PMA (Boulukos et al, 1990; Datta et al, 1991). The involvement of particular ETS family members in Ras-inducible activity of gene expression appears however to also depend upon both promoter context and the particular cell-type involved. For example, whereas the AP1/ETS promoter element is an efficient Ras-responsive element for both ETS-1 and ETS-2 in NIH3T3, only ETS-2 appears to be able to cooperate in these cells with an activated Ras gene in the activation of a reporter gene based on the palindromic ETS element derived from the stromelysin promoter (Yang et al, 1996). Also, in contrast to its lack of activity on the uPA promoter, the restricted expression of ELF-1 to lymphoid cells appears to be an important determinant in the regulated activity of ETS/AP1 motifs present in tissue-specific enhancers. Activation of both the GM-CSF promoter in T-cells in response to PMA or the B-cell-specific intronic enhancer of immunoglobulin heavy chain genes in response to PMA or IgM receptor engagement has been proposed to depend upon the assembly of a complex specifically composed of ELF-1, c-FOS and JUN-B on a composite ETS/AP1 element. Whereas ELF-1 is expressed in quiescent lymphoid cells, transcription of both c-*fos* and *jun-B* is coordinately induced following T- and B-cell activation (Crabtree, 1989; Klemsz et al, 1989). This, together with the cell-cycle, RB-dependent-regulated activity of ELF-1 are proposed to be critical determinants in the tissue-specific activity of these composite AP1/ETS elements. The ELF-1-HMG-I (Y) complex which plays an important role in the tissue-specific expression of the IL2Rα chain promoter (see above), also plays a role in the inducible expression of this promoter (John et al, 1995). This is believed to occur through protein-protein interactions involving specific members of the NF-κB family which are induced to bind, together with SRF on upstream composite element (John et al, 1995). Finally, although PEA3 failed to cooperate with Ras to activate the uPA promoter, both PEA3 and the related ER81 were found to cooperate with activated versions of Ras as well as other components of the Ras signaling pathway to activate other model promoters (Janknecht, 1996).

Serum and Ras-mediated activation of ETS1 and ETS2 depends upon the phosphorylation of a specific threonine residue in a domain conserved in Drosophila PNTP2

Genetic studies have identified PNTP2 as positively acting determinant of photoreceptor cell development in the *sevenless*/Ras/Raf/MEK/ERK signaling pathway (see section "The Lessons of Genetics"). Furthermore, biochemical studies have shown that the function of PNTP2 in eye development and its ability to activate model reporter genes in co-transfection assays critically depend upon the integrity of a single threonine residue (T151) (Brunner et al, 1994; O'Neill et al, 1994) in the *pointed* domain. Since this residue can be phosphorylated by purified Rolled/ERK-A in vitro, these experiments suggested that the transcriptional activity of

PNTP2 and its function in R7 photoreceptor cell development depends upon its direct phosphorylation by Rolled/MAPK in vivo.

Recent experiments have shown that serum stimulation and activation of the Ras signaling pathways result in the phosphorylation of ETS-1 on threonine residue 38 and ETS-2 on threonine residue 72 (Yang et al, 1996; Rabault et al, 1996). These residues are the structural homologs of threonine 151 in PNTP2, although their primary sequence environment is somewhat different. Substitution of both ETS-1 threonine 38 and ETS-2 threonine 72 for alanine suppressed phosphorylation of ETS-1 and ETS-2 on threonine in vivo, their ability to cooperate with Ras in activating transcription of several reporter genes and their ability to rescue the defective mitogenic properties of a mutant CSF-1R, CSF-1R(Y809F), in NIH3T3 cells (Rabault et al, 1996; Yang et al, 1996). These studies also indicate that although ERK-1 and ERK-2 can phosphorylate ETS-1 threonine 38 and ETS-2 threonine 72 in vitro, other members of the MAP kinase family are likely to be responsible for their phosphorylation in vivo. The amino-terminal conserved domains of ETS-1 and ETS-2 does not possess intrinsic transcriptional activation properties but have been shown to modulate the properties of the ETS-1 and ETS-2 activation domains (Schneikert et al, 1992). The carboxy-terminal two-thirds of the *pointed*/B domain of ETS-1, ETS-2 and PNTP2 is shared with other members of the ETS family, including FLI-1, ERG-2, GABPα, TEL in vertebrates and YAN and ELG in *Drosophila* to form a specialized protein-protein interaction interface important to their transcriptional activation properties (see the Introduction). In line with this notion, the cooperation of ETS-1 and PIT-1 to activate the prolactin promoter in response to Ras has been shown to require the integrity of the ETS-1 amino-terminal conserved domain (Bradford et al, 1995). As deletion of this domain does not affect the ability of ETS-1 to localize to the nucleus, nor to bind simple EBS elements in vivo, nor to affect Ras-independent transcription (Boulukos et al, 1989; Gégonne et al, 1992), signal-induced phosphorylation of PNTP2, ETS-1 and ETS-2 in the *pointed*/B domain is likely to affect the interaction of these proteins with specific factors essential to the activation of Ras-responsive promoters.

The ETS proteins with TCF activity harbor a signal-responsive activation domain

The serum response element (SRE) is a promoter element which mediates early gene expression in cells treated with growth factors and cytokines and in response to environmental stress. The human c-*fos* SRE is a bipartite element recognized by serum response factor (SRF) (Treisman, 1986) and by a subfamily of ETS proteins which includes ELK-1, SAP-1a and SAP-2/ERP/NET (Hipskind et al, 1991; Dalton and Treisman, 1992; Giovane et al, 1994; Price et al, 1995). Both elements are required for maximal induction of the SRE in response to extracellular signals (Graham and Gilman, 1991). None of these ETS proteins autonomously bind the c-fos SRE in vitro nor activate the expression of transfected SRE-based reporter genes harboring a mutation in the SRF binding sites in vivo (Dalton and Treisman, 1992; Price et al, 1995). Transcriptional activation of the SRE by these ETS proteins requires their recruitment by SRF at the SRE to form a ternary complex (Hipskind et al, 1991; Janknecht and Nordheim, 1992; Gille et al, 1992; Hill et al, 1993; Janknecht et al, 1994). Since the ability of these proteins to form a ternary complex with SRF at the c-*fos* SRE was recognized before their identification as ETS family members (Shaw et al, 1989), these proteins are usually referred to as ternary complex factors (TCFs).

TCFs share several structural properties which distinguish them from other members of the ETS family. First, unlike other ETS proteins, the ETS domain (domain A: Dalton and Treisman, 1992) forms their amino-terminus. Second, they share three additional and unique regions of homology (homology domains B, C and D: Dalton and Treisman, 1992; Lopez et al, 1994; Giovane et al, 1994; Price et al, 1995). Both the ETS domain and the B domain are required for nuclear localization and recruitment of TCFs to the SRE-bound SRF to form the ternary complex (Dalton and Treisman, 1992; Janknecht and Nordheim, 1992; Hill et al, 1993; Janknecht et al, 1994; Treisman et al, 1992), the B domain forming an autonomous interaction-interface for SRF (Hill et al, 1993; Shore and Sharrocks, 1994). Domain C is essential for transcriptional activation by TCFs and forms part of their signal-regulated activation domains (see below). The function of domain D is presently unknown. SAP-2/ERP/NET is much less efficient than ELK-1 and SAP-1a at forming ternary complexes at the c-*fos* SRE or at binding conventional ETS binding sites in vitro due to a strong intramolecular inhibition by several regions of its carboxy-terminal domain (Lopez et al, 1994; Giovane et al, 1994; Price et al, 1995; Maira et al, 1996). Whether this property reflects a specific difference in the activation properties of SAP-2/ERP/NET as compared to ELK-1 and SAP-1a, e.g., by its interaction with a specific factor (see Price et al, 1995 for a discussion of this point) or its activity as a repressor (Giovane et al, 1994; Maira et al, 1996) remains to be clarified.

Mitogenic stimulation of a variety of cells by serum, growth factors or activated versions of components of the Ras signaling pathway is accompanied by a considerable change in the electrophoretic mobility of the c-*fos* SRE ternary complex which correlates temporally with the induction of c-*fos* promoter activity in vivo (Marais et al, 1993; Zinck et al,

1993; Price et al, 1995). For ELK-1, extensive analyses by deletion and substitution mutagenesis has shown this property to be linked to the signal-induced phosphorylation of several serine and threonine residues in its carboxy-terminal domain (Marais et al, 1993; Hill et al, 1993; Janknecht et al, 1993b; Gille et al, 1995a; Kortenjann et al, 1994). Two-dimensional phosphopeptide analyses, combined with serine-to-alanine substitution mutagenesis has shown that signal-induced phosphorylation of ELK-1 occurs at several serine and threonine residues in the conserved C domain and in residues immediately flanking this domain (Marais et al, 1993; Janknecht et al, 1993b; Gille et al, 1995a). For technical reasons the precise stoechiometry of these modifications has been difficult to establish by two-dimensional phosphopeptide analyses (Marais et al, 1993; Janknecht et al, 1993b; Gille et al, 1995a). The recent development of phosphopeptide-specific antibodies has shown that serine residues 383 and 389 and threonine residues 363, 368 in ELK-1 domain C and threonine 417 are phosphorylated to comparable stoechiometrical levels following growth factor stimulation (F. Cruzalegui and R. Treisman, personal communication). Most of these residues serve as substrates for purified ERK-1 and ERK-2 in vitro and the use of a GST fusion protein containing the 121 carboxy-terminal residues of ELK-1 as substrate in "in gel kinase" assays identified ERK-2 as the major ELK-1 protein kinase in growth factor stimulated NIH3T3 cells (Marais et al, 1993; Gille et al, 1995a). The carboxy-terminal domains of ELK-1, SAP-1a or SAP-2, when fused to the heterologous DNA binding domain of Gal4 or LexA, function as serum, growth factor and Ras/Raf/ERK-activated transcriptional activation domain of Gal4 or LexA responsive reporter genes, respectively (Marais et al, 1993; Janknecht et al, 1994; Gille et al, 1995a; Price et al, 1995; Kortenjann et al, 1994). Activated transcription by these domains correlates with their signal-induced phosphorylation and is dependent upon the integrity of the serine and threonine residues targeted by these phosphorylation events (Marais et al, 1993; Janknecht et al, 1993a; Janknecht et al, 1995; Gille et al, 1995a; Price et al, 1995; Whitmarsh et al, 1995; Kortenjann et al, 1994).

The analysis of the properties of *bona fide* ELK-1, SAP-1a or SAP-2/ERP/NET (Janknecht et al, 1993a; Janknecht et al, 1995; Giovane et al, 1994) and of altered specificity mutants of both SRF and ELK-1 (Hill et al, 1993) on SRE-based reporter genes has shown that serum and Ras/Raf/MAPK-dependent transcriptional activation is dependent upon the same critical residues in domain C as those identified to be required in the activation of Gal4 and LexA fusion proteins. These results show that ELK-1, and probably also SAP-1 and SAP-2/ERP/NET are specifically required and functionally cooperate in the ternary complex to regulate transcription in response to the activation of the Ras/Raf/ERK signaling pathway. SRF does not merely function as a recruitment factor for TCFs but also contributes as co-activator in the regulation of the c-*fos* SRE (Hill et al, 1993). This co-activator function is not constitutive since recent experiments have identified a novel, TCF-independent, signaling pathway at the SRE, controlled by serum components distinct from polypeptidic growth factors. This pathway, which is dependent upon the RhoA family of GTPase, specifically targets SRF by molecular mechanisms which remain to be identified and acts in synergy with the signals that activate the TCFs (Hill et al, 1995).

Recent evidence has shown that, besides the ERK subfamily of protein kinases, other MAP kinase family members including JNKs/SAPKs and p38/MPK2 which are specifically activated in response to proinflammatory cytokines and in response to environmental stress also phosphorylate and activate the activation domain of ELK-1 and SAP-1a (Gille et al, 1995a; Whitmarsh et al, 1995; Zinck et al, 1995; Raingeaud et al, 1996; Price et al, 1996) and ternary complex formation and c-*fos* transcription by these proteins (Price et al, 1996). Signals which specifically activate JNK/SAPK both activate c-*fos* transcription via the SRE and induce phosphorylation of ELK-1 and SAP-1a in vivo. Recombinant ELK-1 and SAP-1a are phosphorylated by JNKs/SAPKs in vitro on specific serine and threonine residues which partially overlaps those phosphorylated by ERKs. The major ELK-1 phosphorylation sites modified upon specific activation of JNK/SAPK include serine 383 (Gille et al, 1995b) but more prominently threonine residues, in particular threonine 363 and 368 (Whitmarsh et al, 1995; Price et al, 1996). Similar to the activation by ERKs and upstream activating components, the transcriptional activity of a Gal4 fusion protein containing the 121 carboxy-terminal residues of ELK-1 is enhanced by signals specific for the activation of JNKs/SAPKs and by a constitutive version of MEKK, a property which requires the integrity of serine 383 (Gille et al, 1995b). ELK-1 is also a substrate for the p38/MPK2 MAP kinase and is phosphorylated at sites which resemble those modified by ERKs, indicating that the stress-inducible JNK/SAPK and p38/MPK2 have different substrate specificities (Price et al, 1996). These data show that ELK-1 and SAP1a are likely to integrate several activating pathways at the SRE, each involving different members of the MAP kinase family. Interestingly, SAP-1a appears to be a poor substrate for JNKs/SAPKs but is phosphorylated as efficiently as ELK-1 by p38/MPK2 (Whitmarsh et al, 1995; Price et al, 1996). Since TCFs appear to be rather ubiquitously expressed (Price et al, 1995), this suggest that one basis of TCF diversity could reside in their preferential response to distinct signaling pathways.

ETS Proteins in Oncogenesis

The v-ets oncogene of avian leukemia virus E26

The naturally occurring E26 retrovirus (Ivanov et al, 1962) induces an acute leukemia in chickens which was initially described as a mixed myeloid/erythroid leukemia (Radke et al, 1982; Moscovici et al, 1983) but has more recently been characterized as a stem cell leukemia (Graf et al, 1992). In tissue culture, E26 transforms myeloblasts and multipotent hematopoietic progenitors (MEPs) able to differentiate along the erythroid, eosinophilic, thrombocytic, granulocytic and monocytic lineages (Graf et al, 1992; Kraut et al, 1994; Frampton et al, 1995). The genome of E26 contains two oncogenes: v-*myb*^{E} which is a truncated version of the c-*myb* proto-oncogene, and v-*ets* which is a truncated and mutated version of c-*ets-1* (Roussel et al, 1979; Leprince et al, 1983; Nunn et al, 1983; Leprince et al, 1988). These oncogenes are expressed as a 135 Kd nuclear fusion protein (Gag-Myb-Ets) that obtains its amino-terminus from a partial retroviral *gag* gene (Beug et al, 1982; Bister et al, 1982). The analysis of the transforming properties of deletion mutants of E26 has shown that both oncogenes are essential to the transforming properties of E26 (Nunn and Hunter, 1989; Metz and Graf, 1991; Domenget et al, 1992). More strikingly, the fusion of the v-Myb^{E} and v-Ets sequences of Gag-Myb-Ets was found to be essential for full transformation of both myeloblasts and MEPs by E26 and for leukemia induction in animals (Metz and Graf, 1991; Metz et al, 1991). These and additional studies using temperature-sensitive mutants in either the v-Myb (E26 ts21) or v-Ets (E26 ts1-1) domains of Gag-Myb-Ets and the analysis of the differentiation properties of E26-transformed MEPs following oncogene excision by a site-specific recombinase have allowed to better define the contribution of both oncogenes. Specifically, v-*ets* was shown to be essential to the ability of E26 to block the differentiation of MEPs along the erythroid lineage (Golay et al, 1988; Domenget et al, 1992; Kraut et al, 1994; Rossi et al, 1996) and also to contribute to the immature phenotype of E26-transformed myeloblasts (Golay et al, 1988; Kraut et al, 1994). Interestingly, a murine retrovirus engineered to express the Gag-Myb-Ets protein of E26 was found to induce a high incidence of erythroleukemia after injection into newborn mice (Ruscetti et al, 1992). The leukemic cells of these animals were found to require erythropoietin (Epo) for their proliferation and to easily generate Epo-dependent cell lines in tissue culture. These leukemic cells are highly blocked in their ability to differentiate into mature erythrocytes as analyzed in standard CFU-E colony assays (Ruscetti et al, 1992). The respective contribution of v-Myb^E and v-Ets to the leukemogenic properties of Gag-Myb-Ets in this system remain to be analyzed.

The lesion in E26 ts 1.1 is a single point mutation resulting in a histidine-to-aspartic acid substitution in the DNA binding domain of v-Ets (Golay et al, 1988) which renders an isolated v-Ets polypeptide – and presumably also Gag-Myb-Ets – temperature sensitive for specific DNA binding to EBSs, suggesting that inhibition of erythroid differentiation by Gag-Myb-Ets requires the DNA protein properties of its v-Ets domain (Kraut et al, 1994). Analyses of the structure of the genomes of E26 viruses recovered from the leukemic cells of animals infected with E26 variants defective in their leukemia-inducing properties also point to a strong – if not absolute – selective pressure for an intact ETS domain for leukemia induction in chickens (Metz et al, 1991). In contrast, the transcriptional activation domain of v-Ets was found to be dispensable for the leukemogenic properties of Gag-Myb-Ets (Metz et al, 1991). This, together with the fact that the v-Myb^E transcriptional activation domain is dispensable for the block of erythroid differentiation (Domenget et al, 1992) suggests that at least part of the activity of Gag-Myb-Ets results from the inhibition of EBS-regulated genes essential to erythroid differentiation. Analysis of the properties of E26 ts1.1 show that the v-Ets domain of Gag-Myb-Ets also contributes to the block of differentiation of MEPs along the myeloid and eosinophilic lineages by a mechanism which may not require v-Ets DNA binding activity (Kraut et al, 1994; Rossi et al, 1996).

E26 ts21 is another temperature-sensitive mutant of E26 which was isolated by virtue of its ability to induce differentiation of transformed myeloblasts into macrophages and granulocytes (Beug et al, 1984; Frampton et al, 1995). The lesion in E26 ts21 is a single base change which modifies a threonine residue into arginine in the v-Myb^E DNA binding domain (Frykberg et al, 1988), an alteration which results in impaired specific DNA binding activity of a v-Myb^E polypeptide at the non permissive temperature (Frampton et al, 1995). Inactivation of v-Myb^E function by shift to the non permissive temperature in MEPs transformed by E26 ts21 specifically induces the differentiation of these cells into thrombocytes, indicating that a major activity of the v-Myb^E domain of Gag-Myb-Ets is to block thrombocytic differentiation of MEPs.

The conditional properties of E26 ts21 and E26 ts1.1 have been used to identify candidate target genes regulated by the Gag-Myb-Ets transforming protein in hematopoietic progenitors. This has led to the identification of *mim*-1, a *bona fide* v-Myb^E target gene which encodes a protein which is unlikely to be related to transformation since it is contained in the granules of both normal and v-Myb^E-transformed promyelocytes (Ness et al, 1989) and of *rem*-1, a putative direct target of the v-Ets domain of Gag-myb-Ets. *Rem* encodes a protein related to EF-hand-containing calcium binding proteins and its expression can confer a growth advantage to E26 ts1.1 transformed MEPs in certain conditions (Kraut et al, 1995).

Friend virus-induced erythroleukemia

The Friend virus complex is composed of a replication-defective spleen focus forming virus (SFFV) and a replication-competent helper virus (F-MLV). Inoculation of adult mice with this complex results first in the polyclonal outgrowth of erythroid progenitors in the spleen of infected animals. These cells are endowed with limited self-renewal capacity and are able to differentiate into mature erythrocytes (for review see Ben-David and Bernstein, 1991). This first step in disease development has been shown to be the direct consequence of the constitutive activation of the erythropoietin receptor (EpoR) as the result of the binding of the 55 Kd SFFV envelope glycoprotein (Hoatlin et al, 1990; Li et al, 1990). Constitutive signaling through the EpoR is sufficient to induce this preleukemic step since an EpoR carrying a single amino acid change in its extracellular domain (EpoR(R129C): Yoshimura et al, 1990), a mutation which results in its constitutive dimerization, is sufficient to induce the first step of the Friend leukemia (Longmore and Lodish, 1991). After several weeks, undifferentiated leukemic cells emerge from the polyclonal population of preleukemic cells, which are characterized by an extended self-renewal capacity and their ability to be established as immortal cell lines in tissue culture (Ben-David et al, 1991). At least two recurrent genetic alterations are associated with this leukemic condition. The first is the overexpression of Spi-1/PU-1 (*S*FFV *P*rovirus *I*ntegration) as the result of insertional mutagenesis by an SFFV provirus (Moreau-Gachelin et al, 1988). The second is the inactivation of the p53 tumor suppressor gene as the result of SFFV proviral insertion, gene deletion and/or point mutations in essential domains of the p53 protein. (Mowat et al, 1985; Chow et al, 1987; Ben-David et al, 1990; Munroe et al, 1990).

A role of p53 inactivation in disease development is suggested by the observation that transgenic mice harboring a mutant p53 progress more rapidly to the leukemic phase than control mice following infection by the Friend virus complex (Lavigueur and Bernstein, 1991).

Several systems have been used to delineate the role of Spi-1 in disease development. Infection of long-term mouse bone marrow cultures with a Spi-1/PU-1 retrovirus results in the proliferation of Epo- and stroma-dependent blast-like cells, suggesting a role of Spi-1/PU-1 in erythroblast proliferation (Schuetze et al, 1993). The use of *Spi-1* antisense oligonucleotide in a Friend tumor cell-line also indicated a role of Spi-1/PU-1 in the control of cell proliferation (Delgado et al, 1994). Expression of Spi-1/PU-1 in primary avian erythroblasts was shown to inhibit the ability of these cells to differentiate terminally in erythrocytes in response to avian Epo (Tran Quang et al, 1995). In this system, expression of a mutant p53 was found to have little effect on erythroid differentiation of CFU-E type progenitors but to be able to relieve the polypeptidic growth factor requirements for long-term proliferation of a rare erythroid progenitor in bone marrow cells (Tran Quang et al, 1995). Similarly, a mouse transgenic line engeneered to express Spi-1/PU-1 was shown to develop erythroleukemia at 2–6 months of age (Moreau-Gachelin et al, 1996). These leukemic erythroblasts were found to be partially blocked in differentiation and to remain Epo-dependent for proliferation.

The contribution of Spi-1/PU-1 is however likely to be more complex as further studies on its effect on the phenotype of avian primary erythroblasts show a role for Spi-1/PU-1 both in protection from apoptosis, blockade of terminal differentiation and induction of proliferation in response to SCF (Kit ligand) (C. Tran Quang, O. Wessely, H. Beug and J. Ghysdael, submitted). Importantly, these properties were found to be dependent upon the cooperation of Spi-1/PU-1 with an activated form of the EpoR.

The F-MLV helper virus can also induce erythroleukemia when injected into susceptible newborn mice (Ben-David et al, 1991). FLI-1 (*F*riend *l*eukemia *i*nduced 1) was found to be overexpressed in about 75% of F-MLV-induced erythroleukemias as the result of a nearby F-MLV provirus insertion at the *FLI-1* locus (Ben-David et al, 1990). As for SFFV-derived cell lines, cell-lines established from F-MLV induced erythroleukemic cells also carry rearrangement, deletion or point mutation in the gene for p53. Subsequent studies have shown that insertional activation of FLI-1 expression is an early event in the course of F-MLV induced leukemias whereas inactivation of p53 is a late event associated with the establishment of erythroleukemic cell-lines in tissue culture (Howard et al, 1996). Inactivation of p53 may not therefore be of central importance in F-MLV induced erythroleukemias – and possibly also SFFV-induced disease – in vivo. The molecular mechanisms which underlie the activity of Spi-1/PU-1 or FLI-1 in leukemia transformation are as yet unknown.

Human solid tumors and leukemias

Ewing sarcomas and the related peripheral neuroepitheliomas are characterized by highly specific and recurrent t(11;22) (q24;q12) or t(21;22) (q22;q12) chromosomal translocation, observed in about 85% and 10% of the cases, respectively (Aurias et al, 1983; Turc-Carel et al, 1983; Zucman et al, 1993b; Sorensen et al, 1993). A rare variant t(7;22) (p22;q12) is also occasionally seen (Jeon et al, 1995). These translocations all result in the fusion on the der(22) chromosome of the 5′ half of the *EWS* gene to the 3′ region of either *FLI-1*, or *ERG-2*, or *ETV-1* (Delattre et al, 1992; Zucman et al, 1993b; Sorensen et al, 1993; Jeon et al, 1995). The EWS-promoter driven expression of these chimeric genes result in the synthesis of EWS-FLI-1, EWS-ERG-2 or EWS-ETV-1 fusion proteins (Figure 2.5) (May et al, 1993a; Zucman et al, 1993b; Bailly et al, 1994). In a similar way, the t(16;21) (p11;q22) translocation found in several types of acute leukemias results in the expression of a chimeric transcript encoding a TLS-ERG-2 hybrid protein containing TLS-derived sequences at its amino-terminal moiety fused to the carboxy-terminal half of ERG-2 (Shimizu et al, 1993; Ichikawa et al, 1994).

EWS and TLS are highly related structurally (Crozat et al, 1993) and form, together with the *D. melanogaster* Cabeza/SARFH protein and the recently characterized hTAFII68 (Bertolotti et al, 1996) a subfamily of protein containing a consensus RNA recognition motif (RRM; RNP-CS) (Burd and Dreyfuss, 1994). This subfamily is characterized by a characteristic RNP-1 box and also includes several clusters of RGG repeats, an RNA binding motif often found in association with other RNA binding domains. Except for SARFH, these proteins also share high similarity in their amino-terminal domains which includes multiple copies of an imperfectly repeated S/G-Y-S/G-Q-Q/S-Q/S/P motif. These proteins are able to bind RNA and single stranded DNA in vitro (Crozat et al, 1993; Ohno et al, 1994; Prasad et al, 1994; Bertolotti et al, 1996) and TLS, EWS and SARFH have been found to be associated in vivo with RNA polymerase II transcripts (Immanuel et al, 1995; Zinszner et al, 1994) with TLS being identical to hnRNPP2 (Calvio et al, 1995). In polytene chromosomes of *D. Melanogaster* salivary glands, SARFH is associated with active transcription units, suggesting a role in RNA polymerase II dependent transcription (Immanuel et al, 1995). Recently, a subpopulation of cellular TLS and hTAFII68 was found to be associated with TBP and core hTAFIIs to form distinct TFIID complexes (Bertolotti et al, 1996). This indicates that TLS and EWS exist as different population in cells and could be involved in several aspects of the control of transcriptional initiation and elongation.

In most of the EWS-FLI-1 and EWS-ERG-2 fusion proteins observed in Ewing sarcoma tumor cell-lines, the amino-terminal domain of FLI-1 or ERG-2 is replaced by the 265 amino-terminal residues forming the SYGQQSP domain of EWS-FLI-1 (type I fusions) (Delattre et al, 1992;

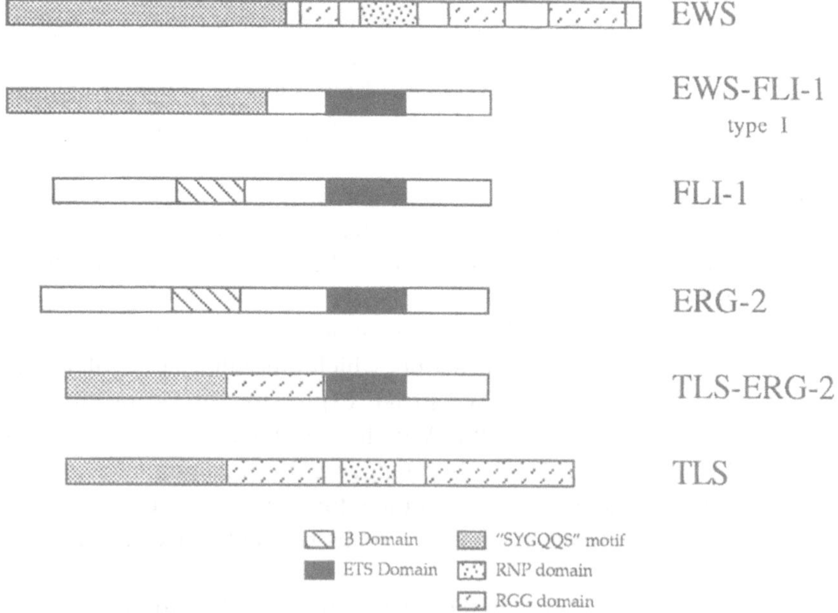

Figure 2.5 Schematic representation of representative fusion proteins involving EWS and TLS in human solid tumors and leukemias. The type I EWS-FLI-1 fusion-protein resulting from the t(11;22) chromosomal translocation of Ewing sarcomas and related tumors (Delattre et al, 1992) is aligned with the parental FLI-1 and EWS proteins. Of note, splice variants of this fusion gene, affecting both the EWS and FLI-1 derived domains, are observed in subset of tumor samples (Zucman et al, 1993b). The TLS-ERG-2 fusion protein resulting from the t(16;21) chromosomal translocation of acute myeloid leukemias is aligned with parental ERG-2 and TLS.

Zucman et al, 1993b). In a minority of cases and in TLS-ERG-2, additional EWS or TLS sequences are present which can include the first RGG motif. EWS-FLI-1, EWS-ERG-2 and TLS-ERG-2 are nuclear proteins which display a similar sequence-specific DNA binding activity as the parental FLI-1 and ERG-2 proteins (May et al, 1993b; Bailly et al, 1994; Prasad et al, 1994; but see also Magnaghi-Jaulin et al, 1996). EWS-FLI-1 and TLS-ERG-2 are sequence-specific transcriptional activators of EBS-driven reporter genes, an activity which is dependent upon the integrity of their EWS and TLS sequences (Ohno et al, 1993; Bailly et al, 1994; Prasad et al, 1994). Both the amino-terminal domains of EWS and TLS can function as strong transcriptional activation domains in the context of Gal4 fusion proteins (May et al, 1993b; Bailly et al, 1994; Zinszner et al, 1994). Expression of EWS-FLI-1 in NIH-3T3 cells induces both morphological transformation and anchorage-independent proliferation in soft agar and confers to these cells the ability to form tumors in nude mice, indicating a central role of EWS-FLI-1 in tumorigenesis (May et al,

1993 a; Zinszner et al, 1994). This activity is dependent upon the integrity of both the EWS and FLI-1 domains of the fusion protein and is specific of EWS-FLI-1 since overexpression of FLI-1 is unable to induce cellular transformation (May et al, 1993 b). This suggests that EWS-FLI-1 and related fusion proteins have diverted one or more aspects of the physiological function(s) of EWS and TLS in transcriptional regulation to generate aberrant transcriptional factors which pervade and deregulate the normal control of gene expression by FLI-1, ERG-2 and/or other members of the ETS family.

Experimental evidence in favor of this model was recently obtained by the identification of several transcripts which are either up-regulated or repressed as the result of EWS-FLI-1 expression in is transformed NIH 3T3 cells (Braun et al, 1995). With the possible exception of stromelysin 1 which contains a tandemly arranged EBS in its promoter region, essential to the Ras-dependent activation of this gene, it is unclear whether these transcripts originate from target genes relevant to Ewing sarcoma tumorigenesis.

Regulatory events other than those controlled by ETS family members appear to be disrupted in other EWS and TLS fusion proteins. Several other chromosomal translocations have been described which involve EWS or TLS and also result in the expression of chimeric transcriptional activators including the fusion of EWS to ATF-1 in malignant melanomas of soft parts (Zucman et al, 1993 a; Brown et al, 1995), to WT-1 in desmoplastic small round cell tumors (Ladanyi and Gerald, 1994), to a member of the NURRI family of orphan nuclear receptors in human myxoid chondrosarcoma (Labelle et al, 1995; Clark et al, 1996) and the fusion of TLS to CHOP in myxoid liposarcoma (Crozat et al, 1993; Rabbitts et al, 1993; Zinszner et al, 1994).

TEL belongs to a subclass of the ETS proteins – which also include ETS-1, ETS-2, FLI-1, ERG-2, GABPα in vertebrates and PNT-P2, YAN and ELG in *Drosophila* – characterized by the presence of a conserved amino-terminal domain of about 65 residues (B domain; *pointed* domain; Figure 2.1 and 2.3). A large spectrum of human leukemias is associated with a rearrangement of the *TEL* gene as the result of specific chromosomal translocations. The translocations involving TEL each result in the expression of specific chimeric transcripts encoding fusion proteins in which different domains of TEL are fused to a variety of partners (Figure 2.6). A subset of chronic myelomonocytic leukemias (CMML) is characterized by a specific and recurrent t(5;12) (q33;p13) balanced translocation. The molecular consequence of this translocation is the expression of a fusion transcript encoding a protein in which the 154 amino-terminal residues of TEL are fused in frame to the transmembrane and tyrosine kinase domain of PDGFRβ (Golub et al, 1994). The same domain of TEL has been shown to be fused to the amino acid sequence encoded by exon 2 of the *cABL* proto-oncogene in a case of childhood ALL with t(9;12) to generate a TEL-ABL fusion protein with elevated tyrosine kinase activity (Papadopoulos et al, 1995). Another TEL-ABL fusion has recently been described in a patient with acute undifferentiated myeloid leukemia presenting a complex t(9;12;14) translocation which result in the fusion of 336 amino acids of TEL to the same domain of ABL (Golub et al, 1996). Consistent with their central role in leukemogenesis, TEL-PDGFRβ and TEL-ABL have been shown to confer growth factor independence for long-term proliferation to the IL3-dependent Ba/F3 hematopoietic cell-line (Golub et al, 1996; Jousset et al, 1997). Furthermore, TEL-ABL has been shown to induce the outgrowth of lymphoid progrenitors from primary bone marrow cells and to transform rat-1 fibroblasts (Golub et al, 1996). Deletion analyses have shown that the conserved amino-terminal domain of TEL (amino acid residues 54–119) is necessary and sufficient to induce oligomerization of TEL itself, TEL-PDGFRβ, TEL-ABL, TEL-AML-1 (Golub et al, 1996; Jousset et al, 1997) and probably also type I TEL-MN1 fusion proteins (see below). This property is unique to TEL since substitution of the amino-terminal conserved domain of TEL in TEL-PDGFRβ by the homologous domain of other vertebrate ETS proteins results in oligomerization-incompetent forms of TEL-PDGFRβ (Jousset et al, 1997), thus providing a rationale for the specific and frequent involvement of TEL in oncogenic fusion proteins.

TEL-induced oligomerization is essential for the constitutive activation of the intrinsic protein kinase activities of the TEL-PDGFRβ and TEL-ABL fusion proteins (Golub et al, 1996; Jousset et al, 1997). Inactivation of the oligomerization properties of TEL-PDGFRβ and TEL-ABL by either deletion of the TEL oligomerization domain or its substitution by the corresponding domains of other ETS proteins and inactivation of their tyrosine protein kinase activity by mutation in their catalytic domains were found to abolish their mitogenic and transforming properties (Golub et al, 1996; Jousset et al, 1997; G. Gilliland, personal communication). Transformation by these fusion proteins appears therefore likely to involve the constitutive phosphorylation of critical substrates. It is also apparent however that transformation by these fusion proteins does not merely recapitulate in a constitutive manner the activation of PDGFRβ and the c-ABL protein kinases, respectively. Indeed, the TEL-ABL and TEL-PDGFRβ fusion proteins localize in subcellular compartment different from those of c-ABL and PDGFRβ, with TEL-ABL localizing to the cytosqueletton of transformed cells (Golub et al, 1996) and TEL-PDGFRβ to the cytosol (Jousset et al, 1997). These oncogenic protein kinases are therefore likely

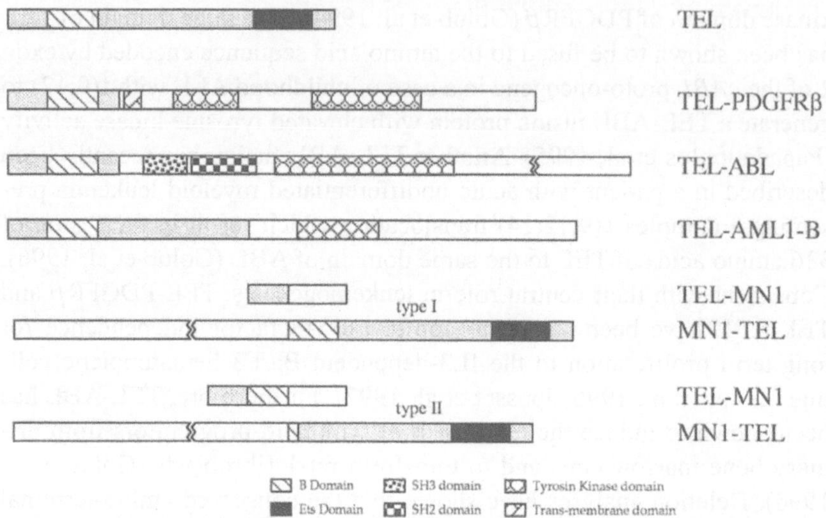

Figure 2.6 Schematic representation of fusion proteins involving TEL in human leukemias. TEL is aligned with the TEL-PDGFRβ and TEL-ABL fusion proteins observed in the leukemias cells of CMML with t(5;12) (Golub et al, 1994) and cALL with t(9;12) (Papadopoulos et al, 1995). Of note, a distinct TEL-ABL fusion is observed in AML with t(9;12) (Golub et al, 1996). Also shown are the TEL-AML1b and TEL-MN1 fusion proteins described in childhood pre B-ALL with t(12;21) and AML with t(12;22) (Golub et al, 1995; Romana et al, 1995a; Buijs et al, 1995).

to access or phosphorylate other protein substrates than those normally involved in c-ABL and PDGFRβ signaling and some of these substrates may play a critical role in their transforming properties.

The TEL-AML-1 fusion was identified in leukemic cells of childhood pre B-ALL as the result of act(12;21) translocation (Golub et al, 1995; Romana et al, 1995b). The consequence of this translocation is the fusion of the 336 amino-terminal residues of TEL (excluding therefore the ETS domain) to most of the sequences of AML-1B, one of the subunit of core binding factor (CBF). The frequency of this chromosomal rearrangement is very high in pediatric cases of ALL, making the TEL-AML-1B fusion the most frequent gene rearrangement in pediatric malignancies (Golub et al, 1995; Romana et al, 1995a; Raynaud et al, 1996). AML-1B, together with the related AML-2 and AML-3 and their mouse homologs form a transcription factor family which display homology to the *D. melanogaster* pair-rule gene *runt*. The *runt* homology domain is responsible for specific binding of AML-1B to the enhancer core sequence (TGT/CGGT), a DNA motif essential to the tissue-specific activity of a number of viral and cellular enhancers. The *runt* domain also mediates interaction with the CBFβ subunit, an association which enhances specific binding to DNA.

The TEL-AML-1B fusion protein contains and intact *runt* domain and consequently was found to interact with CBFβ and to bind core motif sequences (Hiebert et al, 1996). TEL-AML-1B also includes the AML-1B transcriptional activation domain(s) and the TEL-derived oligomerization domain. Yet, TEL-AML-1B failed to activate the expression of a model reporter gene under control of the TCRβ enhancer (Hiebert et al, 1996). Rather, in these assays, TEL-AML-1B functions as a dominant repressor of AML-1B-activated transcription, an activity which requires the integrity of the TEL-derived oligomerization domain (Hiebert et al, 1996).

Embryos in which PEBP2α – the mouse equivalent of AML-1B – is inactivated by homologous recombination lack definitive erythropoiesis of all lineages, suggesting an essential role for AML-1B in the maintenance of stem cells or the proliferation and differentiation control of more mature progenitors (Okuda et al, 1996). Part of the oncogenic properties of TEL-AML-1 could therefore result from dominant inhibition of AML-1B target genes essential to B-cell differentiation and proliferation control.

The occurrence of the t(12;21) translocation is frequently – if not always – associated with the concommitent loss of the second TEL allele (Raynaud et al, 1996), suggesting that the maintenance of TEL expression in leukemic cells could interfere through oligomerization with the oncogenic properties of TEM-AML-1. If this is the case, TEL could define a novel type of tumor suppressor genes displaying specificity toward specific and related oncogenes.

The t(12;22) translocation observed in some patients with myeloid leukemia or a myelodysplastic condition known as refractory anemia with excess blast (RAEB) was found to result in the expression of either of two forms of an MN-1-TEL fusion protein (Buijs et al, 1995). This results in the fusion of the 1259 amino-terminal residues of MN-1, a protein of unknown function, either to amino acid 56 to 452 (end) (type I fusion) or to amino acid 110–452 (type II fusion) of TEL. Both types of fusion proteins maintain the ETS domain of TEL, whereas only type I fusion maintains an intact oligomerization domain. Although the functional properties of MN1-TEL fusions remain to be established, they are likely to act as aberrant transcriptional regulators of TEL-regulated genes.

Conclusions

Although the precise function of the v-*ets* oncogene in E26-induced leukemias is not yet completely understood, the characterization of its cellular homolog has allowed the discovery of a large family of transcriptional regulators which play essential roles in the adaptative response of many cells to a wide variety of extra- and intracellular signals. Importantly, several ETS proteins have now been identified to be essential components in a number of human malignancies and possibly also other diseases (Sumarsono et al, 1996). At present, very little information is available as to the mode of action of these oncogenic forms. There is little doubt however that their study will shed light on their function, not only in disease development but also on the importance of ETS proteins in the normal pathways involved in the control of cell survival, proliferation and differentiation.

Acknowledgments

We thank F. Pio and R. Treisman and members of our laboratory for critical reading of this manuscript and many of our colleagues for sharing unpublished data. Studies in the authors laboratory are supported by funds from the Centre National de la Recherche Scientifique, Institut National de la Recherche Médicale, Institut Curie, Association pour la Recherche sur le Cancer, Ligue Nationale contre le Cancer, Fondation contre la Leucémie, North Atlantic Treaty Organization (NATO) and European Community (EC).

References

Ahne B, Stratling WH (1994): Characterization of a myeloid-specific enhancer of the chicken lysozyme gene. Major role for an Ets transcription factor-binding site. *J Biol Chem* 269: 17794–17801

Albagli O, Soudant N, Ferreira E, Dhordain P, Dewitte F, Begue A, Flourens A, Stehelin D, Leprince D (1994): A model for gene evolution of the ets-1/ets-2 transcription factors based on structural and functional homologies. *Oncogene* 9: 3259–3271

Aurias A, Rimbaut C, Buffe D, Dubousset J, Mazabraud A (1983): Translocation of chromosome 22 in Ewing's sarcoma. *C R Seances Acad Sci III* 296: 1105–1107

Bailly RA, Bosselut R, Zucman J, Cormier F, Delattre O, Roussel M, Thomas G, Ghysdael J (1994): DNA-binding and transcriptional activation properties of the EWS-FLI-1 fusion protein resulting from the t(11;22) translocation in Ewing sarcoma. *Mol Cell Biol* 14: 3230–3241

Beitel GJ, Tuck S, Greenwald I, Horvitz HR (1995): The Caenorhabditis elegans gene lin-1 encodes an ETS-domain protein and defines a branch of the vulval induction pathway. *Genes Dev* 9: 3149–3162

Ben-David Y, Bernstein A (1991): Friend virus-induced erythroleukemia and the multistage nature of cancer. *Cell* 66: 831–834

Ben-David Y, Lavigueur A, Cheong GY, Bernstein A (1990): Insertional inactivation of the p53 gene during friend leukemia: a new strategy for identifying tumor suppressor genes. *New Biol* 2: 1015–1023

Ben-David Y, Giddens EB, Letwin K, Bernstein A (1991): Erythroleukemia induction by Friend murine leukemia virus: insertional activation of a new member of the ets gene family, Fli-1, closely linked to c-ets-1. *Genes Dev* 5: 908–918

Bertolotti A, Lutz Y, Heard D, Chambon P, Tora L (1996): hTAFII68, a novel RNA/ssDNA binding protein with homology to the pro-oncoproteins TLS/FUS and EWS is associated with both TFIID and RNA polymerase II. *EMBO J* 15: 5022–5031

Beug H, Hayman MJ, Graf T (1982): Myeloblasts transformed by the avian acute leukemia virus E26 are hormone-dependent for growth and for the expression of a putative myb-containing protein, p135 E26. *EMBO J* 1: 1069–1073

Beug H, Leutz A, Kahn P, Graf T (1984): Ts mutants of E26 leukemia virus allow transformed myeloblasts, but not erythroblasts or fibroblasts, to differentiate at the nonpermissive temperature. *Cell* 39: 579–588

Bhat NK, Komschlies KL, Fujiwara S, Fisher RJ, Mathieson BJ, Gregorio TA, Young HA, Kasik JW, Ozato K, Papas TS (1989): Expression of ets genes in mouse thymocyte subsets and T-cells. *J Immunol* 142: 672–678

Bhat NK, Thompson CB, Lindsten T, June CH, Fujiwara S, Koizumi S, Fisher RJ, Papas TS (1990): Reciprocal expression of human ETS1 and ETS2 genes during T-cell activation: regulatory role for the protooncogene ETS1. *Proc Natl Acad Sci USA* 87: 3723–3727

Bister K, Nunn M, Moscovici C, Perbal B, Baluda M, Duesberg PH (1982): Acute leukemia viruses E26 and avian myeloblastosis virus have related transformation-specific RNA sequences but different genetic structures, gene products, and oncogenic properties. *Proc Natl Acad Sci USA* 79: 3677–3681

Bories JC, Willerford DM, Grevin D, Davidson L, Camus A, Martin P, Stehelin D, Alt FW (1995): Increased T-cell apoptosis and terminal B-cell differentiation induced by inactivation of the Ets-1 proto-oncogene. *Nature* 377: 635–638

Bosselut R, Duvall JF, Gégonne A, Bailly M, Hemar A, Brady J, Ghysdael J (1990): The product of the c-ets-1 proto-oncogene and the related Ets2 protein act as transcriptional activators of the long terminal repeat of human T-cell leukemia virus HTLV-1. *EMBO J* 9: 3137–3144

Bosselut R, Levin J, Adjadj E, Ghysdael J (1993): A single amino-acid substitution in the Ets domain alters core DNA binding specificity of Ets1 to that of the related transcription factors Elf1 and E74. *Nucleic Acids Res* 21: 5184–5191

Boulukos KE, Pognonec P, Begue A, Galibert F, Gesquiere JC, Stehelin D, Ghysdael J (1988): Identification in chickens of an evolutionarily conserved cellular ets-2 gene (c-ets-2) encoding nuclear proteins related to the products of the c-ets proto-oncogene. *EMBO J* 7: 697–705.

Boulukos KE, Pognonec P, Rabault B, Begue A, Ghysdael J (1989): Definition of an Ets1 protein domain required for nuclear localization in cells and DNA-binding activity in vitro. *Mol Cell Biol* 9: 5718–5721

Boulukos KE, Pognonec P, Sariban E, Bailly M, Lagrou C, Ghysdael J (1990): Rapid and transient expression of Ets2 in mature macrophages following stimulation with cMGF, LPS, and PKC activators. *Genes Dev* 4: 401–409

Bradford AP, Conrad KE, Wasylyk C, Wasylyk B, Gutierrez-Hartmann A (1995): Functional interaction of c-Ets-1 and GHF-1/Pit-1 mediates Ras activation of pituitary-specific gene expression: mapping of the essential c-Ets-1 domain. *Mol Cell Biol* 15: 2849–2857

Braun BS, Frieden R, Lessnick SL, May WA, Denny CT (1995): Identification of target genes for the Ewing's sarcoma EWS/FLI fusion protein by representational difference analysis. *Mol Cell Biol* 15: 4623–4630

Brown AD, Lopez-Terrada D, Denny C, Lee KA (1995): Promoters containing ATF-binding sites are de-regulated in cells that express the EWS/ATF1 oncogene. *Oncogene* 10: 1749–1756

Brown TA, McKnight SL (1992): Specificities of protein-protein and protein-DNA interaction of GABP alpha and two newly defined ets-related proteins. *Genes Dev* 6: 2502–2512

Brunner D, Ducker K, Oellers N, Hafen E, Scholz H, Klambt C (1994): The ETS domain protein pointed-P2 is a target of MAP kinase in the sevenless signal transduction pathway. *Nature* 370: 386–389

Buijs A, Sherr S, van Baal S, van Bezouw S, van der Plas D, Geurts van Kessel A, Riegman P, Lekanne Deprez R, Zwarthoff E, Hagemeijer A, Grosveld G (1995): Translocation (12;22) (p13;q11) in myeloproliferative disorders results in fusion of the ETS-like TEL gene on 12p13 to the MN1 gene on 22q11. *Oncogene* 10: 1511–1519

Burd CG, Dreyfuss G (1994): Conserved structures and diversity of functions of RNA-binding proteins. *Science* 265: 615–621

Burtis KC, Thummel CS, Jones CW, Karim FD, Hogness DS (1990): The *Drosophila* 74EF early puff contains E74, a complex ecdysone-inducible gene that encodes two ets-related proteins. *Cell* 61: 85–99

Calvio C, Neubauer G, Mann M, Lamond AI (1995): Identification of hnRNP P2 as TLS/FUS using electrospray mass spectrometry. *RNA* 1: 724–733

Carter RS, Avadhani NG (1994): Cooperative binding of GA-binding protein transcription factors to duplicated transcription initiation region repeats of the cytochrome c oxidase subunit IV gene. *J Biol Chem* 269: 4381–4387

Carter RS, Bhat NK, Basu A, Avadhani NG (1992): The basal promoter elements of murine cytochrome c oxidase subunit IV gene consist of tandemly duplicated ets motifs that bind to GABP-related transcription factors. *J Biol Chem* 267: 23418–23426

Celada A, Borras F, Soler C, Llobregas J, Klemsz M, Van Beveren C, McKercher S, Maki R (1996): The transcription factor PU.1 is involved in macrophage proliferation. *J Exp Med* 184: 61–69

Chen H, Ray-Gallet D, Zhang P, Hetherington CJ, Gonzalez DA, Zhang DE, Moreau-Gachelin F, Tenen DG (1995a): PU.1 (Spi-1) autoregulates its expression in myeloid cells. *Oncogene* 11: 1549–1560

Chen HM, Zhang P, Voso MT, Hohaus S, Gonzalez DA, Glass CK, Zhang DE, Tenen DG (1995b): Neutrophils and monocytes express high levels of PU.1 (Spi-1) but not Spi-B. *Blood* 85: 2918–2928

Chen JH (1985): The proto-oncogene c-ets is preferentially expressed in lymphoid cells. *Mol Cell Biol* 5: 2993–3000

Chen JH (1988): Complementary DNA clones of chicken proto-oncogene c-ets: sequence divergence from the viral oncogene v-ets. *Oncogene Res* 2: 371–384

Chen T, Bunting M, Karim FD, Thummel CS (1992): Isolation and characterization of five *Drosophila* genes that encode an ets-related DNA binding domain. *Dev Biol* 151: 176–191

Chen ZQ, Kan NC, Pribyl L, Lautenberger JA, Moudrianakis E, Papas TS (1988): Molecular cloning of the ets proto-oncogene of the sea urchin and analysis of its developmental expression. *Dev Biol* 125: 432–440

Chow V, Ben-David Y, Bernstein A, Benchimol S, Mowat M (1987): Multistage Friend erythroleukemia: independent origin of tumor clones with normal or rearranged p53 cellular oncogenes. *J Virol* 61: 2777–2781

Clark J, Benjamin H, Gill S, Sidhar S, Goodwin G, Crew J, Gusterson BA, Shipley J, Cooper CS (1996): Fusion of the EWS gene to CHN, a member of the steroid/thyroid receptor gene superfamily, in a human myxoid chondrosarcoma. *Oncogene* 12: 229–235

Clark KL, Halay ED, Lai E, Burley SK (1993a): Co-crystal structure of the HNF-3/fork head DNA-recognition motif resembles histone H5. *Nature* 364: 412–420

Clark NM, Smith MJ, Hilfinger JM, Markovitz DM (1993b): Activation of the human T-cell leukemia virus type I enhancer is mediated by binding sites for Elf-1 and the pets factor. *J Virol* 67: 5522–5528

Coffer P, de Jonge M, Mettouchi A, Binetruy B, Ghysdael J, Kruijer W (1994): junB promoter regulation: Ras mediated transactivation by c-Ets-1 and c-Ets-2. *Oncogene* 9: 911–921

Conrad KE, Oberwetter JM, Vaillancourt R, Johnson GL, Gutierrez-Hartmann A (1994): Identification of the functional components of the Ras signaling pathway regulating pituitary cell-specific gene expression. *Mol Cell Biol* 14: 1553–1565

Crabtree GR (1989): Contingent genetic regulatory events in T lymphocyte activation. *Science* 243: 355–361

Crozat A, Aman P, Mandahl N, Ron D (1993): Fusion of CHOP to a novel RNA-binding protein in human myxoid liposarcoma. *Nature* 363: 640–644

Dalton S, Treisman R (1992): Characterization of SAP-1, a protein recruited by serum response factor to the c-fos serum response element. *Cell* 68: 597–612

Dalton S, Treisman R (1994): Characterization of SAP-1, a protein recruited by serum response factor to the c-fos serum response element. *Cell* 76: 411

Datta R, Sherman ML, Stone RM, Kufe D (1991): Expression of the jun-B gene during induction of monocytic differentiation. *Cell Growth Differ* 2: 43–49

de la Brousse FC, Birkenmeier EH, King DS, Rowe LB, McKnight SL (1994): Molecular and genetic characterization of GABP beta. *Genes Dev* 8: 1853–1865

Delattre O, Zucman J, Plougastel B, Desmaze C, Melot T, Peter M, Kovar H, Joubert I, de Jong P, Rouleau G, Aurias A, Thomas G (1992): Gene fusion with an ETS DNA-binding domain caused by chromosome translocation in human tumours. *Nature* 359: 162–165

Delgado MD, Hallier M, Meneceur P, Tavitian A, Moreau-Gachelin F (1994): Inhibition of Friend cells proliferation by spi-1 antisense oligodeoxynucleotides. *Oncogene* 9: 1723–1727

Dittmer J, Gégonne A, Gitlin SD, Ghysdael J, Brady JN (1994): Regulation of parathyroid hormone-related protein (PTHrP) gene expression. Sp1 binds through an inverted CACCC motif and regulates promoter activity in cooperation with Ets1. *J Biol Chem* 269: 21428–21434

Domenget C, Leprince D, Pain B, Peyrol S, Li RP, Stehelin D, Samarut J, Jurdic P (1992): The various domains of v-myb and v-ets oncogenes of E26 retrovirus contribute differently, but cooperatively, in transformation of hematopoietic lineages. *Oncogene* 7: 2231–2241

Donaldson L, Petersen J, Graves B, McIntosh L (1996): Solution structure of the ETS domain from murine Ets-1: a winged helix-turn-helix DNA binding motif. *EMBO J* 15: 125–134

Donaldson LW, Petersen JM, Graves BJ, McIntosh LP (1994): Secondary structure of the ETS domain places murine Ets-1 in the superfamily of winged helix-turn-helix DNA-binding proteins. *Proc Natl Acad Sci USA* 33: 13509–13516

Dupriez VJ, Darville MI, Antoine IV, Gégonne A, Ghysdael J, Rousseau GG (1993): Characterization of a hepatoma mRNA transcribed from a third promoter of a 6-phosphofructo-2-kinase/fructose-2,6-bisphosphatase-encoding gene and controlled by ets oncogene-related products. *Proc Natl Acad Sci USA* 90: 8224–8228

Duterque-Coquillaud M, Leprince D, Flourens A, Henry C, Ghysdael J, Debuire B, Stehelin D (1988): Cloning and expression of chicken p54c-ets cDNAs: the first p54c-ets coding exon is located into the 40.0 kbp genomic domain unrelated to v-ets. *Oncogene Res* 2: 335–344

Eichbaum QG, Iyer R, Raveh DP, Mathieu C, Ezekowitz RA (1994): Restriction of interferon gamma responsiveness and basal expression of the myeloid human Fc gamma R1b gene is mediated by a functional PU.1 site and a transcription initiator consensus. *J Exp Med* 179: 1985–1996

Eisenbeis CF, Singh H, Storb U (1993): PU.1 is a component of a multiprotein complex which binds an essential site in the murine immunoglobulin lambda 2–4 enhancer. *Mol Cell Biol* 13: 6452–6461

Eisenbeis CF, Singh H, Storb U (1995): Pip, a novel IRF family member, is a lymphoid-specific, PU.1-dependent transcriptional activator. *Genes Dev* 9: 1377–1387

Eisenmann DM, Kim SK (1994): Signal transduction and cell fate specification during *Caenorhabditis elegans* vulval development. *Curr Opin Genet Dev* 4: 508–516

Ernst P, Hahm K, Smale ST (1993): Both LyF-1 and an Ets protein interact with a critical promoter element in the murine terminal transferase gene. *Mol Cell Biol* 13: 2982–2992

Feinman R, Qiu, WQ, Pearse RN, Nikolajczyk BS, Sen R, Sheffery M, Ravetch JV (1994): PU.1 and an HLH family member contribute to the myeloid-specific transcription of the Fc gamma RIIIA promoter. *EMBO J* 13: 3852–3860

Fisher CL, Ghysdael J, Cambier JC (1991a): Ligation of membrane Ig leads to calcium-mediated phosphorylation of the proto-oncogene product, Ets-1. *J Immunol* 146: 1743–1749

Fisher RJ, Mavrothalassitis G, Kondoh A, Papas TS (1991b): High-affinity DNA-protein interactions of the cellular ETS1 protein: the determination of the ETS binding motif. *Oncogene* 6: 2249–2254

Fletcher JC, Thummel CS (1995): The *Drosophila* E74 gene is required for the proper stage- and tissue-specific transcription of ecdysone-regulated genes at the onset of metamorphosis. *Development* 121: 1411–1421

Frampton J, McNagny K, Sieweke M, Philip A, Smith G, Graf T (1995): v-Myb DNA binding is required to block thrombocytic differentiation of Myb-Ets-transformed multipotent haematopoietic progenitors. *EMBO J* 14: 2866–2875

Frykberg L, Metz T, Brady G, Introna M, Beug H, Vennstrom B, Graf T (1988): A point mutation in the DNA binding domain of the v-myb oncogene of E26 virus confers temperature sensitivity for transformation of myelomonocytic cells. *Oncogene Res* 3: 313–322

Fujiwara S, Koizumi S, Fisher FJ, Bhat NK, Papas TS (1990): Phosphorylation of the ETS-2 protein: regulation by the T-cell antigen receptor-CD3 complex. *Mol Cell Biol* 10: 1249–1253

Galson DL, Hensold JO, Bishop TR, Schalling M, D'Andrea AD, Jones C, Auron PE, Housman DE (1993): Mouse beta-globin DNA-binding protein B1 is identical to a proto-oncogene, the transcription factor Spi-1/PU.1, and is restricted in expression to hematopoietic cells and the testis. *Mol Cell Biol* 13: 2929–2941

Gégonne A, Punyammalee B, Rabault B, Bosselut R, Seneca S, Crabeel M, Ghysdael J (1992): Analysis of the DNA binding and transcriptional activation properties of the Ets1 oncoprotein. *New Biol* 4: 512–519

Gégonne A, Bosselut R, Bailly RA, Ghysdael J (1993): Synergistic activation of the HTLV1 LTR Ets-responsive region by transcription factors Ets1 and Sp1. *EMBO J* 12: 1169–1178

Genuario RR, Kelley DE, Perry RP (1993): Comparative utilization of transcription factor GABP by the promoters of ribosomal protein genes rpL30 and rpL32. *Gene Expr* 3: 279–288

Ghysdael J, Gégonne A, Pognonec P, Dernis D, Leprince D, Stehelin D (1986): Identification and preferential expression in thymic and bursal lymphocytes of a c-ets oncogene-encoded Mr 54,000 cytoplasmic protein. *Proc Natl Acad Sci USA* 83: 1714–1718

Giese K, Kingsley C, Kirshner JR, Grosschedl R (1995): Assembly and function of a TCR alpha enhancer complex is dependent on LEF-1-induced DNA bending and multiple protein-protein interactions. *Genes Dev* 9: 995–1008

Gille H, Sharrocks AD, Shaw PE (1992): Phosphorylation of transcription factor p62TCF by MAP kinase stimulates ternary complex formation at c-fos promoter. *Nature* 358: 414–417

Gille H, Kortenjann M, Thomae O, Moomaw C, Slaughter C, Cobb MH, Shaw PE (1995a): ERK phosphorylation potentiates Elk-1-mediated ternary complex formation and transactivation. *EMBO J* 14: 951–962

Gille H, Strahl T, Shaw PE (1995b): Activation of ternary complex factor Elk-1 by stress-activated protein kinases. *Curr Biol* 5: 1191–1200

Giovane A, Pintzas A, Maira SM, Sobieszczuk P, Wasylyk B (1994): Net, a new ets transcription factor that is activated by Ras. *Gen Dev* 8: 1502–1513

Golay J, Introna M, Graf T (1988): A single point mutation in the v-ets oncogene affects both erythroid and myelomonocytic cell differentiation. *Cell* 55: 1147–1158

Goldberg Y, Treier M, Ghysdael J, Bohmann D (1994): Repression of AP-1-stimulated transcription by c-Ets-1. *J Biol Chem* 269: 16566–16573

Golub T, Goga A, Barker G, Afar D, McLaughlin J, Bohlander S, Rowley J, Witte O, Gilliland D (1996): Oligomerization of the ABL tyrosine kinase by Ets protein TEL in human leukemia. *Mol Cell Biol* 16: 4107–4116

Golub TR, Barker GF, Lovett M, Gilliland DG (1994): Fusion of PDGF receptor beta to a novel ets-like gene, tel, in chronic myelomonocytic leukemia with t(5;12) chromosomal translocation. *Cell* 77: 307–316

Golub TR, Barker GF, Bohlander SK, Hiebert SW, Ward DC, Bray-Ward P, Morgan E, Raimondi SC, Rowley JD, Gilliland DG (1995): Fusion of the TEL gene on 12p13 to the AML1 gene on 21q22 in acute lymphoblastic leukemia. *Proc Natl Acad Sci USA* 92: 4917–4921

Graf T, McNagny K, Brady G, Frampton J (1992): Chicken "erythroid" cells transformed by the Gag-Myb-Ets-encoding E26 leukemia virus are multipotent. *Cell* 70: 201–213

Graham R, Gilman M (1991): Distinct protein targets for signals acting at the c-fos serum response element. *Science* 251: 189–192

Grant PA, Thompson CB, Pettersson S (1995): IgM receptor-mediated transactivation of the IgH 3′ enhancer couples a novel Elf-1-AP-1 protein complex to the developmental control of enhancer function. *EMBO J* 14: 4501–4513

Gugneja S, Virbasius JV, Scarpulla RC (1995): Four structurally distinct, non-DNA-binding subunits of human nuclear respiratory factor 2 share a conserved transcriptional activation domain. *Mol Cell Biol* 15: 102–111

Gunther CV, Graves BJ (1994): Identification of ETS domain proteins in murine T lymphocytes that interact with the Moloney murine leukemia virus enhancer. *Mol Cell Biol* 14: 7569–7580

Gunther CV, Nye JA, Bryner RS, Graves BJ (1990): Sequence-specific DNA binding of the proto-oncoprotein ets-1 defines a transcriptional activator sequence within the long terminal repeat of the Moloney murine sarcoma virus. *Genes Dev* 4: 667–679

Gutman A, Wasylyk B (1990): The collagenase gene promoter contains a TPA and oncogene-responsive unit encompassing the PEA3 and AP-1 binding sites. *EMBO J* 9: 2241–2246

Hagemeier C, Bannister AJ, Cook A, Kouzarides T (1993): The activation domain of transcription factor PU.1 binds the retinoblastoma (RB) protein and the transcription factor TFIID in vitro: RB shows sequence similarity to TFIID and TFIIB. *Proc Natl Acad Sci USA* 90: 1580–1584

Hagman J, Grosschedl R (1992): An inhibitory carboxyl-terminal domain in Ets-1 and Ets-2 mediates differential binding of ETS family factors to promoter sequences of the mb-1 gene. *Proc Natl Acad Sci USA* 89: 8889–8893

Hallier M, Tavitian A, Moreau-Gachelin F (1996): The transcription factor Spi-1/PU.1 binds RNA and interferes with the RNA-binding protein p54nrb. *J Biol Chem* 271: 11 177–11 181

Harrison CJ, Bohm AA, Nelson HC (1994): Crystal structure of the DNA binding domain of the heat shock transcription factor. *Science* 263: 224–227

Henkel G, Brown MA (1994): PU.1 and GATA: components of a mast cell-specific interleukin 4 intronic enhancer. *Proc Natl Acad Sci USA* 91: 7737–7741

Henkel G, McKercher S, Yamamoto H, Anderson K, Oshima R, Maki R (1996): PU.1 but not Ets-2 is essential for macrophage development from embryonic stem cells. *Blood* 88: 2917–2926

Herrera RE, Shaw PE, Nordheim A (1989): Occupation of the c-fos serum response element in vivo by a multi-protein complex is unaltered by growth factor induction. *Nature* 340: 68–70

Hiebert S, Sun W, Davis N, Golub T, Shurtleff S, Buijs A, Downing J, Grosveld G, Roussel M, Gilliland D, Lenny N, Meyers S (1996): The t(12;21) Translocation Converts AML1B from an activator to a Repressor of Transcription. *Mol Cell Biol* 16: 1349–1355

Higashino F, Yoshida K, Fujinaga Y, Kamio K, Fujinaga K (1993): Isolation of a cDNA encoding the adenovirus E1A enhancer binding protein: a new human member of the ets oncogene family. *Nucleic Acids Res* 21: 547–553

Hill CS, Rimmer JM, Green BN, Finch JT, Thomas JO (1991): Histone-DNA interactions and their modulation by phosphorylation of – Ser-Pro-X-Lys/Arg-motifs. *EMBO J* 10: 1939–1948

Hill CS, Marais R, John S, Wynne J, Dalton S, Treisman R (1993): Functional analysis of a growth factor-responsive transcription factor complex. *Cell* 73: 395–406

Hill CS, Wynne J, Treisman R (1995): The Rho family GTPases RhoA, Rac1, and CDC42Hs regulate transcriptional activation by SRF. *Cell* 81: 1159–1170

Hipskind RA, Nordheim A (1991): Functional dissection in vitro of the human c-fos promoter. *J Biol Chem* 266: 19583–19592

Hipskind RA, Rao VN, Mueller CG, Reddy ES, Nordheim A (1991): Ets-related protein Elk-1 is homologous to the c-fos regulatory factor p62TCF. *Nature* 354: 531–534

Ho IC, Bhat NK, Gottschalk LR, Lindsten T, Thompson CB, Papas TS, Leiden JM (1990): Sequence-specific binding of human Ets-1 to the T-cell receptor alpha gene enhancer. *Science* 250: 814–818

Hoatlin ME, Kozak SL, Lilly F, Chakraborti A, Kozak CA, Kabat D (1990): Activation of erythropoietin receptors by Friend viral gp55 and by erythropoietin and down-modulaton by the murine Fv-2r resistance gene. *Proc Natl Acad Sci USA* 87: 9985–9989

Hohaus S, Petrovick MS, Voso MT, Sun Z, Zhang DE, Tenen DG (1995): PU.1 (Spi-1) and C/EBP alpha regulate expression of the granulocyte-macrophage colony-stimulating factor receptor alpha gene. *Mol Cell Biol* 15: 5830–5845

Howard JC, Berger L, Bani MR, Hawley RG, Ben-David Y (1996): Activation of the erythropoietin gene in the majority of F-MuLV-induced erythroleukemias results in growth factor independence and enhanced tumorigenicity. *Oncogene* 12: 1405–1415

Ichikawa H, Shimizu K, Hayashi Y, Ohki M (1994): An RNA-binding protein gene, TLS/FUS, is fused to ERG in human myeloid leukemia with t(16;21) chromosomal translocation. *Cancer Res* 54: 2865–1868

Immanuel D, Zinszner H, Ron D (1995): Association of SARFH (sarcoma-associated RNA-binding fly homolog) with regions of chromatin transcribed by RNA polymerase II. *Mol Cell Biol* 15: 4562–4571

Ivanov X, Mladenov Z, Nedyalkov S, Todorov T (1962): Experimental investigation into avian leukosis. I Transmissions experiments of certain diseases of the leukosis complex found in Bulgaria. *Bull Instl Pathol Comp Anim Domest* 9: 5–36

Jabrane-Ferrat N, Peterlin BM (1994): Ets-1 activates the DRA promoter in B-cells. *Mol Cell Biol* 14: 7314–7321

Janknecht R (1996): Analysis of the ERK-stimulated ETS transcription factor ER81. *Mol Cell Biol* 16: 1550–1556

Janknecht R, Nordheim A (1992): Elk-1 protein domains required for direct and SRF-assited DNA-binding. *Nucleic Acids Res* 20: 3317–3324

Janknecht R, Nordheim A (1996): Regulation of the c-fos promoter by the ternary complex factor Sap-1a and its coactivator CBP. *Oncogene* 12: 1961–1969

Janknecht R, Ernst WH, Houthaeve T, Nordheim A (1993 a): C-terminal phosphorylation of the serum-response factor. *Eur J Biochem* 216: 469–475

Janknecht R, Ernst WH, Pingoud V, Nordheim A (1993 b): Activation of ternary complex factor Elk-1 by MAP kinases. *EMBO J* 12: 5097–5104

Janknecht R, Zinck R, Ernst WH, Nordheim A (1994): Functional dissection of the transcription factor Elk-1. *Oncogene* 9: 1273–1278

Janknecht R, Ernst WH, Nordheim A (1995): SAP1a is a nuclear target of signaling cascades involving ERKs. *Oncogene* 10: 1209–1216

Jeon IS, Davis JN, Braun BS, Sublett JE, Roussel MF, Denny CT, Shapiro DN (1995): A variant Ewing's sarcoma translocation (7;22) fuses the EWS gene to the ETS gene ETV1. *Oncogene* 10: 1229–1234

John S, Reeves RB, I in JX, Child R, Leiden JM, Thompson CB, Leonard WJ (1995): Regulation of cell-type-specific interleukin-2 receptor alpha-chain gene expression: potential role of physical interactions between Elf-1, HMG-I(Y), and NF-kappa B family proteins. *Mol Cell Biol* 15: 1786–1796

Jousset C, Carron C, Boureux A, Tran Quang C, Oury C, Dusanter-Fourt I, Charon M, Levin J, Bernard O, Ghysdael J (1997): A domain of TEL conserved in a subset of Ets protein defines a specific oligomerization interface essential to the transforming function of the TEL-PDGFRβ oncoprotein. *EMBO J* 16: 69–82

Klambt C (1993): The *Drosophila* gene pointed encodes two ETS-like proteins which are involved in the development of the midline glial cells. *Development* 117: 163–176

Klemsz MJ, Justement LB, Palmer E, Cambier JC (1989): Induction of c-fos and c-myc expression during B-cell activation by IL-4 and immunoglobulin binding ligands. *J Immunol* 143: 1032–1039

Klemsz MJ, McKercher SR, Celada A, Van Beveren C, Maki RA (1990): The macrophage and B-cell-specific transcription factor PU.1 is related to the ets oncogene. *Cell* 61: 113–124

Klemsz MJ, Maki RA, Papayannopoulou T, Moore J, Hromas R (1993): Characterization of the ets oncogene family member, fli-1. *J Biol Chem* 268: 5769–5773

Kodandapani R, Pio F, Ni C, Picciali G, Klemsz M, McKercher S, Maki R, Ely K (1996): A new pattern for helix-turn-helix recogniton revealed by the PU.1 ETS-domain-DNA complex. *Nature* 380: 456–460

Koizumi H, Horta MF, Youn BS, Fu KC, Kwon BS, Young JD, Liu CC (1993): Identification of a killer cell-specific regulatory element of the mouse perforin gene: an Ets-binding site-homologous motif that interacts with Ets-related proteins. *Mol Cell Biol* 13: 6690–6701

Kola I, Brookes S, Green AR, Garber R, Tymms M, Papas TS, Seth A (1993): The Ets1 transcription factor is widely expressed during murine embryo development and is associated with mesodermal cells involved in morphogenetic processes such as organ formation. *Proc Natl Acad Sci USA* 90: 7588–7592

Kominato Y, Galson D, Waterman WR, Webb AC, Auron PE (1995): Monocyte expression of the human prointerleukin 1 beta gene (IL 1B) is dependent on promoter sequences which bind the hematopoietic transcription factor Spi-1/PU.1. *Mol Cell Biol* 15: 59–68

Kortenjann M, Thomae O, Shaw PE (1994): Inhibition of v-raf-dependent c-fos expression and transformation by a kinase-defective mutant of the mitogen-activated protein kinase Erk2. *Mol Cell Biol* 14: 4815–4824

Kraulis P (1991): Molscript: a program to produce both detailed and schematic plots of protein structures. *J Appl Crystallogr* 24: 946–950

Kraut N, Frampton J, McNagny KM, Graf T (1994): A functional Ets DNA-binding domain is required to maintain multipotency of hematopoietic progenitors transformed by Myb-Ets. *Genes Dev* 8: 33–44

Kraut N, Frampton J, Graf T (1995): Rem-1, a putative direct target gene of the Myb-Ets fusion oncoprotein in haematopoietic progenitors, is a member of the recoverin family. *Oncogene* 10: 1027–1036

Labelle Y, Zucman J, Stenman G, Kindblom LG, Knight J, Turc-Carel C, Dockhorn-Dworniczak B, Mandahl N, Desmaze C, Peter M, Aurias A, Delattre O, Thomas G (1995): Oncogenic conversion of a novel orphan nuclear receptor by chromosome translocation. *Hum Mol Genet* 4: 2219–2226

Lackner MR, Kornfeld K, Miller LM, Horvitz HR, Kim SK (1994): A MAP kinase homolog, mpk-1, is involved in ras-mediated induction of vulval cell fates in Caenorhabditis elegans. *Genes Dev* 8: 160–173

Ladanyi M, Gerald W (1994): Fusion of the EWS and WT1 genes in the desmoplastic small round cell tumor. *Cancer Res* 54: 2837–2840

Lai ZC, Rubin GM (1992): Negative control of photoreceptor development in *Drosophila* by the product of the yan gene, an ETS domain protein. *Cell* 70: 609–620

LaMarco K, Thompson CC, Byers BP, Walton EM, McKnight SL (1991): Identification of Ets- and notch-related subunits in GA binding protein. *Science* 253: 789–792

Lavigueur A, Bernstein A (1991): p53 transgenic mice: accelerated erythroleukemia induction by Friend virus. *Oncogene* 6: 2197–2201

Leiden JM, Wang CY, Petryniak B, Markovitz DM, Nabel GJ, Thompson CB (1992): A novel Ets-related transcription factor, Elf-1, binds to human immunodeficiency virus type 2 regulatory elements that are required for inducible transactivation in T-cells. *J Virol* 66: 5890–5897

Lelievre-Chotteau A, Laudet V, Flourens A, Begue A, Leprince D, Fontaine F (1994): Identification of two ets related genes in a marine worm, the polychaete annelid Nereis diversicolor. *FEBS Lett* 354: 62–66

Lemarchandel V, Ghysdael J, Mignotte V, Rahuel C, Romeo PH (1993): GATA and Ets cis-acting sequences mediate megakaryocyte-specific expression. *Mol Cell Biol* 13: 668–676

Leprince D, Gégonne A, Coll J, de Taisne C, Schneeberger A, Lagrou C, Stehelin D (1983): A putative second cell-derived oncogene of the avian leukaemia retrovirus E26. *Nature* 306: 395–397

Leprince D, Duterque-Coquillaud M, Li RP, Henry C, Flourens A, Debuire B, Stehelin D (1988): Alternative splicing within the chicken c-ets-1 locus: implications for transduction within the E26 retrovirus of the c-ets proto-oncogene. *J Virol* 62: 3233–3241

Leung S, McCracken S, Ghysdael J, Miyamoto NG (1993): Requirement of an ETS-binding element for transcription of the human lck type I promoter. *Oncogene* 8: 989–997

Levin JM, Garnier J (1988): Improvements in a secondary structure prediction method based on a search for local sequence homologies and its use as a model building tool. *Biochim Biophys Acta* 955: 283–295

Levin JM, Pascarella S, Argos P, Garnier J (1993): Quantification of secondary structure prediction improvement using multiple alignments. *Protein Eng* 6: 849–854

Li JP, D'Andrea AD, Lodish HF, Baltimore D (1990): Activation of cell growth by binding of Friend spleen focus-forming virus gp55 glycoprotein to the erythropoietin receptor. *Nature* 343: 762–764

Liang H, Mao X, Olejniczak ET, Nettesheim DG, Yu L, Meadows RP, Thompson CB, Fesik SW (1994): Solution structure of the ets domain of Fli-1 when bound to DNA. *Nat Struct Biol* 1: 871–875

Lim F, Kraut N, Framptom J, Graf T (1992): DNA binding by c-Ets-1, but not v-Ets, is repressed by an intramolecular mechanism. *EMBO J* 11: 643–652

Lin JX, Bhat NK, John S, Queale WS, Leonard WJ (1993): Characterization of the human interleukin-2 receptor beta-chain gene promoter: regulation of promoter activity by ets gene products. *Mol Cell Biol* 13: 6201–6210

Longmore GD, Lodish HF (1991): An activating mutation in the murine erythropoietin receptor induces erythroleukemia in mice: a cytokine receptor superfamily oncogene. *Cell* 67: 1089–1102

Lopez M, Oettgen P, Akbarali Y, Dendorfer U, Libermann TA (1994): ERP, a new member of the ets transcription factor/oncoprotein family: cloning, characterization, and differential expression during B-lymphocyte development. *Mol Cell Biol* 14: 3292–3309

Magnaghi-Jaulin L, Masutani H, Robin P, Lipinski M, Harel-Bellan A (1996): SRE elements are binding sites for the fusion protein EWS-FLI-1. *Nucl Acids Res* 24: 1052–1058

Maira S, Wurtz J, Wasylyk B (1996): Net (ERP/SAP2), one of the Ras inducible TCFs, has a novel inhibitory domain with ressemblance to the helix-loop-helix motif. *EMBO J* 15: 5849–5865

Mao X, Miesfeldt S, Yang H, Leiden JM, Thompson CB (1994): The FLI-1 and chimeric EWS-FLI-1 oncoproteins display similar DNA binding specificities. *J Biol Chem* 269: 18216–18222

Marais R, Wynne J, Treisman R (1993): The SRF accessory protein Elk-1 contains a growth factor-regulated transcriptional activation domain. *Cell* 73: 381–393

Markiewicz S, Bosselut R, Le Deist F, de Villartay JP, Hivroz C, Ghysdael J, Fischer A, de Saint Basile G (1996): Tissue-specific activity of the gamma chain gene promoter depends upon an Ets binding site and is regulated by GA-binding protein. *J Biol Chem* 271: 14849–14855

Maroulakou IG, Papas TS, Green JE (1994): Differential expression of ets-1 and ets-2 proto-oncogenes during murine embryogenesis. *Oncogene* 9: 1551–1565

Mavrothalassitis G, Fisher RJ, Smyth F, Watson DK, Papas TS (1994): Structural inferences of the ETS1 DNA-binding domain determined by mutational analysis. *Oncogene* 9: 425–435

May WA, Gishizky ML, Lessnick SL, Lunsford LB, Lewis BC, Delattre O, Zucman J, Thomas G, Denny CT (1993a): Ewing sarcoma 11;22 translocation produces a chimeric transcription factor that requires the DNA-binding domain encoded by FLI1 for transformation. *Proc Natl Acad Sci USA* 90: 5752–5756

May WA, Lessnick SL, Braun BS, Klemsz M, Lewis BC, Lunsford LB, Hromas R, Denny CT (1993b): The Ewing's sarcoma EWS/FLI-1 fusion gene encodes a more potent transcriptional activator and is a more powerful transforming gene than FLI-1. *Mol Cell Biol* 13: 7393–7398

McKercher S, Torbett B, Anderson KL, Henkel G, Vestal D, Baribault H, Klemsz M, Feeney A, Wu G, Paige C, Maki R (1996): Targeted disruption of the PU.1 gene results in multiple hematopoietid abnormalities. *EMBO J* 15: 5647–5658

McCracken S, Leung S, Bosselut R Ghysdael J, Miyamoto NG (1994): Myb and Ets related transcription factors are required for activity of the human lck type I promoter. *Oncogene* 9: 3609–3615

Melet F, Motro B, Rossi DJ, Zhang L, Bernstein A (1996): Generation of a novel Fli-1 protein by gene targeting leads to a defect in thymus development and a delay in Friend virus-induced erythroleukemia. *Mol Cell Biol* 16: 2708–2718

Metz T, Graf T (1991): Fusion of the nuclear oncoproteins v-Myb and v-Ets is required for the leukemogenicity of E26 virus. *Cell* 66: 95–105

Metz T, Graf T, Leutz A (1991): Activation of cMGF expression is a critical step in avian myeloid leukemogenesis. *EMBO J* 10: 837–844

Meyer D, Wolff CM, Stiegler P, Senan F, Befort N, Befort JJ, Remy P (1993): Xl-fli, the *Xenopus* homologue of the fli-1 gene, is expressed during embryogenesis in a restricted pattern evocative of neural crest cell distribution. *Mech Dev* 44: 109–121

Monte D, Baert JL, Defossez PA, de Launoit Y, Stehelin D (1994): Molecular cloning and characterization of human ERM, a new member of the Ets family closely related to mouse PEA3 and ER81 transcription factors. *Oncogene* 9: 1397–1406

Moreau-Gachelin F, Tavitian A, Tambourin P (1988): Spi-1 is a putative oncogene in virally induced murine erythroleukaemias. *Nature* 331: 277–280

Moreau-Gachelin F, Ray D, Mattei MG, Tambourin P, Tavitian A (1989): The putative oncogene Spi-1: murine chromosomal localization and transcriptional activation in murine acute erythroleukemias. *Oncogene* 4: 1449–1456

Moreau-Gachelin F, Wendling F, Molina T, Denis N, Titeux M, Grimber G, Briand P, Vainchenker W, Tavitian A (1996): Spi-1/PU.1 transgenic mice develop multistep erythroleukemias. *Mol Cell Biol* 16: 2453–2463

Moscovici MG, Jurdic P, Samarut J, Gazzolo L, Mura CV, Moscovici C (1983): Characterization of the hemopoietic target cells for the avian leukemia virus E26. *Virology* 129: 65–78

Moulton KS, Semple K, Wu H, Glass CK (1994): Cell-specific expression of the macrophage scavenger receptor gene is dependent on PU.1 and a composite AP-1/ets motif. *Mol Cell Biol* 14: 4408–4418

Mowat M, Cheng A, Kimura N, Bernstein A, Benchimol S (1985): Rearrangements of the cellular p53 gene in erythroleukaemic cells transformed by Friend virus. *Nature* 314: 633–636

Munroe DG, Peacock JW, Benchimol S (1990): Inactivation of the cellular p53 gene is a common feature of Friend virus-induced erythroleukemia: relationship of inactivation to dominant transforming alleles. *Mol Cell Biol* 10: 3307–3313

Murakami K, Mavrothalassitis G, Bhat NK, Fisher RJ, Papas TS (1993): Human ERG-2 protein is a phosphorylated DNA-binding protein – a distinct member of the ets family. *Oncogene* 8: 1559–1566

Muthusamy N, Barton K, Leiden JM (1995): Defective activation and survival of T-cells lacking the ETS-1 transcription factor. *Nature* 377: 639–642

Nelsen B, Tian G, Erman B, Gregoire J, Maki R, Graves B, Sen R (1993): Regulation of lymphoid-specific immunoglobulin μ heavy chain gene enhancer by ETS-domain proteins. *Science* 261: 82–86

Ness SA, Marknell A, Graf T (1989): The v-myb oncogene product binds to and activates the promyelocyte-specific mim-1 gene. *Cell* 59: 1115–1125

Nunn MF, Hunter T (1989): The ets sequence is required for induction of erythroblastosis in chickens by avian retrovirus E26 *J Virol* 63: 398–402

Nunn MF, Seeburg PH, Moscovici C, Duesberg PH (1983): Tripartite structure of the avian erythroblastosis virus E26 transforming gene. *Nature* 306: 391–395

Nye JA, Petersen JM, Gunther CV, Jonsen MD, Graves BJ (1992): Interaction of murine ets-1 with GGA-binding sites establishes the ETS domain as a new DNA-binding motif. *Genes Dev* 6: 975–990

Ohno T, Rao VN, Reddy ES (1993): EWS/Fli-1 chimeric protein is a transcriptional activator. *Cancer Res* 53: 5859–5863

Ohno T, Ouchida M, Lee L, Gatalica Z, Rao VN, Reddy ES (1994): The EWS gene, involved in Ewing family of tumors, malignant melanoma of soft parts and desmoplastic small round cell tumors, codes for an RNA binding protein with novel regulatory domains. *Oncogene* 9: 3087–3097

Okuda T, van Deursen J, Hiebert SW, Grosveld G, Downing JR (1996): AML1, the target of multiple chromosomal translocations in human leukemia, is essential for normal fetal liver hematopoiesis. *Cell* 84: 321–330

O'Neill EM, Rebay I, Tjian R, Rubin GM (1994): The activities of two Ets-related transcription factors required for *Drosophila* eye development are modulated by the Ras/MAPK pathway. *Cell* 78: 137–147

O'Prey J, Ramsay S, Chambers I, Harrison PR (1993): Transcriptional upregulation of the mouse cytosolic glutathione peroxidase gene in erythroid cells is due to a tissue-specific 3′ enhancer containing functionally important CACC/GT motifs and binding sites for GATA and Ets transcription factors. *Mol Cell Biol* 13: 6290–6303

Pahl HL, Scheibe RJ, Zhang DE, Chen HM, Galson DL, Maki RA, Tenen DG (1993): The proto-oncogene PU.1 regulates expression of the myeloid-specific CD11b promoter. *J Biol Chem* 268: 5014–5020

Papadopoulos P, Ridge SA, Boucher CA, Stocking C, Wiedemann LM (1995): The novel activation of ABL by fusion to an ets-related gene, TEL. *Cancer Res* 55:34–38

Paul R, Schuetze S, Kozak SL, Kozak CA, Kabat D (1991): The Sfpi-1 proviral integration site of Friend erythroleukemia encodes the ets-related transcription factor PU.1. *J Virol* 65: 464–467

Petersen JM, Skalicky JJ, Donaldson LW, McIntosh LP, Alber T, Graves BJ (1995): Modulation of transcription factor Ets-1 DNA binding: DNA-induced unfolding of an alpha helix. *Science* 269: 1866–1869

Pio F, Ni CZ, Mitchell RS, Knight J, McKercher S, Klemsz M, Lombardo A, Maki RA, Ely KR (1995): Co-crystallization of an ETS domain (PU.1) in complex with DNA. Engineering the length of both protein and oligonucleotide. *J Biol Chem* 270: 24258–24263

Pio F, Kodandapani R, Ni C, Shepard W, Klemsz M, McKercher S, Maki R, Ely K (1996): New insights on the DNA recognition by ets proteins from the crystal structure of the PU.1 ETS domain-DNA complex. *J Biol Chem* 271: 23329–23337

Pognonec P, Boulukos KE, Gesquiere JC, Stehelin D, Ghysdael J (1988): Mitogenic stimulation of thymocytes results in the calcium-dependent phosphorylation of c-ets-1 proteins. *EMBO J* 7: 977–983

Pognonec P, Boulukos KE, Ghysdael J (1989): The c-ets-1 protein is chromatin associated and binds to DNA in vitro. *Oncogene* 4: 691–697

Pongubala JM, Nagulapalli S, Klemsz MJ, McKercher SR, Maki RA, Atchison ML (1992): PU.1 recruits a second nuclear factor to a site important for immunoglobulin κ3′ enhancer activity. *Mol Cell Biol* 12: 368–378

Pongubala JM, Van Beveren C, Nagulapalli S, Klemsz MJ, McKercher SR, Maki RA, Atchison ML (1993): Effect of PU.1 phosphorylation on interaction with NF-EM5 and transcriptional activation. *Science* 259: 1622–1625

Prasad DD, Ouchida M, Lee L, Rao VN, Reddy ES (1994): TLS/FUS fusion domain of TLS/FUS-erg chimeric protein resulting from the t(16;21) chromosomal translocation in human myeloid leukemia functions as a transcriptional activation domain. *Oncogene* 9: 3717–3729

Pribyl LJ, Watson DK, McWilliams MJ, Ascione R, Papas TS (1988): The *Drosophila* ets-2 gene: molecular structure, chromosomal localization, and developmental expression. *Dev Biol* 127: 45–53

Pribyl LJ, Watson DK, Schulz RA, Papas TS (1991): D-elg, a member of the *Drosophila* ets gene family: sequence, expression and evolutionary comparison. *Oncogene* 6: 1175–1183

Price MA, Rogers AE, Treisman R (1995): Comparative analysis of the ternary complex factors Elk-1, SAP-1a and SAP-2 (ERP/NET). *EMBO J* 14: 2589–2601

Price M, Cruzalegui F, Treisman R (1996): The p38 and ERK MAP kinase pathways co-operate to activate Ternary Complex Factors and c-fos transcription in response to UV light. *EMBO J* 15: 6552–6563

Prosser HM, Wotton D, Gégonne A, Ghysdael J, Wang S, Speck NA, Owen MJ (1992): A phorbol ester response element within the human T-cell receptor beta-chain enhancer. *Proc Natl Acad Sci USA* 89: 9934–9938

Punyammalee B, Crabeel M, de Lannoy C, Perbal B, Glansdorff N (1991): Two c-myb proteins differing by their aminotermini exhibit different transcriptional transactivation activities (yeast/reporter-effector system). *Oncogene* 6: 11–19

Qi S, Chen ZQ, Papas TS, Lautenberger JA (1992): The sea urchin erg homolog defines a highly conserved erg-specific domain. *DNA Seq* 3: 127–129

Rabault B, Ghysdael J (1994): Calcium-induced phosphorylation of ETS1 inhibits its specific DNA binding activity. *J Biol Chem* 269: 28143–28151

Rabault B, Roussel M, Tran Quang C, Ghysdael J (1996). Phosphorylation of Ets1 regulates the complementation of a CSF-1 receptor impaired in mitogenesis. *Oncogene* 13: 877–881

Rabbitts T, Forster A, Larson R, Nathan P (1993): Fusion of the dominant negative transcription regulator CHOP with a novel gene FUS by translocation t(12;16) in malignant liposarcoma. *Nat Genet* 4: 175–180

Radke K, Beug H, Kornfeld S, Graf T (1982): Transformation of both erythroid and myeloid cells by E26, an avian leukemia virus that contains the myb gene. *Cell* 31: 643–653

Raingeaud J, Whitmarsh AJ, Barrett T, Derijard B, Davis RJ (1996): MKK3- and MKK6-regulated gene expression is mediated by the p38 mitogen-activated protein kinase signal transduction pathway. *Mol Cell Biol* 16: 1247–1255

Ramakrishnan V, Finch JT, Graziano V, Lee PL, Sweet RM (1993): Crystal structure of globular domain of histone H5 and its implications for nucleosome binding. *Nature* 362: 219–223

Rao VN, Papas TS, Reddy ES (1987): erg, a human ets-related gene on chromosome 21: alternative splicing, polyadenylation, and translation. *Science* 237: 635–639

Rao VN, Huebner K, Isobe M, ar-Rushdi A, Croce CM, Reddy ES (1989): elk, tissue-specific ets-related genes on chromosomes X and 14 near translocation breakpoints. *Science* 244: 66–70

Rao VN, Ohno T, Prasad DD, Bhattacharya G, Reddy ES (1993): Analysis of the DNA-binding and transcriptional activation unctions of human Fli-1 protein. *Oncogene* 8: 2167–2173

Ray D, Culine S, Tavitain A, Moreau-Gachelin F (1990): The human homologue of the putative proto-oncogene Spi-1: characterization and expression in tumors. *Oncogene* 5: 663–668

Ray D, Bosselut R, Ghysdael J, Mattei MG, Tavitian A, Moreau-Gachelin F (1992): Characterization of Spi-B, a transcription factor related to the putative oncoprotein Spi-1/PU.1. *Mol Cell Biol* 12: 4297–4304

Ray-Gallet D, Mao C, Tavitian A, Moreau-Gachelin F (1995): DNA binding specificities of Spi-1/PU.1 and Spi-B transcription factors and identification of a Spi-1/Spi-B binding site in the c-fes/c-fps promoter. *Oncogene* 11: 303–313

Raynaud S, Cave H, Baens M, Bastard C, Cacheux V, Grosgeorge J, Guidal-Giroux C, Guo C, Vilmer E, Marynen P, Grandchamp B (1996): The 12;21 translocation involving TEL and deletion of the other TEL allele: two frequently associated alterations found in childhoed acute lymphoblastic leukemia. *Blood* 87: 2891–2899

Rebay I, Rubin GM (1995): Yan functions as a general inhibitor of differentiation and is negatively regulated by activation of the Ras1/MAPK pathway. *Cell* 81: 857–866

Reddy ES, Rao VN (1988): Structure, expression and alternative splicing of the human c-ets-1 proto-oncogene. *Oncogene Res* 3: 239–246

Reddy ES, Rao VN, Papas TS (1987): The erg gene: a human gene related to the ets oncogene. *Proc Natl Acad Sci USA* 84: 6131–6135

Rivera RR, Stuiver MH, Steenbergen R, Murre C (1993): Ets proteins: new factors that regulate immunoglobulin heavy-chain gene expression. *Mol Cell Biol* 13: 7163–7169

Romana SP, Mauchauffe M, Le Coniat M, Chumakov I, Le Paslier D, Berger R, Bernard OA (1995a) The t(12;21) of acute lymphoblastic leukemia results in a tel-AML1 gene fusion. *Blood* 85: 3662–3670

Romana SP, Poirel H, Leconiat M, Flexor MA, Mauchauffe M, Jonveaux P, MacIntyre EA, Berger R, Bernard OA (1995b): High frequency of t(12;21) in childhood B-lineage acute lymphoblastic leukemia. *Blood* 86: 4263–4269

Romano-Spica V, Georgiou P, Suzuki H, Papas TS, Bhat NK (1995): Role of ETS1 in IL-2 gene expression. *J Immunol* 154: 2724–2732

Rosmarin AG, Caprio DG, Kirsch DG, Handa H, Simkevich CP (1995): GABP and PU.1 compete for binding, yet cooperate to increase CD18 (beta 2 leukocyte integrin) transcription. *J Biol Chem* 270: 23627–23633

Rossi F, McNagny M, Smith G, Frampton J, Graf T (1996): Lineage commitment of transformed haematopoietic progenitors is determined by the level of PKC activity. *EMBO J* 15: 1894–1901

Roussel M, Saule S, Lagrou C, Rommens C, Beug H, Graf T, Stehelin D (1979): Three new types of viral oncogene of cellular origin specific for haematopoietic cell transformation. *Nature* 281: 452–455

Ruscetti S, Aurigemma R, Yuan CC, Sawyer S, Blair DG (1992): Induction of erythropoietin responsiveness in murine hematopoietic cells by the gag-myb-ets-containing ME26 virus. *J Virol* 66: 20–26

Salmon P, Giovane A, Wasylyk B, Klatzmann D (1993): Characterization of the human CD4 gene promoter: transcription from the CD4 gene core promoter is tissue-specific and is activated by Ets proteins. *Proc Natl Acad Sci USA* 90: 7739–7743

Schneikert J, Lutz Y, Wasylyk B (1992): Two independent activation domains in c-Ets-1 and c-Ets-2 located in non-conserved sequences of the ets gene family. *Oncogene* 7: 249–256

Schuetze S, Stenberg PE, Kabat D (1993): The Ets-related transcription factor PU.1 immortalizes erythroblasts. *Mol Cell Biol* 13: 5670–5678

Schulz RA, The SM, Hogue DA, Galewsky S, Guo Q (1993): Ets oncogene-related gene Elg functions in *Drosophila* oogenesis. *Proc Natl Acad Sci USA* 90: 10076–10080

Schwarzenbach H, Newell JW, Matthias P (1995): Involvement of the Ets family factor PU.1 in the activation of immunoglobulin promoters. *J Biol Chem* 270: 898–907

Scott EW, Simon MC, Anastasi J, Singh H (1994): Requirement of transcription factor PU.1 in the development of multiple hematopoietic lineages. *Science* 265: 1573–1577

Seth A, Robinson L, Thompson DM, Watson DK, Papas TS (1993): Transactivation of GATA-1 promoter with ETS1, ETS2 and ERGB/Hu-FLI-1 proteins: stabilization of the ETS1 protein binding on GATA-1 promoter sequences by monoclonal antibody. *Oncogene* 8: 1783–1790

Seth A, Robinson L, Panayiotakis A, Thompson DM, Hodge DR, Zhang XK, Watson DK, Ozato K, Papas TS (1994): The EndoA enhancer contains multiple ETS binding site repeats and is regulated by ETS proteins. *Oncogene* 9: 469–477

Sgouras DN, Athanasiou MA, Beal GJ Jr, Fisher RJ, Blair DG, Mavrothalassitis GJ (1995): ERF: an ETS domain protein with strong transcriptional repressor activity, can suppress ets-associated tumorigenesis and is regulated by phosphorylation during cell cycle and mitogenic stimulation. *EMBO J* 14: 4781–4793

Shapiro LH (1995): Myb and Ets proteins cooperate to transactivate an early myeloid gene. *J Biol Chem* 270: 8763–8771

Shaw PE, Frasch S, Nordheim A (1989): Repression of c-fos transcription is mediated through p67SRF bound to the SRE. *EMBO J* 8: 2567–2574

Shimizu K, Ichikawa H, Tojo A, Kaneko Y, Maseki N, Hayashi Y, Ohira M, Asano S, Ohki M (1993): An ets-related gene, ERG, is rearranged in human myeloid leukemia with t(16;21) chromosomal translocation. *Proc Natl Acad Sci USA* 90: 10280–10284

Shin MK, Koshland ME (1993): Ets-related protein PU.1 regulates expression of the immunoglobulin J-chain gene through a novel Ets-binding element. *Genes Dev* 7: 2006–2015

Shore P, Sharrocks AD (1994): The transcription factors Elk-1 and serum response factor interact by direct protein-protein contacts mediated by a short region of Elk-1. *Mol Cell Biol* 14: 3283–3291

Siddique HR, Rao VN, Lee L, Reddy ES (1993): Characterization of the DNA binding and transcriptional activation domains of the erg protein. *Oncogene* 8. 1751–1755

Sieweke MH, Tekotte H, Frampton J, Graf T (1996): MafB is an interaction partner and repressor of Ets-1 that inhibits erythroid differentiation. *Cell* 85: 49–60

Skalicky J, Donaldson L, Petersen J, Graves B, McIntosh L (1996): Structural coupling of the inhibitory regions flanking the ETS domain of murine Ets-1. *Protein Sci* 5: 296–309

Sorensen PH, Liu XF, Delattre O, Rowland JM, Biggs CA, Thomas G, Triche TJ (1993): Reverse transcriptase PCR amplification of EWS/FLI-1 fusion transcripts as a diagnostic test for peripheral primitive neuroectodermal tumors of childhood. *Diagnost Mol Pathol* 2: 147–157

Speck NA, Renjifo B, Hopkins N (1990): Point mutations in the Moloney murine leukemia virus enhancer identify a lymphoid-specific viral core motif and 1,3-phorbol myristate acetate-inducible element. *J Virol* 64: 543–550

Stiegler P, Wolff CM, Baltzinger M, Hirtzlin J, Senan F, Meyer D, Ghysdael J, Stehelin D, Befort N, Remy P (1990) Characterization of *Xenopus laevis* cDNA clones of the c-ets-1 proto-oncogene. *Nucleic Acids Res* 18: 5298

Sumarsono SH, Wilson TJ, Tymms MJ, Venter DJ, Corrick CM, Kola R, Lahoud MH, Papas TS, Seth A, Kola I (1996): Down's syndrome-like skeletal abnormalities in Ets2 transgenic mice. *Nature* 379: 534–537

Tei H, Nihonmatsu I, Yokokura T, Ueda R, Sano Y, Okuda T, Sato K, Hirata K, Fujita SC, Yamamoto D (1992): pokkuri, a *Drosophila* gene encoding an E-26-specific (Ets) domain protein, prevents overproduction of the R7 photoreceptor. *Proc Natl Acad Sci USA* 89: 6856–6860

The SM, Xie X, Smyth F, Papas TS, Watson DK, Schulz RA (1992): Molecular characterization and structural organization of D-elg, an ets proto-oncogene-related gene of Drosophila. *Oncogene* 7: 2471–2478

Thomspon CB, Wang CY, Ho IC, Bohjanen PR, Petryniak B, June CH, Miesfeldt S, Zhang L, Nabel GJ, Karpinski B, Leiden JM (1992): cis-acting sequences required for inducible interleukin-2 enhancer function bind a novel Ets-related protein, Elf-1. *Mol Cell Biol* 12: 1043–1053

Thompson CC, Brown TA, McKnight SL (1991): Convergence of Ets- and notch-related structural motifs in a heteromeric DNA binding complex. *Science* 253: 762–768

Thompson JD, Higgins DG, Gibson TJ (1994): CLUSTAL W: improving the sensitivity of progressive multiple sequence alignment through sequence weighting, positionspecific gap penalties and weight matrix choice. *Nucleic Acids Res* 22: 4673–4680

Thummel CS, Burtis KC, Hogness DS (1990): Spatial and temporal patterns of E74 transcription during *Drosophila* development. *Cell* 61: 191–111

Tran Quang C, Pironin M, von Lindern M, Beug H, Ghysdael J (1995): Spi-1 and mutant p53 regulate different aspects of the proliferation and differentiation control of primary erythroid progenitors. *Oncogene* 11: 1229–1239

Treier M, Bohmann D, Mlodzik M (1995): JUN cooperates with the ETS domain protein pointed to induce photoreceptor R7 fate in the *Drosophila* eye. *Cell* 83: 753–760

Treismann R (1986): Identification of a protein-binding site that mediates transcriptional respone of the c-fos gene to serum factors. *Cell* 46: 567–574

Treismann R (1992): The serum respone element. *Trends Biochem Sci* 17: 423–426

Treismann R, Marais R, Wynne J (1992): Spatial flexibility in ternary complexes between SRF and its accessory proteins. *EMBO J* 11: 4631–4640

Turc-Carel C, Philip I, Berger MP, Philip T, Lenoir G (1983): Chromosomal translocation (11;22) in cell lines of Ewing's sarcoma. *C R Seances Acad Sci III* 296: 1101–1103

Urness LD, Thummel CS (1990): Molecular interactions within the ecdysone regulatory hierarchy: DNA binding properties of the *Drosophila* ecdysone-inducibile E74A protein. *Cell* 63: 47–61

Vandenbunder B, Pardanaud L, Jaffredo T, Mirabel MA, Stehelin D (1989): Complementary patterns of expression of c-ets 1, c-myb and c-myc in the blood-forming system of the chick embryo. *Development* 107: 265–274

Virbasius JV, Scarpulla RC (1991): Transcriptional activation through ETS domain binding sites in the cytochrome c oxidase subunit IV gene. *Mol Cell Biol* 11: 5631–5638

Virbasius CA, Virbasius JV, Scarpulla RC (1993a): NRF-1, an activator involved in nuclear-mitochondrial interactions, utilizes a new DNA-binding domain conserved in a family of developmental regulators. *Genes Dev* 7: 2431–2445

Virbasius JV, Virbasius CA, Scarpulla RC (1993b): Identity of GABP with NRF-2, a multisubunit activator of cytochrome oxidase expression, reveals a cellular role for an ETS domain activator of viral promoters. *Genes Dev* 7: 380–392

Wang CY, Petryniak B, Ho IC, Thompson CB, Leiden JM (1992): Evolutionarily conserved Ets family members display distinct DNA binding specificities. *J Exp Med* 175: 1391–1399

Wang CY, Petryniak B, Thompson CB, Kaelin WG, Leiden JM (1993): Regulation of the Ets-related transcription factor Elf-1 by binding to the retinoblastoma protein. *Science* 260: 1330–1335

Wang CY, Bassuk AG, Boise LH, Thompson CB, Bravo R, Leiden JM (1994): Activation of the granulocyte-macrophage colony-stimulating factor promoter in T-cells requires cooperative binding of Elf-1 and AP-1 transcription factors. *Mol Cell Biol* 14: 1153–1159

Wasylyk B, Wasylyk C, Flores P, Begue A, Leprince D, Stehelin D (1990): The c-ets proto-oncogenes encode transcription factors that cooperate with c-Fos and c-Jun for transcriptional activation. *Nature* 346: 191–193

Wasylyk C, Flores P, Gutman A, Wasylyk B (1989): PEA3 is a nuclear target for transcription activation by non-nuclear oncogenes. *EMBO J* 8: 3371–3378

Wasylyk C, Gutman A, Nicholson R, Wasylyk B (1991): The c-Ets oncoprotein activates the stromelysin promoter through the same elements as several non-nuclear oncoproteins. *EMBO J* 10: 1127–1134

Wasylyk C, Kerckaert JP, Wasylyk B (1992) A novel modulator domain of Ets transcription factors. *Genes Dev* 6: 965–974

Watanabe H, Sawada J, Yano K, Yamaguchi K, Goto M, Handa H (1993): cDNA cloning of transcription factor E4TF1 subunits with Ets and notch motifs. *Mol Cell Biol* 13: 1385–1391

Watson DK, McWilliams MJ, Lapis P, Lautenberger JA Schweinfest CW, Papas TS (1988a): Mammalian ets-1 and ets-2 genes encode highly conserved proteins. *Proc Natl Acad Sci USA* 85: 7862–7866

Watson DK, McWilliams MJ, Papas TS (1988b): A unique amino-terminal sequence predicted for the chicken proto-ets protein. *Virology* 167: 1–7

Watson DK, Mavrothalassitis GJ, Jorcyk CL, Smyth FE, Papas TS (1990): Molecular organization and differential polyadenylation sites of the human ETS2 gene. *Oncogene* 5: 1521–1517

Watson DK, Smyth FE, Thompson DM, Cheng JQ, Testa JR, Papas TS, Seth A (1992): The ERGB/Fli-1 gene: isolation and characterization of a new member of the family of human ETS transcription factors. *Cell Growth Differ* 3: 705–713

Werner M, Clore G, Fisher C, Fisher R, Trinh L, Shiloach J, Gronenborn A (1995): The solution structure of the human ETS1-DNA complex reveals a novel mode of binding and true side chain intercalation. *Cell* 83: 761–771

Werner M, Clore G, Fisher C, Fisher R, Trinh L, Shiloach J, Gronenborn A (1996): The solution structure of the human ETS-1-DNA Complex Reveals a Novel Mode of Binding and True Side Chain Interaction. *Cell* 87: 357 (Erratum)

Whitmarsh AJ, Shore P, Sharrocks AD, Davis RJ (1995): Integration of MAP kinase signal transduction pathways at the serum response element. *Science* 269: 403–407

Wolff CM, Stiegler P, Baltzinger M, Meyer D, Ghysdael J, Stehelin D, Befort N, Remy P (1991): Cloning, sequencing, and expression of two *Xenopus laevis* c-ets-2 protooncogenes. *Cell Growth Differ* 2: 447–456

Woods DB, Ghysdael J, Owen MJ (1992): Identification of nucleotide preferences in DNA sequences recognised specifically by c-Ets-1 protein. *Nucleic Acids Res* 20: 699–704

Wotton D, Ghysdael J, Wang S, Speck NA, Owen MJ (1994): Cooperative binding of Ets-1 and core binding factor to DNA. *Mol Cell Biol* 14: 840–850

Wu H, Moulton K, Horvai A, Parik S, Glass CK (1994): Combinatorial interactions between AP-1 and ets domain proteins contribute to the developmental regulation of the macrophage scavenger receptor gene. *Mol Cell Biol* 14: 2129–2139

Wurster AL, Siu G, Leiden JM, Hedrick SM (1994): Elf-1 binds to a critical element in a second CD4 enhancer. *Mol Cell Biol* 14: 6452–6463

Xin JH, Cowie A, Lachance P, Hassell JA (1992): Molecular cloning and characterization of PEA3, a new member of the Ets oncogene family that is differentially expressed in mouse embryonic cells. *Genes Dev* 6: 481–496

Yang BS, Hauser CA, Henkel G, Colman MS, Van Beveren C, Stacey KJ, Hume DA, Maki RA, Ostrowski MC (1996): Ras-mediated phosphorylation of a conserved threonine residue enhances the transactivation activities of c-Ets1 and c-Ets2. *Mol Cell Biol* 16: 538–547

Yokomori N, Kobayashi R, Moore R, Sueyoshi T, Negishi M (1995): A DNA methylation site in the male-specific P450 (Cyp 2d-9) promoter and binding of the heteromeric transcription factor GABP. *Mol Cell Biol* 15: 5355–5362

Yoshimura A, Longmore G, Lodish HF (1990): Point mutation in the exoplasmic domain of the erythropoietin receptor resulting in hormone-independent activation and tumorigenicity. *Nature* 348: 647–649

Zhang DE, Hetherington CJ, Chen HM, Tenen DG (1994): The macrophage transcription factor PU.1 directs tissue-specific expression of the macrophage colony-stimulating factor receptor. *Mol Cell Biol* 14: 373–381

Zhang L, Lemarchandel V, Romeo PH, Ben-David Y, Greer P, Bernstein A (1993): The Fli-1 proto-oncogene, involved in erythroleukemia and Ewing's sarcoma, encodes a transcriptional activator with DNA-binding specificities distinct from other Ets family members. *Oncogene* 8: 1621–1630

Zinck R, Hipskind RA, Pingoud V, Nordheim A (1993): c-fos transcriptional activation and repression correlate temporally with the phosphorylation status of TCF. *EMBO J* 12: 2377–2387

Zinck R, Cahill MA, Kracht M, Sachsenmaier C, Hipskind RA, Nordheim A (1995): Protein synthesis inhibitors reveal differential regulation of mitogen-activated protein kinase and stress-activated protein kinase pathways that converge on Elk-1. *Mol Cell Biol* 15: 4930–4938

Zinszner H, Albalat R, Ron D (1994): A novel effector domain from the RNA-binding protein TLS or EWS is required for oncogenic transformation by CHOP. *Genes Dev* 8: 2513–1526

Zipursky S, Rubin G (1994): Determination of neuronal cell fate: lessons from the R7 neuron of *Drosophila*. *Annu Rev Neurosci* 17: 373–397

Zucman J, Delattre O, Desmaze C, Epstein AL, Stenman G, Speleman F, Fletchers CD, Aurias A, Thomas G (1993a) EWS and ATF-1 gene fusion induced by t(12;22) translocation in malignant melanoma of soft parts. *Nat Genet* 4: 341–345

Zucman J, Melot T, Desmaze C, Ghysdael J, Plougastel B, Peter M, Zucker JM, Triche TJ, Sheer D, Turc-Carel C, Ambros P, Combaret V, Lenoir G, Aurias A, Thomas G, Delattre O (1993b): Combinatcrial generation of variable fusion proteins in the Ewing family of tumours. *EMBO J* 12: 4481–4487

3

Oncogenes as Transcriptional Regulators
Vol. 1: Retroviral Oncogenes
ed. by M. Yaniv and J. Ghysdael

myb Proto-Oncogene Product as a Transcriptional Regulator

SHUNSUKE ISHII

Introduction

The v-*myb* oncogene was originally identified as an oncogene carried by two actuely leukemogenic avian retroviruses, avian myeloblastosis virus (AMV) and E26 (for review see Graf, 1992). AMV transforms myeloid cells and causes a monoblastic leukemia, whereas E26 transforms myeloid and erythroid cells in vitro and causes predominantly an erythroleukemia (Radke et al, 1992; Moscovici et al, 1983). In 1988–1989, the c-*myb* proto-oncogene product (c-Myb) was demonstrated to function as a transcriptional activator that binds to specific DNA sequences. In this review, I shall address the structure of each functional domain in c-Myb, regulation of c-Myb activity, and mechanism of oncogenic activation of c-*myb*.

Physiological Role of c-*myb*

c-*myb* expression is linked to the differentiation state of the hematopoietic cells. Early experiments had demonstrated that c-*myb* mRNA levels were relatively low in more mature hematopoietic cells compared with immature cells (Westin et al, 1982; Sheiness and Gardinier, 1984; Gonda and Metcalf, 1984; Duprey and Boettinger, 1985; Thompson et al, 1986; Liebermann and Hoffman-Liebermann, 1989). This down-regulation of c-*myb* expression is necessary for differentiation of hematopoietic cells, since sustained expression of c-*myb* blocks the induced differentiation of immature erythroleukemia and myeloid leukemia cells (Clarke et al, 1988; McMahon et al, 1988; Todokoro et al, 1988; Yanagisawa et al, 1991). Similarly, AMV and E26 also interfere with monocyte-macrophage differentiation (Beug et al, 1987; Ness et al, 1987; Introna et al, 1990; Metz and Graf, 1991; Graf et al, 1992). Conversely, ablation of c-*myb* expression by the use of c-*myb*-specific antisense oligonucleotides greatly reduces colony formation by bone marrow and the growth of myeloid leukemia cell lines (Gewirtz and Calabretta, 1988; Anfossi et al, 1989). Furthermore, homozygous c-*myb* mutant mice generated by homologous recombination in embryonic stem cells displayed a specific failure of fetal hepatic hematopoiesis (Mücenski et al, 1990). These results all indicate a role for c-*myb* in maintaining the proliferative state of hematopoietic progenitor cells.

However, c-*myb* plays an important role for not only immature hematopoietic cells, but also some types of mature cells. c-*myb* expression is induced after mitogenic stimulation of both T-lymphocytes and smooth muscle cells, in parallel to entry into S phase (Stern and Smith, 1986; Brown et al, 1992). Expression of c-*myb* in the mitogen- or antigen-stimulated T-lymphocytes peaks as the activated cells undergo G1 progression, and c-*myb*-specific antisense oligonucleotides specifically block cells in late G1 or early S phase of cell cycle (Gewirtz et al, 1989), suggesting that c-*myb* is required for G1/S transition in normal T-lymphocytes. Inhibition of c-*myb* function by dominant negative forms blocks thymopoiesis and proliferation of mature T-cells (Badiani et al, 1994), whereas overexpression of c-*myb* induces degeneration of skeltal and cardiac muscles and cardiomegaly (Furuta et al, 1993).

A number of the *myb* homologues and the *myb*-related genes were identified in non-vertebrates, and the physiological roles of some of these genes were also characterized. The *Drosophila myb* gene which has a high homology with the vertebrate *myb* genes in the DNA-binding domain was identified (Katzen et al, 1985; Peters et al, 1987). The homozygous mutant for *Drosophila myb* is lethal, indicating that *Drosophila myb* has an essential function (Bishop et al, 1991). In plant, many *myb*-related genes were identified. All of them have the striking homology with vertebrate *myb* in the DNA-binding domain. The *GL1* gene product of *Arabidopsis thaliana* is required for leaf trichome differentiation (Oppenheimer et al, 1991). The *C1* and *P* genes found in maize, *Arabidopsis thaliana*, and *Nicotiana tabacum* are involved in the regulation of flower color by producing pigment (Paz-Ares et al, 1987; Lloyd et al, 1992). The *P* gene product controls phlobaphene pigmentation by directly activating a flavonoid biosynthetic gene subset (Grotewold et al, 1994). The *mixta* gene product of *Antirrhinum majus* is also involved in the regulation of flower color intensity by controlling the cell shape (Noda et al, 1994). In the filamentous fungus *Aspergillus nidulans*, the Myb-related transcription factor encoded by the *flbD* gene regulates the coordinate initiation of conidiophore development (Wieser and Adams, 1995). Yeast homologue BAS1 controls histidine biosynthesis (Tice-Baldwin et al, 1989). Thus, quite different functions have been demonstrated for different members of the *myb* family.

Binding to Specific DNA Sequences

Sequence specificity for DNA binding

Both v-*myb* and c-*myb* encode nuclear phosphoproteins that bind DNA directly (Moelling et al, 1985; Klempnauer and Sippel, 1987). The function of c-Myb as a transcriptional activator was demonstrated using various systems (Nishina et al, 1989; Weston and Bishop; 1989; Ness et al, 1989; Klempnauer et al, 1989; Nakagoshi et al, 1990; Ibanez and Lipsick, 1990; Kalkbrenner et al, 1990). The Myb-recognition element was originally defined as YAACKG, where K stands for G or T, as derived from comparison of isolated chicken DNA-binding sites for v-Myb (Biedenkapp et al, 1988). Then, two extended consensus sequences, the 9-bp YGRCGTTR motif (Howe and Watson, 1991), where Y and R denote pyrimidines and purines, respectively, and the 8-bp YAACKGHH motif (Weston, 1992), where H denotes A, C, or T (i.e. not G), were obtained from binding-site selection protocols. In order to characterize quantitatively the sequence specificity in c-Myb binding, an extensive binding analysis by using the Myb-binding site, MBS-I, in SV40 was performed (Tanikawa et al, 1993). This mutational analyses showed that the specific interactions are not uniformly distributed in the TAACTGAC region of MBS-I; the 2nd A, the 4th C and the 6th G are involved in very specific interactions with Myb, whereas the interactions at the 3rd A and the 8th C are less specific. Thus, mammalian c-Myb specifically recognizes the sequence AACNG (Figure 3.1 d).

Three members of vertebrate *myb* family (c-Myb, A-Myb, and B-Myb) can bind to the consensus sequence AACNG. However, some plant Myb-related proteins have a different sequence specificity for DNA binding. The *Am305* gene product from *Antirrhinum* and *P* protein from maize recognize a sequence ACCNACC (Grotewold et al, 1994; Sablowski et al, 1994). Myb.Ph3 from *Petunia* recognizes two types of DNA sequences: AACNG and AACNAAC (Solano et al, 1995). The mechanism by which these multiple sequences are recognized is not clear at present.

Structure of DNA-binding domain

c-Myb is comprised of three domains responsible for DNA binding, transcriptional activation, and negative regulation (Figure 3.1 a) (Gonda et al, 1985; Sakura et al, 1989). The DNA-binding domain is located at the N-terminal side and consists of three homologous tandem repeats of 51 or 52 amino acids (Figure 3.1) (designated as R1, R2, and R3 from the

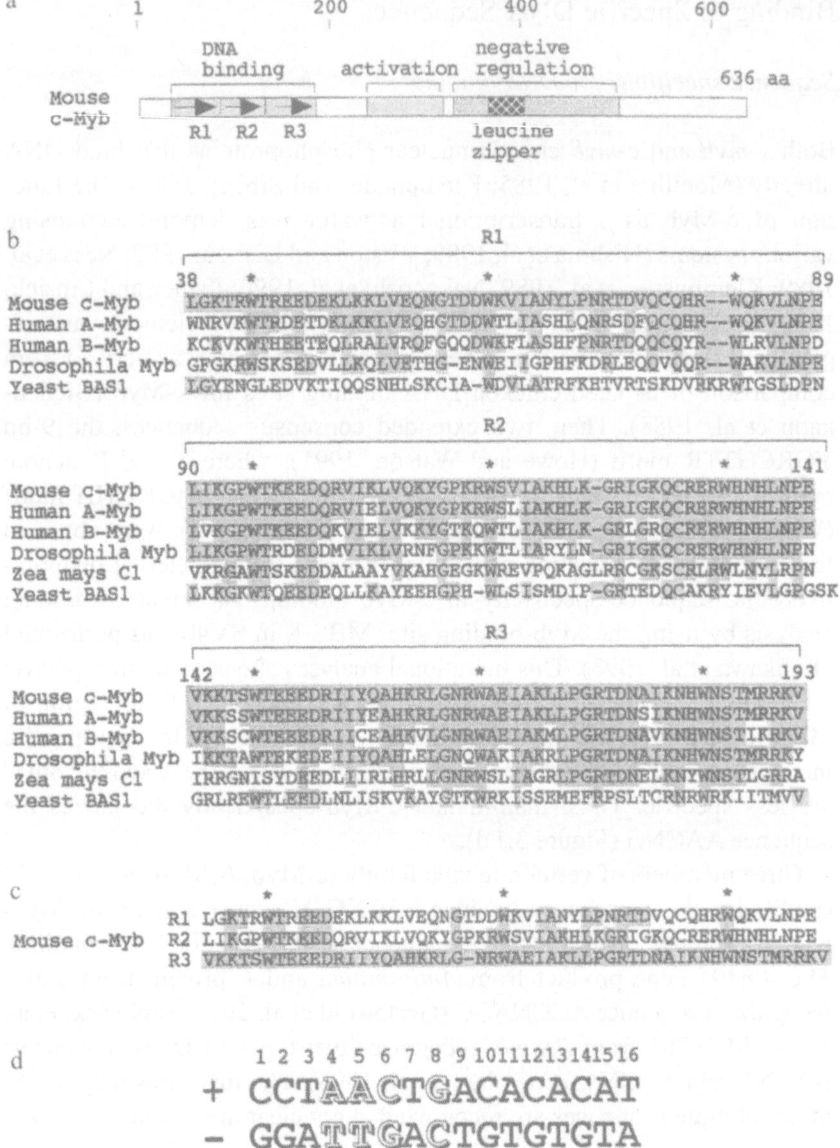

Figure 3.1 Structure and recognition sequence of c-Myb. (a) Schematic representation of functional domains in mouse c-Myb. The three functional domains, which are responsible for DNA binding, transcriptional activation, and negative regulation, respectively, are shown. Arrows represent the three fold tandem repeats of 51 and 52 amino acids. (b) Conservation of the structure of Myb DNA-binding domain. The amino acid sequences of the DNA-binding domain from various members of the *myb* family are shown. The identical amino acids are indicated by green. The conserved tryptophans are shown by asterisks. (c) Comparison of three repeats in the DNA-binding domain. The amino acid sequences of each repeat in the DNA-binding domain of mouse c-Myb are shown. (d) The Myb-binding site, MBS-I, found in the SV40 enhancer. The four bases which are specifically in contact with c-Myb are indicated by bold letters.

N terminus) (Klempnauer and Sippel, 1987; Sakura et al, 1989; Howe et al, 1990). Each repeat has three conserved tryptophans spaced 18 or 19 residues apart (Figure 3.1 c) (Anton and Frampton, 1988). This repeat structure with conserved tryptophans seems to be a general motif utilized in many different transcription factors found in a wide spectrum of eukaryotes, including vertebrates, insects, yeast, cellular slime mold, and higher plants (Figure 3.1 b) (Frampton et al, 1989; Katzen et al, 1985: Tice-Baldwin et al, 1989; Paz-Ares et al, 1987; Boulukos et al, 1988). Among the three repeats, R1 can be deleted without significant loss of DNA-binding activity (Sakura et al, 1989; Howe et al, 1990), and plays a minor role in sequence recognition. The two repeats R2 and R3 are necessary and sufficient for the recognition of specific DNA sequences.

Based on model building and mutational analysis, structures of the DNA-binding domain of c-Myb were proposed (Kanei-Ishii et al, 1990; Saikumar et al, 1990; Frampton et al, 1991; Gabrielsen et al, 1991). In all these structures, each repeat had a helix-turn-helix (HTH) motif, and the conserved tryptophans played an important role in generating the hydrophobic core. Consistent with this, the NMR analysis of the DNA-binding domain revealed that the three repeats have very similar folding architecture, containing three well-defined helices (Ogata et al, 1992; Ogata et al, 1995) (Figure 3.2). The second and third helices in each repeat form a HTH variation motif that contains a longer turn than the corresponding turn in the prototypical HTH motif. The three helices in each repeat are maintained by a hydrophobic core that includes the three conserved tryptophans, indicating that the tryptophan residues, conserved in many transcription factors, form the hydrophobic core. This is consistent with the fact that any single or multiple mutations of tryptophans in R2 and R3 to hydrophilic residues or alanine abolished or greatly reduced the DNA-binding activity (Kanei-Ishii et al, 1990; Saikumar et al, 1990). The NMR analysis of R2R3-DNA complex indicated that the most C-terminal helix in each of R2 and R3 functions as a recognition helix (Figure 3.3) (Ogata et al, 1994). R2 and R3 are closely packed in the major groove, so that the two recognition helices contact each other directly to bind to the specific base sequence, AACNG cooperatively. The first A, the G in the opposite strand corresponding to the third C, and the fifth G are interacting with Asn-183 (R3), Lys-182 (R3), and Lys-128 (R2), respectively. The homologous R2 and R3 repeats, which are directly connected in tandem, bind to the major groove of DNA continuously. In this sense, the binding of R2 and R3 is similar to that of transcription factor IIIA (TFIIIA)-type Zn fingers. Unlike the case of the TFIIIA-type Zn fingers, however, the recognition helices of R2 and R3 are more closely packed together in the major groove. So far, this type of direct interaction between the recognition

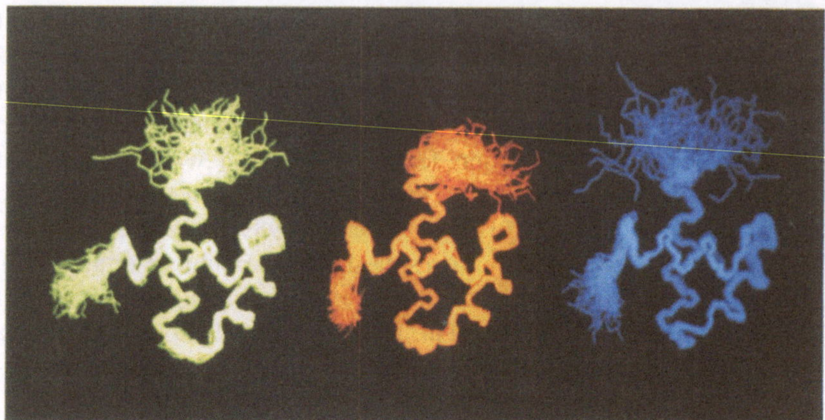

Figure 3.2 Structure of three repeats of Myb DNA-binding domain. Best-fit superpositions of the 50 structures determined by NMR are shown, and the backbone atoms of the peptide fragments of R1, R2, and R3 are shown in yellow, red, and blue, respectively.

helices from different DNA-binding units appears to be unique among many DNA-binding proteins. In contrast to R2 and R3, R1 does not bind tightly to DNA and may fluctuate between free and bound state at a fast rate with only a minor population in the DNA-bound state.

In spite of the structural similarities of the three repeats, their thermal stabilities are remarkably different from each other in the DNA-free state. R1 and R3 are thermodynamically stable (Tms of 61 °C for R1 and 57 °C for R3), whereas R2 is more unstable (Tm of 43 °C) (Sarai et al, 1993). In addition, R2 has slow conformational fluctuations. This is due to a cavity inside the hydrophobic core of R2 but not in R1 and R3, because a cavity-filling mutant does not have the conformational fluctuations (Ogata et al, 1996). The cavity-filling mutation significantly reduces specific DNA-binding activity, indicating that the conformational fluctuation in R2 is important for DNA binding. Comparison of the structures between the free and DNA-complexed states indicates that the indole ring of Trp-95 is significantly shifted toward the cavity of R2 on DNA binding (Figure 3.4). Thus, a cavity inside a hydrophobic core in R2 plays an important role to facilitate a conformational change.

In the v-Myb protein derived from AMV, Ile-91, Leu-106, and Val-117 in R2 are replaced with Asn, His, and Asp, respectively. It is interesting that a phenotype of the v-*myb*-transformed cell is altered by these point mutations (Introna et al, 1990). c-Myb and v-Myb encoded by the virus E26, where no mutations in these three positions are observed, can activate the *mim-1* gene, which is one of the *myb* target genes (Ness et al, 1989), but v-Myb encoded by AMV cannot. The three point mutations in R2 of v-Myb encoded by AMV result, therefore, in a loss in the activating ability

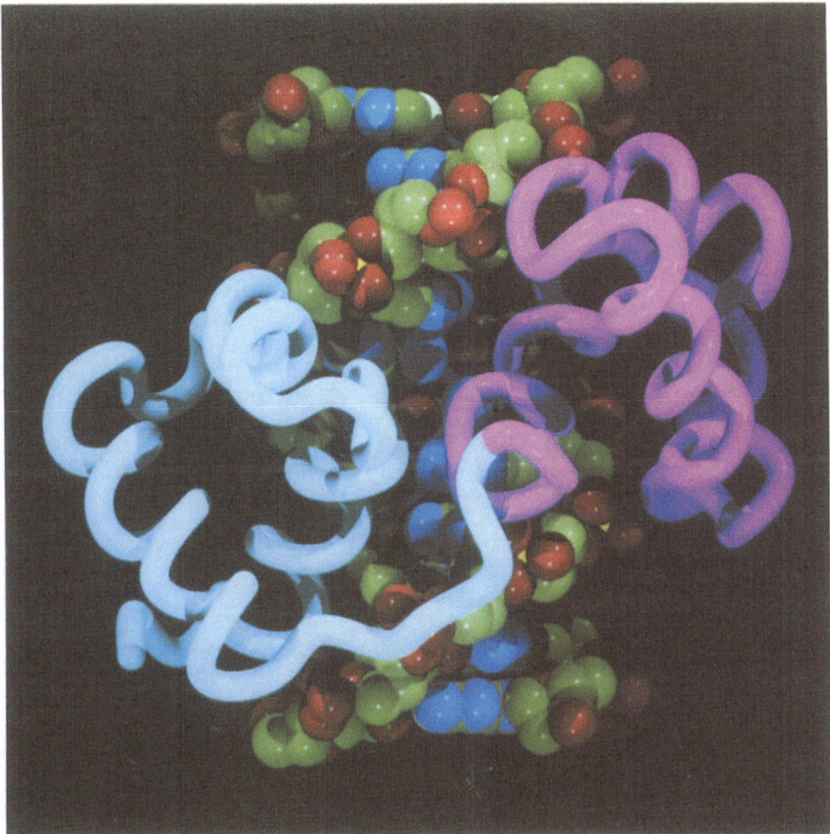

Figure 3.3 Solution structure of R2R3-DNA complex determined by NMR. The backbone atoms in the R2 region of R2R3 are shown in purple, and those in the R3 region are shown in blue.

of the *mim-1* gene. Furthermore, a myeloid-specific transcription factor NF-M (also called C/EBPβ or NF-IL6) is required for the expression of the *mim-1* gene (Ness et al, 1993; Burk et al, 1993). The structure of R2R3-DNA complex indicates that these three amino acids, together with their adjacent amino acids, form a hydrophobic patch on the surface in the DNA-complexed form (Ogata et al, 1995). Since the three mutated residues in the R2 of AMV-derived v-Myb protein are involved in this hydrophobic patch, it may mediate an interaction between Myb and NF-M.

Interaction with Coactivator CBP

The transcriptional activation domain of c-Myb, which is rich in acidic amino acids, is adjacent to the DNA-binding domain. The transcriptional

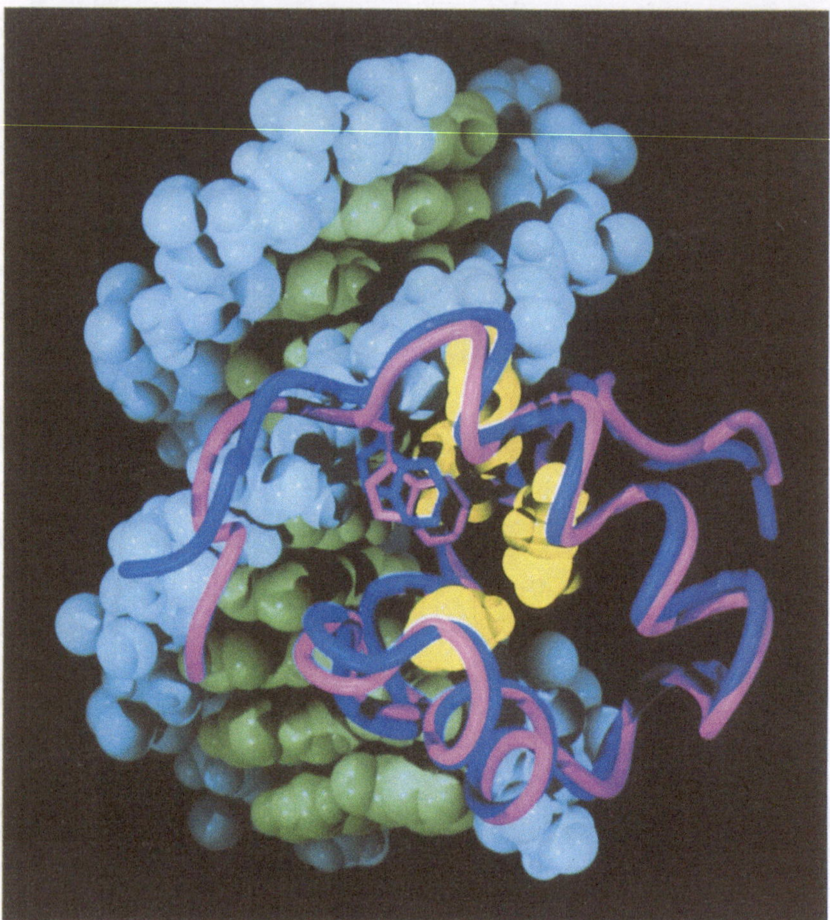

Figure 3.4 Role of the conformational flexibilty caused by the cavity in the hydrophobic core. The superimposed average structures of backbone of R2 and also the side chain of Trp-95 in the free (blue) and DNA-complexed forms (purple) are shown. The van der Waals surfaces of Val-103, Cys-130, and Arg-133 side chains in the DNA-complexed form are shown in yellow. Comparison of two structures indicates that the indole ring of Trp-95 enters into the cavity on DNA binding.

coactivator CBP, which was originally identified as a CREB (cAMP response element-binding protein)-binding protein (Chrivia et al, 1993), binds to this activation domain of c-Myb and mediate c-Myb-induced transcriptional activation (Dai et al, 1996; Oelgeschläger et al, 1996). An increase in intracellular cAMP concentration leads to the activation of the cAMP-dependent protein kinase A (PKA) that directly phosphorylates CREB at Ser-133 in the transcriptional activation domain (Gonzarez and Montminy, 1989). CBP binds to CREB only when Ser-133 is phosphorylated and functions as a coactivator (Kwok et al, 1994). However,

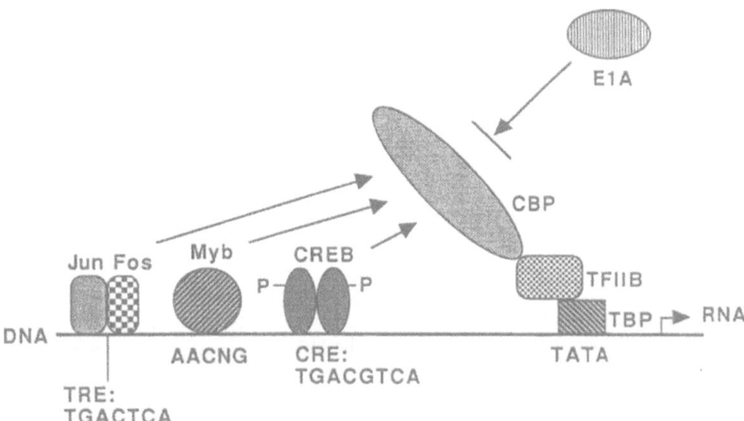

Figure 3.5 Interaction of c-Myb and coactivator CBP. Coactivator CBP binds to multiple transcription factors including CREB, AP-1, and c-Myb. CBP mediates the transcriptional activation by acting as a bridge between TFIIB and these transcription factors. Binding of E1A to CBP blocks the transcriptional activation by these factors.

c-Myb binds to CBP in a phosphorylation-independent manner. Since CBP interacts with TFIIB, which contacts TBP and also functions in the recruitment to the promoter of RNA polymerase II, CBP mediates CREB- or c-Myb-dependent transcriptional activation as a bridge between phosphorylated CREB or c-Myb and the RNA polymerase II complex. Recently, CBP was demonstrated to function as a transcriptional co-activator for many other transcriptional activators including c-Jun and c-Fos (Figure 3.5) (Arias et al, 1994: Bannister and Kouzareides, 1995). Interestingly, mutations in the human CBP gene cause Rubinstein-Taybi syndrome through haploinsufficiency (Petrij et al, 1995), suggesting that the amount of CBP in cells is not excessive and a 50% decrease in the amount affects normal development. Therefore, cross-talk between the c-Myb, cAMP, and AP-1 pathways through competition for binding to CBP is possible.

Regulation of c-Myb Activity

Role of negative regulatory domain (NRD)

The C-terminal region of c-Myb contains the functional domains which are important for modulation of c-Myb activity. A series of sequential C-terminal deletions was used to define a region spanning c-Myb amino acids 326–500 that is referred to as the negative regulatory domain (NRD)

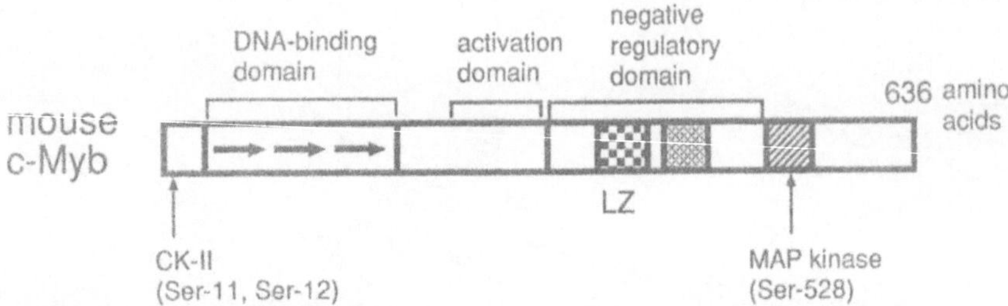

Figure 3.6 Modulation of c-Myb activity. Three elements in the carboxyl region of the c-Myb protein, which were shown to negatively regulate c-Myb activity, are indicated by shaded boxes. The CK-II site in the amino-terminal region and the MAP kinase site in the carboxyl region are shown by arrows.

(Figure 3.6). Removal of NRD results in increase in both *trans*-activating and transformin capacities (Sakura et al, 1989; Hu et al, 1990). This is partly because a C-truncated Myb protein binds to a Myb-recognition sequence with a higher affinity than the full-length form (Ramsay et al, 1991; Ramsay et al, 1992). A potential leucine zipper structure is located in the N-terminal portion of NRD (Figure 3.6). This region was predicted to form an amphipathic α-helix and contains characteristic hydrophobic residues at every seventh position. Disruption of this leucine zipper by site-directed mutagenesis markedly increases both trans-activating and transforming capacities (Kanei-Ishii et al, 1992). Since the leucine zipper motif was originally identified as a mediating dimerization of several DNA-binding proteins (Landschulz et al, 1988), these results indicate that c-Myb activity is negatively regulated through the leucine zipper and imply the presence of an inhibitor(s) of c-Myb. One protein that may interact with the leucine zipper motif of the c-Myb NRD is c-Myb itself, because c-Myb is capable of forming homodimers via the leucine zipper (Nomura et al, 1993). However, it has not been possible to demonstrate leucine zipper-mediated crosslinking of full-length c-Myb dimers in solution (our unpublished results, and R. Ramsay, personal communication). Recently, two proteins, p67 and p160, that bound to GST-NRD proteins containing the wild type leucine zipper, but not to those carrying a mutated version, were identified (Favier and Gonda, 1994). These two proteins are closely related as shown by peptide mapping with V8 protease. However, their distribution differs in that p160 is found in all the murine cell lines examined while p67 is found in a subset of immature myeloid cell lines. In addition to the leucine zipper, Dubendorf et al, (1993) identified a small region (amino acids 425–464) in the C-terminal region of NRD that suppressed c-Myb *trans*-activating activity (Figure 3.6). Thus, the NRD appears to have multiple subdomains.

Other domain in the C-terminal region

Kalkbrenner et al (1990) identified a region, amino acids 496–640, of the human c-Myb that also negatively influenced the ability of c-Myb to *trans*-activate. This is immediately C-terminal to the NRD, suggesting that two non-overlapping regions in the C-terminal portion of c-Myb might play a role in regulating c-Myb activity. Consistent with this, Dubendorf et al (1993) also identified a small region, amino acids 499–558, that negatively regulates c-Myb activity (Figure 3.6). This is in the region defined by Kalkbrenner et al (1990). Both groups reported that this region inhibits transcriptional activation by c-Myb in trans, possible through interacting with the transcriptional activation domain (Dubendorf et al, 1992; Vorbrueggen et al, 1994). This region may directly interact with the transcriptional activation domain, or alternatively the putative factor that binds to this region may modulate the interaction between CBP and the transcriptional activation domain.

Regulation of c-Myb activity by phosphorylation

c-Myb is hyperphosphorylated during mitosis in an avian lymphoma cell line, suggesting that changes in the state of c-Myb phosphorylation may alter c-Myb activity during the cell cycle (Lüscher und Eisenman, 1992). Casein kinase II phosphorylates c-Myb in vitro at Ser-11 and Ser-12 in the N-terminal region (Figure 3.6), and phosphorylation of these residues inhibits the DNA binding activity of c-Myb and the transcriptional cooperativity with NF-M (Lüscher et al, 1990; Oelgeschläger et al, 1995). In contrast to this, however, phosphatase treatment of the in vitro translated c-Myb was demonstrated to inhibit DNA binding, and substitution of Ser-11 and Ser-12 with Glu and Ala in *E. coli*-expressed Myb demonstrated that these amino terminal residues influence the negative effect on DNA binding by the leucine zipper in NRD (Ramsay et al, 1995). This discrepancy could be due to the expression system of c-Myb using mammalian cells, insect cells, or bacteria. A similar discrepancy was also observed for the effect of NRD on DNA binding (Ramsay et al, 1991; Krieg et al, 1995).

The 42-kDA MAP kinase also phosphorylates c-Myb at the C-terminal region which is lacking in v-Myb proteins. The phosphopeptide analysis identified Ser-528, which is located in the region identified by Kalkbrenner et al (1990) and Dubendorf et al (1993), as a phosphorylation site by MAP kinase (Figure 3.6) (Aziz et al, 1995). Replacement of Ser-528 with alanine results in significant increase in the ability of c-Myb to *trans*-activate a synthetic promoter containing the Myb-binding sites. Thus, changes in the state of phosphorylation serve to regulate c-Myb function.

Redox regulation

The Cys-130 in R2 of the DNA-binding domain is well conserved in various members of the *myb* gene family. Site-directed mutagenesis of this Cys-130 to serine almost completely abolishes DNA binding and transformation (Grässer et al, 1992; Guehmann et al, 1992). An early preliminary NMR analysis indicated that the third helix of R2 was disordered in the DNA-free condition (Jamin et al, 1993). However, the more extensive NMR analysis showed that R2 has the structure similar to R1 and R3 in the presence of high concentration reducing reagent (Ogata et al, 1995). R2 is thermally unstable and has slow conformational fluctuation due to the presence of a cavity inside the hydrophobic core, and its structure is stabilized on binding to DNA (Ogata et al, 1996). Since the Cys-130 is involved in the hydrophobic core in the DNA-complexed state, this cysteine residue cannot be oxidized in the DNA-complexed state. On the other hand, this cysteine residue is liable to be oxidized in the DNA-free condition due to the conformational fluctuation in R2. In fact, the Cys-130 was demonstrated to be accessible to alkylating reagent in the DNA-free state (Myrset et al, 1993). Thus, the Cys-130 in R2 could act as a molecular sensor for redox regulation. Consistent with these results, the redox regulator, Ref-1 (redox factor-1), which was originally identified as a positive regulator of AP-1 DNA binding, stimulates the DNA-binding activity of c-Myb in vitro (Xanthoudakis et al, 1992).

Oncogenic Activation and Functional Domains

Analysis of various oncogenically activated *myb* genes suggested that truncation of the N- or C-terminus of c-Myb can cause oncogenic activation (Figure 3.7). Both v-Myb proteins encoded by chicken leukemia viruses AMV and E26 are N- and C-terminally truncated versions of c-Myb (Roussel et al, 1979; Klempnauer et al, 1982; Nunn et al, 1983). Furthermore, integration of chronically transforming retroviruses into the c-*myb* locus can also result in truncation of the N- or C-terminus of c-Myb, inducing myelogenous diseases in mice and B-cell lymphomas in chicken (Shen-Ong et al, 1986; Weinstein et al, 1986; Shen-Ong et al, 1987; Gonda et al, 1987; Kanter et al, 1988; Pizer and Humphries, 1989; Wolff et al, 1991; Mukhopadhyaya and Wolff, 1992). Thus, there are two types of mode of oncogenic activation, N-terminal truncation and C-terminal truncation of c-Myb protein.

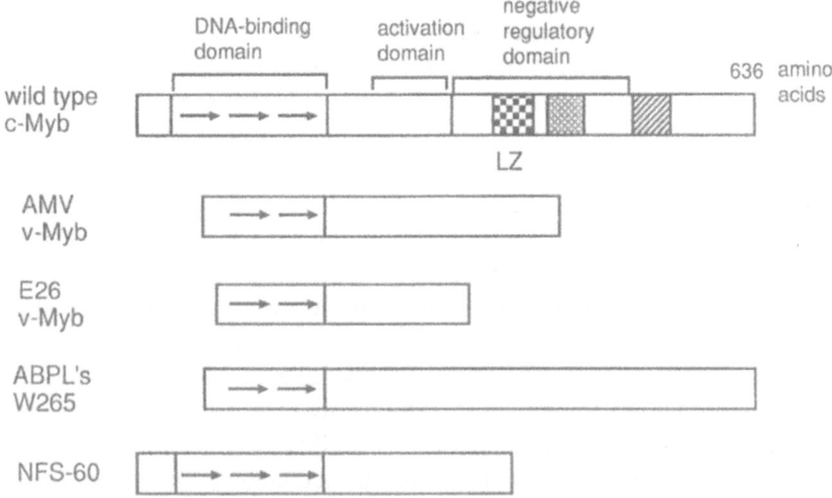

Figure 3.7 Oncogenic activation by N- or C-terminal truncation. At the top, the functional domains, in c-Myb are indicated. Two v-Myb proteins encoded by AMV and E26, and two Myb proteins encoded by the c-*myb* genes activated by the retroviral insertion are shown below.

Oncogenic activation by C-terminal truncation

The transforming capacity by C-truncated forms is well correlated with the *trans*-activating capacity (Gonda et al, 1989; Hu et al, 1991; Grässer et al, 1991), supporting the view that C-truncated forms of Myb transform by increasing the expression of targct genes. Removal of the NRD is responsible for oncogenic activation of C-truncated forms of Myb (Gonda et al, 1989; Hu et al, 1991; Grasser et al, 1991). v-Myb encoded by AMV lacks the C-proximal region in NRD, while v-Myb encoded by E26 lacks most of the NRD including the leucine zipper (Figure 3.7). In addition, a disruption of the leucine zipper alone also results in oncogenic activation of c-*myb* (Kanei-Ishii et al, 1992). Thus, the NRD contains multiple subdomains, and deletion of either of them may cause oncogenic activation of c-*myb*.

Oncogenic activation by N-terminal truncation

In contrast to the case of C-terminal truncation, the mechanism of oncogenic activation by N-terminal truncation is not clear. Both v-Myb proteins encoded by AMV and E26 lack the N-terminal 77 and 85 amino acids which contain the N-terminal region upstream of the R1 and about half of

R1 (Figure 3.7). Based on the observations that phosphorylation of serines 11 and 12 of c-Myb by casein kinase II can inhibit the binding of c-Myb to DNA, it was speculated that loss of the casein kinase II (CK-II) phosphorylation site by N-terminal truncation could uncouple c-Myb activity from its normal physiological regulators (Lüscher et al, 1990). This was further supported by the finding that proviral integration associated with murine promonocytic leukemias leads to loss of only 20 amino acids from the N-terminus of c-Myb which contains the CK-II phosphorylation sites (Mukkopadhyaya et al, 1992). However, it was demonstrated in the chicken system that truncation of R1 is sufficient for transformation, while deletion of the CK-II phosphorylation site is not (Dini et al, 1993). R1 is not necessary for recognition of the specific sequence, but it stabilizes the Myb-DNA complex (Sakura et al, 1989; Howe et al, 1990; Tanikawa et al, 1993). Truncation of R1 decreases the ability of Myb to bind DNA, indicating that the mechanism of oncogenic activation by N-terminal truncation may be different from that by C-terminal truncation. One hypothesis proposed is that this decreased affinity for DNA results in the regulation of a subset of c-Myb-regulated genes which control proliferation but not terminal differentiation (Dini et al, 1993). This implies that the promoters of proliferation genes regulated by Myb may possess Myb-binding sites with higher affinity or greater number than do differentiation genes. Another possibility is that N-terminal truncated c-Myb fails to repress transcription of some of the target genes due to the lack of a part of the DNA-binding domain (see below). This hypothesis suggests that the target genes, of which transcription is repressed by c-Myb, are critical for proliferation control of hematopoietic progenitor cells.

Target Genes

Multiple target genes activated by c-Myb

Identification of the target genes is one of the most important issues in understanding the molecular mechanism of cellular transformation by nuclear oncogenes. So far several potential target genes of c-Myb have been identified. Ness et al (1989) used differential hybridization to screen for v-Myb-regulated genes in cells transformed by a ts mutant of v-*myb* and identified the *mim-1* gene. This gene encodes a secretable protein contained in the granules of promyelocytes. Several groups found that the promoter of c-*myc* gene also contains multiple Myb-binding sites and its activity is stimulated by c-Myb (Evans et al, 1990; Zobel et al, 1991; Nakagoshi et al, 1992; Cogswell et al, 1993). In addition, c-Myb binds to the promoter regions of the *cdc2* and c-*myb* gene itself, and activates transcription (Ku et al, 1993; Nicolaides et al, 1991). Other genes which were identified as potential target genes include DNA polymerase α gene (Venturelli et al, 1990; Valtieri et al, 1991), lysozyme, MD-1 (Burk and Klempnauer, 1991), gene encoding PR264-splicing factor (Sureau et al, 1992), CD4 (Siu et al, 1992), CD34 (Melotti et al, 1994), and T-cell receptor δ (Hernandez-Munain and Krangel, 1994), as well as the long terminal repeat (LTR) promoters of the human immunodeficiency virus type-1 (HIV-1) (Dasgupta et al, 1992) and the human T-lymphotropic virus type-I (HTLV-I) (Dasgupta et al, 1992). The results of extensive differential screenings suggest that c-Myb may activate a larger number of genes than expected previously (Burk and Klempnauer, 1991; Nakano and Graf, 1992). In fact, a screening for c-Myb binding sites using a protocol which was developed from the systematic binding data obtained from measurement of binding affinity for the Myb-binding sites revealed a number of potential target genes including GM-CSF and T-cell receptor (Deng et al, 1996). However, it is not clear at present whether the increase and/or deregulation of the expression of specific target genes alone is sufficient for cellular transformation.

Synergistic action of c-Myb with other transcription factors

c-Myb can cooperate with other transcription factors in transcriptional activation of specific target genes. Although Ets-2 alone, which is also a nuclear oncogene, does not activate the *mim-1* promoter, it enhances the *mim-1* transcriptional activation by c-Myb (Dudek et al, 1992). c-Myb also synergistically activates the *mim-1* promoter with a myeloid-specific transcription factor NF-M (also called NF-IL6 or C/EBPβ) (Ness et al, 1993; Burk et al, 1993). Since the binding sites for c-Myb and NF-M are very close in the *mim-1* promoter, a direct interaction between these factors was suggested. The third transcription factor which cooperates with c-Myb is AML1 (acute myeloid leukemia 1, also called PEBP2 or CBFβ). AML1 is a member of a family of transcription factors with homology to the *Drosophila* pair-rule gene *runt*, and binds to various promoters and enhancers (Kagoshima et al, 1993; see also the chapter by Yto and Bae). The AML1-mediated transcriptional activation of the T-cell receptor genes is dependent on the c-Myb-binding site near the AML1 site (Hernandez-Munain and Krangel, 1995). Interestingly, the phenotype of mouse embryos with homozygous mutations in *AML1* closely resemble that of the c-*myb*-deficient mice (Okuda et al, 1996). Homozygous loss of c-*myb* or *AML1* results in normal yolk sac-derived erythropoiesis, but a severe defect in fetal liver hematopoiesis. This strongly suggest that c-Myb and AML1 coordinately regulate partially overlapping sets of target genes.

Other target genes

Using the artificial promoter linked to c-Myb-binding sites, c-Myb was also demonstrated to repress transcription in some cases (Nakagoshi et al, 1989). In fact, c-Myb represses the human c-*erbB*-2 promoter activity (Mizuguchi et al, 1995). The c-*erbB*-2 proto-oncogene (also called neu or HER2) encodes a 185-kDa transmembrane glycoprotein that has significant structural similarity to the EGF receptor. Among multiple Myb-binding sites in the c-*erbB*-2 promoter, two Myb-binding siters are critical for transcriptional repression by c-Myb. Myb represses the c-*erbB*-2 promoter activity by competing with positive regulators involving TFIID. The c-Myb mutant, which lacks the N-terminal 76 amino acids including about half of R1 in the DNA-binding domain, is a much weaker repressor than normal c-Myb. If some target genes, which are repressed by c-Myb, are critical for proliferation control, the N-terminal truncated c-Myb lacking R1 would cause transformation by deregulating these target genes.

Interestingly, c-Myb can activate the *hsp70* promoter without direct binding to its promoter region (Klempnauer et al, 1989; Kanei-Ishii et al, 1994). In this case, two distinct mechanisms seem to occur, since two elements in the *hsp70* promoter, the heat shock element and the TATA box, were identified as c-Myb responsive *cis*-elements (Foos et al, 1993; Kanei-Ishii et al, 1994). Since c-Myb bind to neither of these two elements, these results suggest that c-Myb can activate transcription of specific target genes by interacting with unidentified *trans*-acting factors. The biological significance of this phenomena should be clarified, in order to understand the physiological role of c-Myb in such cases.

Structure and Function of A-Myb and B-Myb

In addition to c-*myb*, the vertebrate *myb* gene family contains two other members A-*myb* and B-*myb* genes (Figure 3.8) (Nomura et al, 1988). Although c-*myb* is expressed at high levels in immature hematopoietic cells, it is also expressed in non-hematopoietic tissues: colon and small intestine at moderated levels and testis, ovary, and lung at low level. Tissue distribution of the B-*myb* mRNA is broader than c-*myb*, since it is also expressed in spleen, placenta, and pancreas in addition to the tissues mentioned above (Tashiro et al, 1995). Like in the case of c-*myb*, expression of B-*myb*

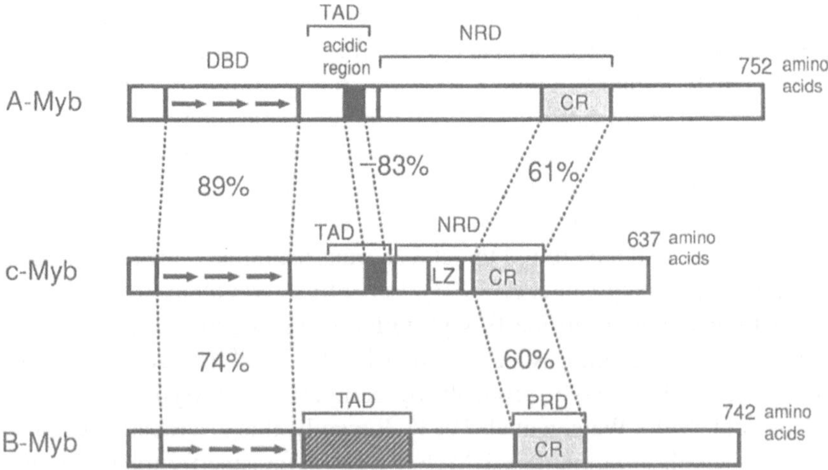

Figure 3.8 Comparison of functional domains between three members of the human *myb* gene family. The three tandem repeats in the DNA-binding domain are shown by arrows. The closed box indicates the region that is rich in acidic amino acids. The region conserved between three members (CR) is indicated by shaded box. DBD, DNA-binding domain; TAD, transcriptional activation domain; NRD, negative regulatory domain, LZ, leucine zipper structure. Percentages indicate percent identity in each region emcompassed by lines between human c-Myb and human A-Myb or between human c-Myb and human B-Myb.

correlates with cellular proliferation. B-*myb* mRNA is not expressed in resting cells but is induced late in G1 to maximal levels that are maintained through S phase (Reiss et al, 1992; Golay et al, 1991). B-*myb* mRNA levels decrease when HL60 or U937 cells are induced to differentiate (Reiss et al, 1991; Arsura et al, 1992). Inhibition of B-*myb* expression by introduction of a B-*myb* antisense construct diminished cell proliferation of hematopoietic cells and fibroblasts, whereas constitutive expression of B-*myb* induced a transformed phenotype (Arsura et al, 1992; Sala et al, 1992). These results suggest that B-Myb is a positive regulator for proliferation like c-Myb. In contrast to c-*myb* and B-*myb*, A-*myb* is expressed only in testis and peripheral blood leukocytes (Takahashi et al, 1995). A-*myb* is also expressed in resting T-cells, and its levels gradually decrease after mitogenic stimulation (Golay et al, 1991). These facts suggest that physiological role of A-Myb is distinct from that of c-Myb and B-Myb. These different tissue distributions of mRNA also suggest different physiological roles of three members.

Among the three members, A-Myb has the strongest *trans*-activating capacity (Trauth et al, 1994; Golay et al, 1994; Takahashi et al, 1995). The functional domains of A-Myb are similar to those of c-Myb except for the leucine zipper (Figure 3.8). The A-Myb activity is negatively regulated through the NRD that is located downstream of the transcriptional activation domain. There was a discrepancy for the *trans*-activating capacity of B-Myb. Originally, human b-Myb was reported to be a transcriptional activator, and to activate the artificial promoter containing multiple Myb-binding sites (MBS-I) (Mizuguchi et al, 1990) and the human c-*myc* promoter (Nakagoshi et al, 1992). Furthermore, the region located downstream of the DNA-binding domain in the B-Myb protein, which is rich in acidic amino acid was found to act as a transcriptional activation domain (Figure 3.8) (Nakagoshi et al, 1993). In contrast, two other groups reported that chicken and murine B-Myb failed to *trans*-activate the promoter containing Myb-binding sites, and that B-Myb inhibited *trans*-activation of promoters by c-Myb, probably by competing for Myb-binding sites (Foos et al, 1992; Watson et al, 1993). Recently, this discrepancy was clarified by an observation that B-Myb is a cell type-specific transcriptional activator (Tashiro et al, 1995). B-Myb functions as a transcriptional activator in CV-1 and HeLa cells, but not in NIH3T3 cells. Deletion analyses of B-Myb demonstrated that the region conserved between three members of the *myb* gene family (CR) is necessary for *trans*-activation by B-Myb (Figure 3.8). An in vivo competition assay suggests that regulatory factor(s) that binds to the CR of B-Myb is required for *trans*-activation. Analyses using affinity resin showed that multiple proteins bind to the CR of B-Myb and that the CR-binding proteins in CV-1 and HeLa cells are different from those in NIH3T3 cells. Thus, the CR-binding cofactor(s) may be critical for the cell type-specific *trans*-activation by B-Myb.

Conclusions and Perspectives

The proto-oncogene c-*myb* plays a pivotal role in growth control, differentiation, and tumorigenesis. Toward understanding how these cellular processes are regulated, it is essential to understand the role of c-Myb. Identifying the multiple c-*myb* target genes involved in cellular transformation, and how specific target genes are regulated by the -*myb* family of transcription factors is crucial to achieving these goals. Many observations indicate that the regulation of c-Myb activity by the carboxyl region of the protein is complex, and probably results from the action of several distinct elements. To clarify the role of the carboxy terminal region, it is necessary to characterize the leucine zipper-binding proteins, to determine the structure of the intact molecule, and to analyze the phosphorylation of this region. The structural analysis solved the unique structure of the DNA-binding domain of c-Myb, and we can now understand the molecular interaction between Myb and specific DNA sequence at the atomic level.

However, the role of R1 is not clear at present. Analyses of the structure of R1 and the further upstream N-terminal region is undoubtedly necessary to understand the mechanism of oncogenic activation of c-*myb* by N-terminal truncation of the protein. One of the obvious questions remaining is how growth signal(s) can be transduced from growth factor(s) to the Myb protein in nuclei. A c-*myb* deficient homozygous mutant does not develop fetal blood cells, indicating a similarity between the disrupted c-*myb* phenotype and the dominant white-spotting and steel mutants which affect the c-kit receptor and its ligand SCF. These results imply that c-Myb could be the nuclear target of the c-kit signaling pathway. The inability of normal c-Myb to transform primary hematopietic cells could therefore be due to its failure to escape normal posttranslational regulation by such cytokine receptor pathways. Further analysis of the mechanism by which the c-Myb activity is modulated will give us the answer to this question.

Acknowledgments

I thank Akinori Sarai, Kazuhiro Ogata, and Haruki Nakamura for the preparation of the figures.

References

Anfossi G, Gewirtz A, Calabretta B (1989): An oligomer complementary to c-*myb*-encoded mRNA inhibits proliferation of human myeloid leukemia cells. *Proc Natl Acad Sci USA* 86: 3379–3383

Anton IA, Frampton J (1988): Tryptophans in *myb* proteins. *Nature* 336: 719

Arias J, Alberts AS, Brindle P, Claret FX, Smeal T, Karin M, Feramisco J, Montminy MR (1994): Activation of cAMP and mitogen responsive genes relies on a common nuclear factor. *Nature* 370: 226–229

Arsura M, Introna M, Passerini F, Mantovani A, Golay, J (1992): B-*myb* antisense oligonucleotides inhibit proliferation of human hematopoietic cell lines. *Blood* 79: 2708–2716

Aziz N, Miglarese MR, Hendrickson RC, Shabanowitz J, Sturgill TW, Hunt DF, Bender T (1995): Modulation of c-Myb-induced transcription activation by a phosphorylation site near the negative regulatory domain. *Proc Natl Acad Sci USA* 92: 6429–6433

Badiani P, Corabella P, Kioussies D, Marvel J, Weston K (1994): Dominant interfering alleles difine a role for c-Myb in T-cell development. *Genes Dev* 8: 770–782

Bannister AJ, Kouzarides T (1995): CBP-induced stimulation of c-Fos activity is abrogated by E1A, *EMBO J* 14: 4758–4762

Beug H, Blundell P, Graf T (1987): Reversibility of differentiation and proliferative capacity in avian myelomonocytic cells transformed by ts E26 leukemia virus. *Genes Dev* 1: 277–286

Biedenkapp H, Borgmeyer U, Sippel AE, Klempnauer KH (1988): Viral *myb* oncogene encodes a sequence-specific DNA binding activity. *Nature* 335: 835–837

Bishop JM, Eilers M, Katzen AL, Kornberg T, Ramsay G, Schim S (1991): *myb* and *myc* in the cell cycle. Cold Spring Harbor Symposia on Quantitative Biology. Vol. LVI: 99–107

Boulukos KE, Pognonec P, Begue A, Galibert F, Gesquiere JC, Stehelin D, Ghysdael J (1988): Identification in chickens of an evolutionarily conserved cellular *ets*-2 gene (c-*ets-2*) encoding nuclear proteins related to the products of the c-*ets* proto-oncogene. *EMBO J* 7: 697–705

Brown KE, Kindy MS, Sonenshein GE (1992): Expression of the c-*myb* proto-oncogene in bovine vascular smooth muscle cells. *J Biol Chem* 267: 4625–4630

Burk O, Klempnauer KH (1991): Estrogen-dependent alterations in differentiation state of myeloid cells caused by a v-Myb/estrogen receptor fusion protein. *EMBO J* 10: 3713–3719

Burk O, Mink S, Ringwald M, Klempnauer KH (1993): Synergistic activation of the chicken *mim-1* gene by v-*myb* and C/EBP transcription factors. *EMBO J* 12: 2027–2038

Chrivia JC, Kwok RPS, Lamb N, Hagiwara M, Montminy MR, Goorman RH (1993): Phosphorylated CREB bind specifically to the nuclear protein CBP. *Nature* 365: 855–859

Clarke MF, Kukowska-Latallo JF, Westin E, Smith M, Prochownik EU (1988): Constitutive expression of a c-*myb* cDNA blocks Friend murine erythroleukemia cell differentiation. *Mol Cell Biol* 8: 884–892

Cogswell JP, Cogswell PC, Kuehl WM, Cuddihy AM, Bender TM, Engelke U, Marcu KB, Ting JPY (1993): Mechanism of c-*myc* regulation by c-Myb in different cell lineages. *Mol Cell Biol* 13: 2858–2869

Dai P, Akimaru H, Tanaka Y, Hou DX, Yasukawa T, Kanei-Ishii C, Takahashi T, Ishii S (1996): CBP as a transcriptional coactivator of c-Myb. *Genes Dev* 10: 528–540

Dasgupta P, Saikumar P, Reddy CD, Reddy EP (1990): Myb protein binds to human immunodeficiency virus 1 long terminal repeat (LTR) sequences and transactivates LTR-mediated transcription. *Proc Natl Acad Sci USA* 87: 8090–8094

Dasgupta P, Reddy CD, Saikumar P, Reddy EP (1992): The cellular proto-oncogene product Myb acts as transcriptional activator of the long terminal repeat of human T-lymphotropic virus type I. *J Virol* 66: 270–276

Deng QL, Ishii S, Sarai A (1996): Binding site analysis of c-Myb: screening of potential binding sites by using the mutation matrix derived from systematic binding affinity measurements. *Nucleic Acids Res* 24: 766–774

Dini PW, Lipsick JS (1993): Oncogenic truncation of the first repeat of c-Myb decreases DNA binding in vitro and in vivo. *Mol Cell Biol* 13: 7334–7348

Dubendorff JW, Whittaker LJ, Eltman JT, Lipsick JS (1992): Carboxyl-terminal elements of c-Myb negatively regulate transcriptional activation in *cis* and in *trans*. *Genes Dev* 6: 2524–2535

Dudek H, Tantravahi RV, Rao VN, Reffy ESP, Reddy EP (1992): Myb and Ets proteins cooperate to transcriptional activation of the *mim-1* promoter. *Proc Natl Acad Sci USA* 89: 1291–1295

Duprey SP, Boettiger D (1985): Developmental regulation of c-*myb* in normal myeloid progenitor cells. *Proc Natl Acad Sci USA* 82: 6937–6941

Evans JT, Moore TL, Kuehl WM, Bender T, Ting JPY (1990): Functional analysis of c-Myb protein in T-lymphocytic cell lines shows that in *trans*-activates the c-*myc* promoter. *Mol Cell Biol* 10: 5747–5752

Favier D, Gonda TJ (1994): Detection of proteins that bind to the leucine zipper motif of c-Myb. *Oncogene* 9: 305–311

Foos G, Grimm S, Klempnauer KH (1992): Functional antigonism between members of the *myb* family: B-*myb* inhibits v-*myb*-induced gene activation. *EMBO J* 11: 4619–4629

Foos G, Natour S, Klempnauer KH (1993): TATA-box dependent *trans*-activation of the human HSP70 promoter by Myb proteins. *Oncogene* 8: 1775–1782

Frampton J, Leutz A, Gibson TJ, Graf T (1989): DNA-binding domain ancestry. *Nature* 342: 134

Frampton J, Gibson TJ, Ness SA, Döderlein G, Graf T (1991): Proposed structure for the DNA-binding domain of the Myb oncoprotein based on model building and mutational analysis. *Protein Engng* 4: 891–901

Furuta Y, Aizawa S, Suda Y, Ikawa Y, Nakagoshi H, Nishina Y, Ishii S (1993): Degeneration of skeletal and cardiac muscles in c-*myb*-transgenic mice. *Transgenic Res* 2: 199–207

Gabrielsen OS, Sentenac A, Fromageot P (1991): Specific DNA binding by c-Myb: evidence for a double helix-turn-helix-related motif. *Science* 253: 1140–1143

Gewirtz A, Anfossi G, Venturelli D, Valpreda S, Sims R, Calabretta B (1989): G1/S transition in normal human T-lymphocytes requires the nuclear protein encoded by c-*myb*. *Science* 245: 180–183

Gewirtz AM, Calabretta B (1988): A c-*myb* antisense oligodeoxynucleotide inhibits normal human hematopoiesis in vitro. *Science* 242: 1303–1306

Golay J, Capucci A, Arsura M, Casettelano M, Rizzo V, Introna M (1991): Expression of c-*myb* and B-*myb*, but not A-*myb*, correlates with proliferation in human hematopoietic cells. *Blood* 7: 149–158

Golay JL, Loffarelli M, Luppi M, Castellano M, Introna M (1994): The human A-*myb* protein is a strong activator of transcription. *Oncogene* 9: 2469–2479

Gonda TJ, Metcalf D (1984): Expression of *myb*, *myc*, and *fos* proto-oncogenes during the differentiation of a murine myeloid leukaemia. *Nature* 310: 249–251

Gonda TJ, Gough NM, Dunn AR, de Blaquiere J (1985): Nucleotide sequence of cDNA clones of the murine *myb* proto-oncogene. *EMBO J* 4: 2003–2008

Gonda TJ, Cory S, Sobieszczul P, Holtzman D, Adams JM (1987): Generation of altered transcripts by retroviral insertion within the c-*myb* gene in two murine monocytic leukemias. *J Virol* 61: 2754–2763

Gonda TJ, Buckmaster C, Ramsay RG (1989): Activation of c-*myb* by carboxy-terminal truncation: Relationship to transformation of murine haemopoietic cells in vitro. *EMBO J* 8: 1777–1783

Gonzalez GA, Montminy MR (1989): Cyclic AMP stimulates somatostatin gene transcription by phosphorylation of CREB at serine 133. *Cell* 59: 675–680

Graf T (1992): Myb: A transcriptional activator linking proliferation and differentiation in hematopoietic cells. *Curr Opin Gen Dev* 2: 249–255

Graf T, McNagny K, Brady G, Frampton J (1992): Chicken "erythroid" cells transformed by the gag-*myb*-ets-encoding E26 leukemia virus are multipotent. *Cell 70*: 201–213

Grässer FA, Graf T, Lipsick JS (1991): Protein truncation is required for the activation of the c-*myb* protoconcogene. *Mol Cell Biol* 11: 3987–3996

Grässer FA, LaMontagne K, Whittaker L, Stohr S, Lipsick JS (1992): A highly conserved cystein in the v-Myb DNA-binding domain is essential for transformation and transcriptional *trans*-activation. *Oncogene* 7: 1005–1009

Grotewold E, Drummond BJ, Bowen B, Peterson T (1994): The *myb*-homologous *P* gene controls phlobaphene pigmentation in maize floral organs by directly activating a flavonoid biosynthetic gene subset. *Cell* 76: 543–553

Guehmann S, Vorbrueggen G, Kalkbrenner F, Moelling K (1992): Reduction of a conserved Cys is essential for Myb DNA-binding. *Nuclei Acids Res* 20: 2279–2286

Hernandez-Munain C, Krangel MS (1994): Regulation of the T-cell receptor δ enhancer by funtional cooperation between c-Myb and core-binding factors. *Mol Cell Biol* 14: 473–483

Hernandez-Munain C, Krangel MS (1995): c-Myb and core-binding factor/PEBP-2 display synergy but bind independently to adjacent sites in the T-cell receptor enhancer. *Mol Cell Biol* 15: 3090–3099

Howe KM, Watson RJ (1991): Nucleotide preferences in sequence-specific recognition of DNA by c-*myb* protein. *Nucleic Acids Res* 19: 3913–3919

Howe KM, Reakes CFL, Watson RJ (1990): Characterization of the sequence-specific interaction of mouse c-*myb* protein with DNA. *EMBO J* 1: 161–169

Hu Y, Ramsay RG, Kanei-Ishii C, Ishii S, Gonda TJ (1991): Transformation by carboxyl-deleted Myb reflects increased transactivating capacity and disruption of a negative regulatory domain. *Oncogene* 6: 1549–155

Ibanez CE, Lipsick JS (1990): *trans*-Activation of gene expression by v-*myb*. *Mol Cell Biol* 10: 2285–2293

Introna M, Golay J, Frampton J, Nakano T, Ness SA, Graf T (1990): Mutations in v-*myb* alter the differentiation of myelomonocytic cells transformed by the oncogene. *Cell* 63: 1287–1297

Jamin N, Gabrielsen OS, Gilles N, Lirsac PN, Toma F (1993): Secondary structure of the DNA-binding domain of the c-Mybc oncoprotein in solution. *Eur J Biochem* 216: 147–154

Kagoshima H, Shigesada K, Satake M, Ito Y, Miyoshi H, Ohki M, Pepling M, Gergen P (1993): The runt domain identifies a new family of heteromeric transcriptional regulators. *Trends Genet* 9: 338–341

Kalkbrenner F, Guehmann S, Moeling K (1990): Transcriptional activation by human c-*myb* and v-*myb* genes. *Oncogene* 5: 657–661

Kanei-Ishii, Sarai A, Sawazaki T, Nakagoshi H, He DN, Ogata K, Nishimura Y, Ishii S (1990): The tryptophan cluster: a hypothetical structure of the DNA-binding domain of the *myb* protooncogene product. *J Biol Chem* 265: 19990–19995

Kanei-Ishii, MacMillan EM, Nomura T, Sarai A, Ramsay RG, Aimoto S, Ishii S, Gonda TJ (1992): Transactivation and transformation by Myb are negatively regulated by a leucine-zipper structure. *Proc Natl Acad Sci USA* 89: 3088–3092

Kanei-Ishii, Yasukawa T, Morimoto RI, Ishii S (1994): c-Myb-induced *trans*-activation mediated by heat shock elements without sequence-specific DNA binding of c-Myb. *J Biol Chem* 269: 15768–15775

Kanter MR, Smith RE, Hayward WS (1988): Rapid induction of B-cell lymphomas: insertional activation of c-*myb* by avian leukosis virus. *J Virol* 62: 1423–1432

Katzen AL, Korngerg TB, Bishop JM (1985): Isolation of the proto-oncogene c-*myb* from *D. melanogaster. Cell* 41: 449–456

Klempnauer KH, Sippel AE (1987): The highly conserved amino-terminal region of the protein encoded by the v-*myb* oncogene functions as a DNA-binding domain. *EMBO J* 6: 2719–2725

Klempnauer KH, Gonda TJ, Bishop JM (1982): Nucleotide sequence of the retroviral leukemia gene v-*myb* and its cellular progenitor c-*myb*: the architecture of a transduced oncogene. *Cell* 31: 453–463

Klempnauer KH, Arnold H, Biedenkapp H (1989): Activation of transcription by v-*myb*; evidence for two different mechanisms. *Genes Dev* 3: 1582–1589

Krieg J, Oelgeschläger M, Janknecht R, Lüscher B (1995): High affinity DNA binding of native full length c-Myb and differential proteolytic sensitivity of its N- and C-terminal domains. *Oncogene* 10: 2211–2228

Ku DH, Wen SC, Engelhard A, Nicolaides NC, Lipson KE, Marino TA, Calabretta B (1993): c-*myb* transactivates *cdc*2 expression via Myb binding sites in the 5′-flanking region of the human *cdc*2 gene. *J Biol Chem* 268: 2255–2259

Kwok RPS, Lundblad JR, Chrivia JC, Richards JP, Bachinger HP, Brennan RG, Roberts SGR, Green MR, Goodman RH (1994): Nuclear protein CBP is a coactivator for the transcription factor CREB. *Nature* 370: 223–226

Landschulz WH, Johnson PF, McKnight SL (1988): The leucine zipper, a hypothetical structure common to a new class of DNA-binding proteins. *Science* 240: 1759–1766

Liebermann DA, Hoffman-Liebermann B (1989): Proto-oncogene expression and dissection of the myeloid growth to differentiation development cascade. *Oncogene* 4: 583–592

Lloyd AM, Walbot V, Davis RW (1992): *Arabidopsis and Nicotiana* anthocyanin production activated by maize regulators *R* and *C1*. *Science* 258: 1773–1775

Lüscher B, Eisenmann RN (1992): Mitosis-specific phosphorylation of the nuclear oncoproteins Myb and Myc. *J Cell Biol* 118: 775–784

Lüscher B, Christenson E, Litchfield DW, Krebs EG, Eisenman RN (1990): *Myb* DNA binding inhibited by phosphorylation at a site deleted during oncogenic activation. *Nature* 344: 517–522

McMahon J, Howe KM, Watson RJ (1988): The induction of Friend erythroleukaemia differentiation is markedly affected by expression of a transfected c-*myb* cDNA. *Oncogene* 3: 717–720

Melotti P, Ku DH, Calabretta B (1994): Regulation of the expression of the hematopoietic stem cell antigen CD34: role of c-myb *J Exp Med* 179: 1023–1028

Metz T, Graf T (1991): v-*myb* and v-*ets* transform chicken erythroid cells and cooperate both in *trans* and in *cis* to distinct differentiation phenotypes. *Genes Dev* 5: 369–380

Mizuguchi G, Nakagoshi H, Nagase T, Nomura N, Date T, Ueno Y, Ishii S (1990): DNA-binding activity and transcriptional activator function of the human B-*myb* protein compared with c-*myb*. *J Biol Chem* 265: 9280–9284

Mizuguchi G, Kanei-Ishii C, Takahashi T, Yasukawa T, Nagase T, Horikoshi M, Yamamoto T, Ishii S (1995): c-Myb repression of c-*erbB*-2 transcription by direct binding of the c-*erbB*-2 promoter. *J Biol Chem* 270: 9384–9389

Moelling K, Pfaff E, Beug H, Beimling P, Bunte T, Schaller HE, Graf T (1985): DNA-binding activity is associated with purified Myb proteins from AMV and E26 viruses and is temperature-sensitive for E26 ts mutants. *Cell* 40: 983–990

Moscovici MG, Jurdic P, Samarut J, Gazzolo L, Mura CV, Moscovici C (1983): Characterization of the hemopoietic target cells for the avian leukemia virus E26. *Virology* 129: 65–78

Mücenski ML, McLain K, Kier AB, Swerdlow SH, Schereiner CM, Miller TA, Pietryga DW, Scott WJ, Potter SS (1991): A functional c-*myb* gene is required for normal murine fetal hepatic hematopoiesis. *Cell* 65: 677–689

Mukhopadhyaya R, Wolff L (1992): New sites of proviral integration associated with murine promonocytic leukemias and evidence for alternate modes of c-*myb* activation. *J Virol* 66: 6035–6044

Myrset AH, Bostat A, Jamin N, Lirsac PN, Toma F, Gabrielsen OS (1993): DNA and redox state induced conformational changes in the DNA-binding domain of the Myb oncoprotein. *EMBO J* 12: 4625–4633

Nakagoshi H, Nagase T, Ueno Y, Ishii S (1989): Transcriptional *trans*-repression by the c-*myb* proto-oncogene product. *Nucleic Acids Res* 17: 7315–7324

Nakagoshi H, Nagase T, Kanei-Ishii C, Ueno Y, Ishii S (1990): Binding of the c-*myb* proto-oncogene product to the simian virus 40 enhancer stimulates transcription. *J Biol Chem* 265: 3479–3483

Nakagoshi H, Kanei-Ishii C, Sawazaki T, Mizuguchi G, Ishii S (1992): Transcriptional activation of the c-*myc* gene by the c-*myb* and B-*myb* gene products. *Oncogene* 7: 1233–1240

Nakagoshi H, Takemoto Y, Ishii S (1993): Functional domains of the human B-*myb* gene product. *J Biol Chem* 268: 14161–14167

Nakano T, Graf T (1992): Identification of genes differentially expressed in two types of v-*myb*-transformed avian myelomonocytic cells. *Oncogene* 7: 527–534

Ness SA, Beug H, Graf T (1987): v-*myb* dominance over c-*myc* in doubly transformed chick myelomonocytic cells. *Cell* 51: 41–50

Ness SA, Marknell A, Graf T (1989): The v-*myb* oncogene product binds to and activates the promyelocyte-specific *mim-1* gene. *Cell* 59: 1115–1125

Ness SA, Kownez-Leutz E, Casini T, Graf T, Leutz A (1993): Myb and NF-M: combinatorial activators of myeloid genes in heterologous cell types. *Genes Dev* 7: 749–759

Nicolaides NC, Gualdi R, Casadevall C, Manzella L, Calabretta B (1991): Positive autoregulation of c-*myb* expression via Myb binding sites in the 5′ flanking region of the human c-*myb* gene. *Mol Cell Biol* 11: 6166–6176

Nishina Y, Nakagoshi H, Imamoto F, Gonda TJ, Ishii S (1989): *trans*-Activation by the c-*myb* proto-oncogene. *Nucleic Acids Res* 17: 107–117

Noda K, Glover BJ, Linstead P, Martin C (1994): Flower colour intensity depends on specialized cell shape controlled by a Myb-related transcription factor. *Nature* 369: 661–664

Nomura N, Takahashi M, Matsui M, Ishii S, Date T, Sasamoto S, Ishizaki R (1988): Isolation of human cDNA clones of *myb*-related genes, A-*myb* and B-*myb*. *Nucleic Acids Res* 16: 11075–11089

Nomura T, Sakai N, Sarai A, Sudo T, Kanei-Ishii C, Ramsay RG, Favier D, Gonda TJ, Ishii S (1993): Negative autoregulation of c-Myb activity by homodimer formation through the leucine zipper. *J Biol Chem* 268: 21914–21923

Nunn MF, Seeburg PH, Moscovici C, Duesberg PH (1983): Tripartite structure of the avian erythroblastosis virus E26 transforming gene. *Nature* 306: 391–395

Oelgeschläger M, Krieg J, Lüscher-Firzlaff JM, Lüscher B (1995): Casein kinase II phosphorylation site mutations in c-Myb affect DNA binding and transcriptional cooperativity with NF-M. *Mol Cell Biol 15*: 5966–5974

Oelgeschläger M, Janknecht R, Krieg J, Schreek S, Lüscher B (1996): Interaction of the coactivator CBP with Myb proteins: effects on Myb-specific transactivation and on the cooperativity with NF-M. *EMBO J* 15: 2771–2780

Ogata K, Hojo H, Aimoto S, Nakai T, Nakamura H, Sarai A, Ishii S, Nishimura Y (1992): Solution structure of a DNA-binding unit of Myb: a helix-turn-helix-related motif with conserved tryptophans forming a hydrophobic core. *Proc Natl Acad Sci USA* 89: 6428–6432

Ogata K, Morikawa S, Nakamura H, Sekikawa A, Inoue T, Kanai H, Sarai A, Ishii S, Nishimura Y (1994): Solution structure of a specific DNA complex of the Myb DNA-binding domain with cooperative recognition helixes. *Cell* 79: 639–648

Ogata K, Morikawa S, Nakamura H, Hojo H, Yoshimura S, Zhang R, Aimoto S, Ametani Y, Hirata Z, Sarai A, Ishii S, Nishimura Y (1995): Comparison of the free and DNA complexed forms of the DNA-binding domain from c-Myb. *Nature Struct Biol* 2: 309–320

Ogata K, Kanei-Ishii C, Sasaki M, Hatanaka H, Nagadoi A, Enari M, Nakamura H, Nishimura Y, Ishii S, Sarai A (1996): The cavity in the hydrophobic core of Myb DNA-binding domain is reserved for DNA recognition and *trans*-activation. *Nature Struct Biol* 3: 178–187

Okuda T, van Deursen J, Hiebert SW, Grosveld G, Downing JR (1996): AML1, the target of multiple chromosomal translocations in human leukemia, is essential for normal fetal liver hematopoiesis. *Cell* 84: 321–330

Oppenhaimer DG, Herman PL, Sivakmaran S, Esch J, Marks MD (1991): A *myb* gene required for leaf trichome differentiation in *Arabidopsis* is expressed in stipules. *Cell* 67: 483–493

Paz-Ares J, Ghosal D, Wienand U, Peterson PA, Saedler H (1987): The regulatory *c1* locus of *Zea mays* encodes a protein with homology to *myb* proto-oncogene products and with structural similarities of transcriptional activators. *EMBO J* 6: 3553–3558

Peters CW, Sippel AE, Vingron M, Klempnauer KH (1987): *Drosophila* and vertebrate *myb* proteins share two conserved regions, one of which functions as a DNA-binding domain. *EMBO J* 6: 3085–3090

Petrij F, Giles RH, Dauwerse HG, Saris JJ, Hennekam RCM, Masuno M, Tommerup N, van Ommen GB, Goodman RH, Peters DJM, Breuning MH (1995): Rubinstein-Taybi syndrome caused by mutations in the transcriptional co-activator CBP. *Nature* 376: 348–351

Pizer E, Humphries EH (1989): RAV-1 insertional mutagenesis: Disruption of the c-*myb* locus and development of avian B-cell lymphomas. *J Viol* 63: 1630–1640

Radke K, Beug H, Kornfeld S, Graf T (1982): Transformation of both erythroid and myeloid cells by E26, an avian leukemia virus that contains the *myb* gene. *Cell* 31: 643–653

Ramsay RG, Ishii S, Gonda TJ (1991): Increase in specific DNA binding by carboxyl truncation suggests a mechanism for activation of Myb. *Oncogene* 6: 1875–1879

Ramsay RG, Ishii S, Gonda TH (1992): Interaction of the Myb protein with specific DNA binding sites. *J Biol Chem* 267: 5656–5662

Ramsay RG, Morrice N, Van Eden P, Kanagasundaram V, Nomura T, De Blaquiere J, Ishii S, Wettenhall R (1995): Regulation of c-Myb through protein phosphorylation and leucine zipper interaction. *Oncogene* 11: 2113–2120

Reiss K, Travali S, Calabretta B, Baserga R (1991): Growth regulated expression of B-*myb* in fibroblasts and hematopoietic cells. *J Cell Physiol* 148: 338–343

Roussel M, Saule S, Lagrou C, Rommens C, Beug H, Graf T, Stehelin D (1979): Three new types of viral oncogenes of cellular origin specific for haematopoietic cell transformation. *Nature* 281: 452–455

Sablowski RWM, Moyano E, Culianez-Macia FA, Schuch W, Martin C, Bevan M (1994): A flower-specific Myb protein activates transcription of phenylpropanoid biosynthetic genes. *EMBO J* 13: 128–137

Saikumar P, Murali R, Reddy EP (1990): Role of tryptophan repeats and flanking amino acids in Myb-DNA interactions. *Proc Natl Acad Sci USA* 87: 8452–8456

Sakura H, Kanei-Ishii C, Nagase T, Nakagoshi H, Gonda TJ, Ishii S (1989): Delineation of three functional domains of the transcriptional activator encoded by the c-*myb* proto-oncogene. *Proc Natl Acad Sci USA* 86: 5758–5762

Sala A, Calabretta B (1992): Regulation of BALB c3T3 fibroblast proliferation by B-*myb* is accompanied by selective activation of *cdc*2 and cyclin D1 expression. *Proc Natl Acad Sci USA* 89: 10415–10419

Sarai A, Uedaira H, Morii H, Yasukawa T, Ogata K, Nishimura Y, Ishii S (1993): Thermal stability of the DNA-binding domain of the Myb oncoprotein. *Biochemistry* 32: 7759–7764

Sheiness D, Gardinier M (1984): Expression of a proto-oncogene (proto-*myb*) in hemopoietic tissues of mice. *Mol Cell Biol* 4: 1206–1212

Shen-Ong GLC, Wolff L (1987): Moloney murine leukemia virus-induced myeloid tumors in adult BALB/c mice: requirement of c-*myb* activation but lack of v-*abl* involvement. *J Virol* 61: 3721–3725

Shen-Ong GLC III, Morse HC, Potter M, Mushinski JF (1986): Two modes of c-*myb* activation in virus-induced mouse myeloid tumors. *Mol Cell Biol* 6: 380–392

Siu G, Wurster AL, Lipsick JS, Hedrick SM (1992): Expression of the CD4 gene requires a Myb transcriptional factor. *Mol Cell Biol* 12: 1592–1604

Solano R, Nieto C, Avila J, Canas L, Diaz I, Paz-Ares J (1995): Dual DNA binding specificity of a petal epidermis-specific MYB transcription factor (MYB.Ph3) from *Petunia hybrida*. *EMBO J* 14: 1773–1784

Stern J, Smith KA (1986): Interleukin-2 induction of T-cell G1 progression and c-*myb* expression. *Science* 233: 203–206

Sureau A, Soret J, Vellard M, Crochet J, Perbal B (1992): The PR264/c-*myb* connection: expression of a splicing factor modulated by a nuclear protooncogene. *Proc Natl Acad Sci USA* 89: 11683–11687

Takahashi T, Nakagoshi H, Sarai A, Nomura N, Yamamoto T, Ishii S (1995): Human A-*myb* gene encodes a transcriptional activator containing the negative regulatory domains. *FEBS Lett.* 358: 89–96

Tanikawa J, Yasukawa T, Enari M, Ogata K, Nishimura Y, Ishii S, Sarai A (1993): Recognition of specific DNA sequences by the c-*myb* proto-oncogene product-role of three repeat units in the DNA-binding domain. *Proc Natl Acad Sci USA* 90: 9320–9324

Tashiro S, Takemoto Y, Handa H, Ishii S (1995): Cell type-specific *trans*-activation by the B-*myb* gene product: requirement of the putative cofactor binding to the C-terminal conserved domain. *Oncogene* 10: 1699–1707

Thompson CB, Challoner PB, Neiman PE, Groudine M (1986): Expression of the c-*myb* proto-oncogene during cellular proliferation. *Nature* 319: 374–380

Tice-Baldwin K, Fink GR, Arndt KT (1989): BAS1 has a *myb* motif and activates *HIS4* transcription only in combination with BAS2. *Science* 246: 931–935

Todokoro K, Watson RJ, Higo H, Amanuma H, Kuramochi S, Yanagisawa H, Îkawa Y (1988): Down-regulation of c-*myb* gene expression is a prerequisite for erythropoietin-induced erythroid differentiation. *Proc Natl Acad Sci USA* 85: 8900–8904

Trauth K, Mutschler B, Jenkins NA, Gilbert DJ, Copeland NG, Klempnauer KH (1994): Mouse A-*myb* encodes a *trans*-activator and is expressed in mitotically active cells of the developing central nervous system, adult testis and B lymphocytes. *EMBO J* 13: 5994–6005

Valtieri M, Venturelli D, Care A, Fossati C, Pelosi E, Labbaye C, Mattia G, Gewirtz AM, Calabretta B, Peschle C (1991): Antisense *myb* inhibition of purified erythroid progenitors in development and differentiation is linked to cycling activity and expression of DNA polymerase α. *Blood* 77: 1181–1190

Venturelli D, Travali S, Calabretta B (1990): Inhibition of T-cell proliferation by a MYB antisense oligomer is accompanied by selective down-regulation of DNA polymerase α expression. *Proc Natl Acad Sci USA* 87: 5963–5967

Vorbrueggen G, Kalkbrenner F, Guehmann S, Moelling K (1994): The carboxy-terminus of human c-myb protein stimulates activated transcription in *trans*. *Nucleic Acids Res* 22: 2466–2475

Watson RJ, Robinson C, Lam EW (1993): Transcription regulation by murine B-*myb* is distinct from that by c-*myb*. *Nucleic Acids Res* 21: 267–272

Weinstein Y, Ihle JN, Lavu S, Reddy P (1986): Truncation of the c-*myb* gene by a retroviral integration in an interleukin-3-dependent myeloid leukemia cell line. *Proc Natl Acad Sci USA* 83: 5010–5014

Westin EH, Gallo GC, Arya SK, Eva A, Souza LM, Baluda MA, Aaronson SA, Wong-Staal F (1982): Differential expression of the *amv* gene in human hematopoietic cells. *Proc Natl Acad Sci USA* 79: 2194–2198

Weston K (1992): Extension of the DNA binding consensus of the chicken c-Myb and v-Myb protiens. *Nucleic Acids Res* 20: 3043–3049

Weston K, Bishop JM (1989): Transcriptional activation by the v-*myb* oncogene and its cellular progenitor, c-*myb*. *Cell* 58: 85–93

Wieser J, Adams TH (1995): *flbD* encodes a Myb-like DNA-binding protein that coordinates initiation of *Aspergillus nidulans* conidiophore development. *Genes Dev* 9: 491–502

Wolff L, Koller R, Davidson W (1991): Acute myeloid leukemia induction by amphotropic murine retrovirus (4070A): Clonal integrations involve c-*myb* in some but not all leukemias. *J Virol* 65: 3607–3616

Xanthoudakis S, Miao G, Wang F, Pan YCE, Curran T (1992): Redox activation of Fos-Jun DNA binding activity is mediated by a DNA repair enzyme. *EMBO J* 11: 3323–3335

Yanagisawa H, Nagasawa T, Kuramochi S Abe T, Ikawa Y, Todokoro K (1991): Constitutive expression of exogenous c-*myb* gene causes maturation block in monocyte-macrophage differentiation. *Biochim Biophys Acta* 1988: 380–384

Zobel A, Kalkbrenner F, Guehmann S, Nawrath M, Vorbrueggen G, Moelling K (1991): Interaction of the v- and c-Myb proteins with regulatory sequences of the human c-*myb* gene. *Oncogene* 6: 1397–1407

4

Oncogenes as Transcriptional Regulators
Vol. 1: Retroviral Oncogenes
ed. by M. Yaniv and J. Ghysdael

The v-erbA Oncogene

ANNE RASCLE, OLIVIER GANDRILLON, GÉRARD CABELLO AND JACQUES SAMARUT

Introduction

The v-erbA oncogene has been identified as one of the two oncogenes of the Avian Erythroblastosis Virus (AEV), a natural field-isolated avian acute leukemia retrovirus. When injected intravenously into newborn chicks, it induces acute and fatal erythroleukemia. When injected subcutaneously it also induces sarcomas, but the erythroleukemia usually quickly takes over and kills the animal before the sarcoma has extended. This observation could also explain that no other pathology has yet been described in chickens infected by AEV.

In vitro, AEV transforms erythrocytic progenitors from bone marrow, embryonic yolk sac and early embryonic blastoderm. The target cells for transformation are very early erythrocytic progenitors which become blocked in their differentiation as they pass through the differentiation window between the BFU-E and CFU-E stages (Samarut and Gazzolo, 1982). The leukemic cells enter long-term self renewal and sometimes can be established as cell lines (Beug et al, 1982).

The virus also transforms primary chicken embryo fibroblasts in culture. These cells make typical foci and are able to develop colonies in soft agar. The in vitro-transformed fibroblasts develop tumour nodules when grafted onto the chorioallantoic membrane of chicken embryo (Gandrillon et al, 1989).

Like all acute leukemia retroviruses, AEV owes its oncogenic properties to the expression of viral oncogenes that the virus has transduced from cellular proto-oncogenes. AEV carries two viral oncogenes in its genome, respectively v-erbA and v-erbB. The v-erbB oncogene originates from the c-erbB proto-oncogene which encodes the membrane tyrosine kinase receptor for EGF and TGFα. As a result of many rearrangements during the transduction process of the c-erbB gene, the v-erbB oncogene encodes a highly mutated product which represents a membrane protein devoid of most of the extracellular domain and containing point mutations and deletions in its carboxy terminal intracellular domain. As a consequence of these alterations, the v-erbB protein delivers in the infected cells a constitutive and amplified tyrosine kinase activity which generates a permanent mitogenic signal.

The v-erbA oncogene originates from the c-erbAα proto-oncogene and encodes the p75 gag-v-erbA as a unique product (see below).

Expression of both oncogenes is necessary for the virus to induce erythroleukemia in vivo and to generate fully tumorigenic sarcomas (Frykberg et al, 1983; Gandrillon et al, 1987). AEV has been one of the first examples illustrating the cooperation of oncogenes in neoplastic transformation (Graf and Beug, 1983). In this cooperation the v-erbB product essentially induces a constitutive growth factor-independent mitogenic activation of the transformed cells (Gandrillon et al, 1989; Pain et al, 1991).

Biological Effects of the v-erbA Oncogene

Erythrocytic cells

The v-erbA oncogene is responsible for blocking the differentiation program of the early CFU-E (Gandrillon et al, 1989). However the blocked cells are strictly dependent upon exogenous growth factors, mainly TGFα, for their growth in vitro (Pain et al, 1991). A mutant AEV which expresses only the v-erbA oncogene is unable to induce an erythroleukemia in young chicks, which shows that acquisition of growth factor independence and mitogenic stimulation provided by v-erbB is necessary for leukemogenic transformation in vivo (Frykberg et al, 1983; Gandrillon et al, 1989). Recently, an engineered retrovirus vector carrying the v-erbA oncogene only and allowing production of high titre recombinant virus was shown to induce a leukemia. It is not yet understood how increasing the infectivity of a v-erbA-expressing virus might be sufficient to induce the leukemia (Casini and Graf, 1995).

Chicken embryo fibroblasts (CEF)

The first identified autonomous activity of the v-erbA oncoprotein was its ability to alter the growth control of CEF (Gandrillon et al, 1987). When grown in vitro CEF expressing v-erbA as the only oncogene exhibit long-term growth, decreased growth factor requirement and limited anchorage-independent growth as they are able to develop minute colonies in soft agar, but they are not tumorigenic. These CEF also show a strongly enhanced sensitivity to the mitogenic effect of EGF (Khazaie et al, 1991). All these effects are likely to result from induction by v-erbA of the resistance of the CEF to the growth inhibitory effect of retinoic acid and/or T3 which are present either as such or as derivatives in the serum and other biological fluids (Desbois et al, 1991a). Abrogation by v-erbA of the down modulation of AP-1 activity by retinoic acid and T3 (Desbois et al, 1991 b) is likely the basic mechanism at the origin of the growth activation by the oncogene (see below).

Myogenic cells

The influence of v-erbA has been studied in secondary cultures of embryonic quail myoblasts and in the quail myogenic line QM7. In these cells, activation of the proliferation induced by v-erbA is associated with a potent stimulation of myoblast terminal differentiation (Cassar-Malek et al, 1994). These data indicate that v-erbA displays an unexpected myogenic activity. Expression of chimeric c-erbA/v-erbA proteins in these cells demonstrated that the DNA binding domain of the oncoprotein plays a crucial role in this process. In particular, changing the DNA binding domain of the c-erbAα1 receptor for the DNA binding domain of v-erbA confers to this chimeric receptor the ability to enhance myoblast differentiation in absence of T3. Additional experiments indicated that substitution of serine 61 located in the first zinc finger of v-erbA by the corresponding glycine of c-erbAα1 abrogates the myogenic potency of the oncoprotein, thus underlining the importance of this natural point mutation in the biological effect of the oncoprotein (Cassar-Malek et al, 1994).

In myoblasts, the proliferative influences of v-erbA and v-erbB are not additive since expression of v-erbB abrogates the myogenic activity of v-erbA and efficiently inhibits myoblast differentiation. As a consequence of this lack of cooperation, co-expression of the two oncogenes does not transform myoblasts into tumour cells (I. Cassar-Malek et al, unpublished data).

Neuronal cells

The v-erbA oncogene induces the proliferation of chicken neuroretina cells and cooperates with various oncogenes to transform these cells. It renders the cells highly sensitive to the mitogenic effect of FGF (Garrido et al, 1993).

The effect of v-erbA on neuronal differentiation has been examined in the rat pheochromocytoma cell line PC12. PC12 cells can differentiate into either neuronal cells or chromaffin cells upon induction with NGF or glucocorticoids respectively. When expressed in these cells, the v-erbA oncogene inhibits NGF-induced neuronal differentiation and induces an aberrant interlineage phenotype (Munoz et al, 1993).

The v-erbA oncogene also promotes some transformation features (increased survival, invasiveness and anchorage-independent growth) in cultured glial cells via the activation of the PDGF/c-sis gene (Iglesias et al, 1995; Llanos et al, 1996).

Thyroid cells

To investigate the potential oncogenic role of v-erbA in epithelial cells, the oncogene was introduced into cells of a rat thyroid cell line, PC Cl 3 which exhibits several markers of thyroid differentiation. The cells expressing v-erbA show inhibition of iodide uptake, acquires thyrotropin-independent growth, but expression of other differentiation markers is not altered and the cells do not become tumorigenic (Trapasso et al, 1996).

Other tissues

Transgenic mice carrying the v-erbA oncogene downstream from a human β-actin promoter have been generated (Barlow et al, 1994). The effects of v-erbA in these animals are pleiotropic, tissue-specific and dose dependent. Mice had breeding disorders, reduced brown and white adipose tissues, hypothyroidism with normal levels of TSH and TRH, and enlarged seminal vesicles in males. Animals of both sexes were sterile. In the thyroid, the follicular structures producing thyroid hormone precursors were disorganised. These observations suggest that v-erbA perturbs the functionality of the hypothalamic-pituitary-thyroid axis. In addition, most of the males developed hepatocellular carcinomas after 9 months thereby demonstrating that the oncogene can promote neoplasia in mammals.

The c-erbA Product and the Nuclear Receptors

Two c-erbA genes (α and β), respectively located on human chromosomes 17 and 3, encode the α and β forms of the nuclear receptor for thyroid hormone (Sap et al, 1986; Weinberger et al, 1986). The c-erbA products belong to the vast superfamily of nuclear receptors for lipophilic ligands (steroids, thyroid hormones, retinoids and vitamin D3). This superfamily can be broadly divided into four classes based on their dimerization and DNA binding properties. Class I receptors include steroid hormone receptors. Class II encloses the receptors for respectively thyroid hormone (TR), *all-trans* retinoic acid (RAR), 9-*cis*-retinoic acid (RXR), vitamin D3 (VDR), the peroxysome proliferator activated receptor (PPAR) and some orphan receptors. For the purpose of this review, we will further refer to this class as the c-erbA family. Classes III and IV contain receptors which bind to DNA as monomers and which are mostly orphan receptors awaiting identification of ligands or which are true ligand-independent transcription regulators (review in Mangelsdorf et al, 1995; Gronemeyer and Laudet, 1995).

General structure of receptors of the c-erbA family

The nuclear receptors are modular proteins composed of six distinct domains referred to as A to F which exhibit different evolutionary sequence conservation (Figure 4.1). The structure of the receptors for thyroid hormone and for retinoids have been studied in most detail to date (review in Evans 1988; Leid et al, 1992a). Recently tri-dimensional analyses have considerably increased our knowledge about the fine structures of TR, RAR and RXR (Rastinejad et al, 1995; Renaud et al, 1995; Wagner et al, 1995; Bourguet et al, 1995).

The amino terminal domain A/B is poorly conserved among receptors. It contains a constitutive transactivating function AF-1 which is cell and promoter specific. The length of this domain is quite variable among the receptors (24 to 603 amino acids) It is relatively short in the thyroid hormone receptors and does not present significant homology between the α and β forms.

The C domain is highly conserved and is the DNA binding domain which targets the receptors to specific DNA sequences known as hormone response elements (HRE). It is composed of two highly conserved zinc fingers (Figure 4.1). The first (amino terminal or proximal) finger specifies the recognition of the nucleotide sequence of the target HRE through a motif located at its C-terminus edge (P box). Interestingly, the five amino acids which define the P box are totally conserved among receptors of the c-erbA family, thus predicting that they bind to common DNA sequences (Figure 4.2). The second finger (distal) is directly involved in the dimerization of the receptors on their HRE through a short amino acid sequence localised at its N-terminus edge (D box). Both fingers are separated by a helical domain (helix C1) which includes the P box. The last four amino acids of the distal finger together with the downstream eight

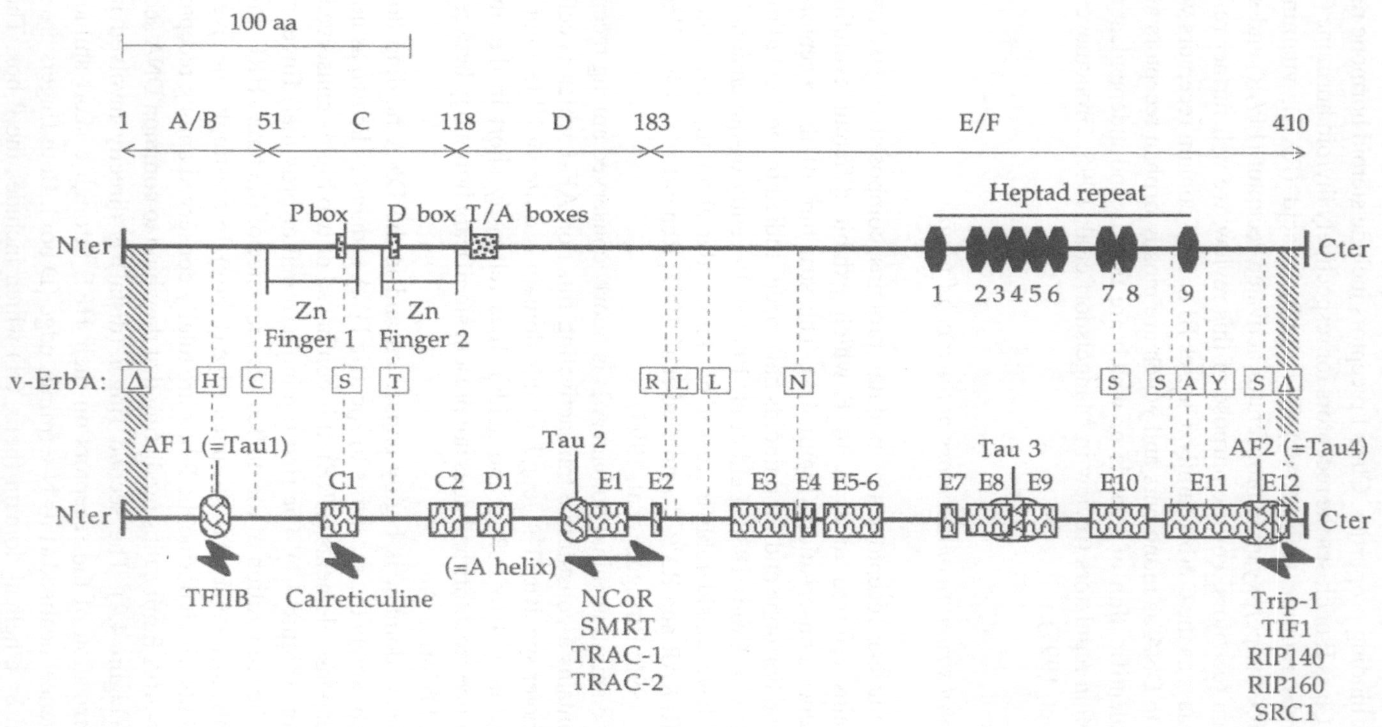
100 aa
1 A/B 51 C 118 D 183 E/F 410
Heptad repeat
P box
D box
T/A boxes
Nter
Cter
1 2 3 4 5 6 7 8 9
Zn Finger 1
Zn Finger 2
v-ErbA:
Δ H C S T R L L N S S A Y S Δ
AF 1 (=Tau1)
Tau 2
Tau 3
AF2 (=Tau4)
C1 C2 D1 E1 E2 E3 E4 E5-6 E7 E8 E9 E10 E11 E12
Nter
Cter
(=A helix)
TFIIB
Calreticuline
NCoR
SMRT
TRAC-1
TRAC-2
Trip-1
TIF1
RIP140
RIP160
SRC1

Receptor	P box sequence
ER	EGCKA
TRα	EGCKG
TRβ	EGCKG
RARα	EGCKG
RARβ	EGCKG
RARγ	EGCKG
RXRα	EGCKG
RXRβ	EGCKG
VDR	EGCKG
Revα	EGCKG
PPAR	EGCKG
v-erbA	EGCKS
COUP	EGCKS

Figure 4.2 Aminoacid sequences of the P boxes of different members of the c-erbA family. For comparison purposes are also shown the P box sequences of the oestrogen receptor and of the orphan COUP.

residues constitute helix C2. Helix C1 and helix C2 pack together at right angles to form the core of the DNA binding domain (Figure 4.3). A further downstream helix defines the T-A box involved in contacting the minor grove of DNA.

The D domain has been considered for a long time only as a flexible hinge without any specific function. Recently, this domain has been shown to exhibit silencing activity and to bind transcriptional repressors (see below). In addition, it contains an amino acid sequence governing the nuclear import of the receptor (Lee and Mahdavi, 1993).

Figure 4.1 Schematic representation of the c-ErbA molecule showing the region that are either structurally or functionally defined. For clarity the c-ErbA molecule has been depicted twice. The upper part displays the localisation of:
(1) the zinc finger regions;
(2) the P, D and T/A boxes ( ), and
(3) the 9 heptad repeats ( ).
The lower part displays the localisation of: (1) the four regions (Tau 1–4;) known to play important functions in the transactivating/transrepressing functions of the c-ErbA molecules; (2) the regions known to form an α-helical structure ( ). They have been labelled according to their regional location (C, D or E regions) and their position within this region. The D1 helix is known in the literature as the A helix (Rastinejad et al, 1995), and (3) the regions known to interact with identified cofactors ( ). The names of the factors are indicated below the line. See text for details on the nature and functions of the mentioned cofactors. In between the two c-ErbA representations are positioned the aminoacids differences (letters in squares) between v-erbA and c-ErbA. Δ represents deletion.

A

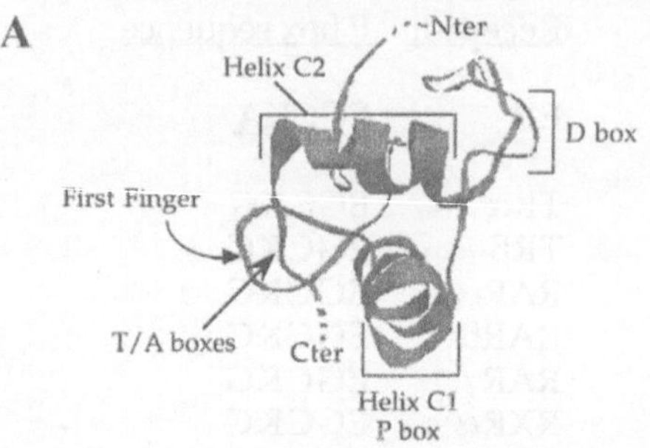
Nter
Helix C2
D box
First Finger
T/A boxes
Cter
Helix C1
P box

B

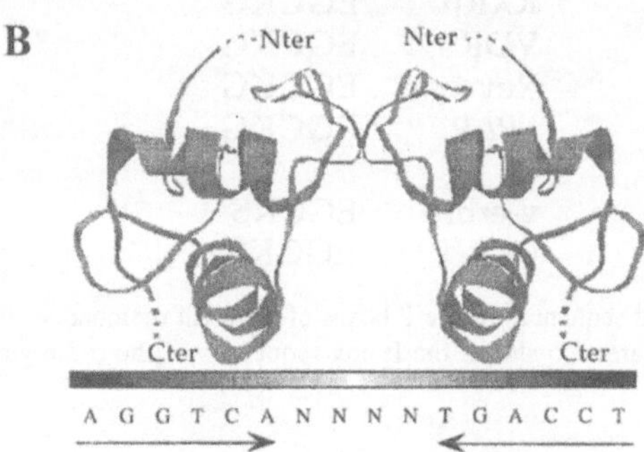
Nter
Nter
Cter
Cter
A G G T C A N N N N T G A C C T

C

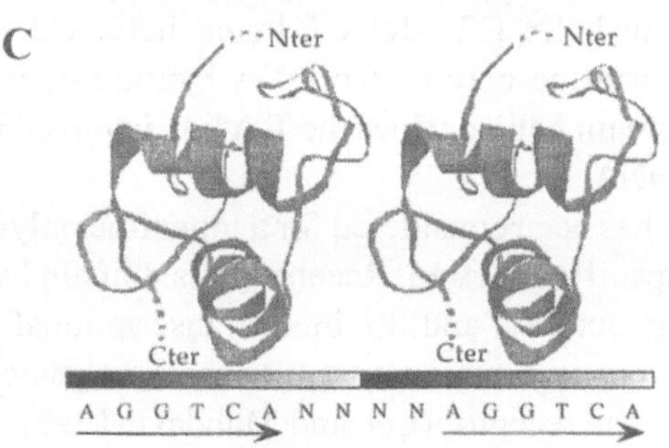
Nter
Nter
Cter
Cter
A G G T C A N N N N A G G T C A

D

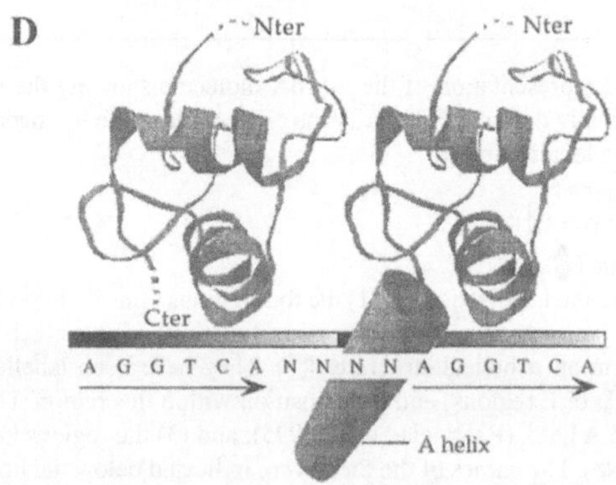
Nter
Nter
Cter
A G G T C A N N N N A G G T C A
A helix

The E domain is the ligand binding domain which also includes homo- and heterodimerization interface and a hormone-dependent transcriptional activation function (AF-2). It mediates dimerization of the receptors in solution and probably stabilizes the dimers on their HRE but it does not play any role in HRE selection.

The E domain is made of several successive helices (11 or 12 according to receptors) which are grouped into three layers making an anti parallel α-helical sandwich. This intramolecular architecture is stabilised by the presence of nine heptad repeats distributed from helix E 6 to helix E 10 (Forman and Samuels, 1990). This structure generates a pocket in which the ligand is buried. The last helix encloses the AF-2 domain, a highly conserved domain containing the specific motif $\Phi\Phi$XE$\Phi\Phi$ (Φ = hydrophobic amino acid). Comparison of the 3D structures of apo- and holo-receptors suggests that upon binding of the ligand, the AF-2 domain flips over and becomes exposed for the binding of regulatory cofactors (see below) (Renaud et al, 1995).

The most carboxy-terminal domain, domain F, has not yet been assigned any specific function. It length is quite variable among receptors. It is virtually absent in TR and RXR.

The receptors for thyroid hormone

From chicken and human cDNA libraries, c-erbA cDNA homologous to the v-erbA oncogene were identified and shown to belong to two independent genes (Sap et al, 1986; Weinberger et al, 1986). These two genes were found to encode two isotypes of the thyroid hormone receptor, respectively α and β and hence referred to as TRα or c-erbAα and TRβ or c-erbAβ. Both are conserved in all vertebrate species and have been cloned now in many species from fish to human. The two isotypes of receptors bind triiodothyronine T3 with a high affinity and T4 at a much lower affinity in vitro. In vivo there is no evidence that T4 can activate the function of TRα

Figure 4.3 Schematic representation of the three-dimensional structure of (A) the DNA-binding region of the human RARβ (Brookhaven ID: 1 HRA; Knegtel et al, 1993); (B) and (C): representation of the binding of two RAR molecules on a palindromic sequence with a four nucleotide-long spacer (B) or to a direct repeat (DR-4; C). (D) Schematic representation of the binding of a heterodimer between an RXR (left molecule) and a TR (right molecule) to a DR-4. We schematised the position of a protruding A-helix that prevents binding of such a heterodimer to a response element with less than four bases-long spacer. These schemes were redrawn after Rastinejad et al, (1995), Glass (1994), and Gronemeyer and Laudet (1995).

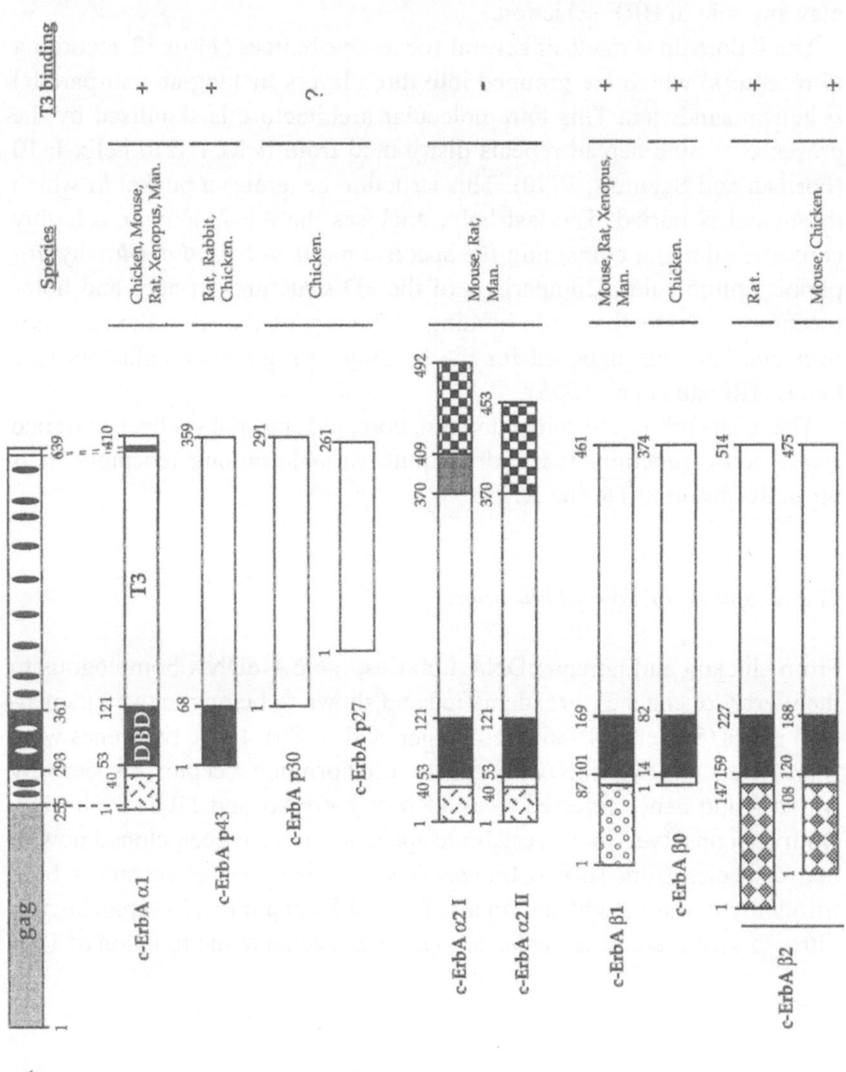
Species
T3 binding
v-ErbA
gag
DBD
T3
c-ErbA α1
Chicken, Mouse, Rat, Xenopus, Man.
c-ErbA p43
Rat, Rabbit, Chicken.
c-ErbA p30
Chicken.
c-ErbA p27
c-ErbA α2 I
c-ErbA α2 II
Mouse, Rat, Man.
c-ErbA β1
Mouse, Rat, Xenopus, Man.
c-ErbA β0
Chicken.
c-ErbA β2
Rat.
Mouse, Chicken.

and TRβ, but experimental proofs are difficult to provide, due to production of T3 by deiodination of T4 within the cells. The v-erbA oncogene is the oncogenic form of TRα.

Both c-erbAα and c-erbAβ genes encode several isoforms of the respective receptor (Figure 4.4). In human, mouse and rat the c-erbAα gene encodes the regular receptor TRα1 and two isoforms, TRα2 (or c-erbAα2-I) and TRα3 (or c-erbAα2-II), resulting from alternative splicing of the c-erbAα mRNA. The TRα2 and TRα3 isoforms differ from TRα1 at the carboxyterminus of the E domain. Nothing is known about TRα3, but as demonstrated for TRα2 it probably does not bind T3 and does not activate transcription. When overexpressed, TRα2 inhibits transcription activated by TRα1 and TRβ (Mitsuhashi et al, 1988; Izumo and Mahdavi, 1988; Koenig et al, 1989). No TRα2 isoform has yet been identified in chicken and amphibians. In chicken erythrocytic cells, truncated forms of TRα have been identified (Bigler and Eisenman, 1988). One of these isoforms, p43, is located in the mitochondrial matrix where it binds to specific sequences of the mitochondrial genome (Wrutniak et al, 1995). At the nuclear level, shorter forms devoid of the domain C do not activate transcription and inhibit transcriptional activity of the TRα1 isoform and RAR (Bigler et al, 1992).

In human, rat, mouse and chicken two isoforms of TRβ, respectively TRβ1 (TRβ0 in chicken) and TRβ2 have been isolated (Weinberger et al, 1986; Hodin et al, 1989; Forrest et al, 1990a, Wood et al, 1991; Sjoberg et al, 1992). These two isoforms result from promoter choice in the TRβ gene and differ in their A/B domains. They both activate transcription upon T3 binding. The homology between the TRα1 and TRβ1 isotypes is quite high. The C-ter of E domains are 100% identical and the homology is 80% in the rest of the E domain and 90% in the C domain.

TRα1 are present in almost all tissues. TRα2 are also widely distributed with highest expression in brain, testis, muscle, adipocytes and fibro-

Figure 4.4 Structural comparison between different c-ErbA proteins. The c-ErbAα1 molecule has been chosen as the comparison point. On top of it is shown the gag-v-ErbA protein. Underneath are shown in the following order: (1) a group of small proteins that are derived by alternative translation initiation codon usage from the c-erbAα sequence (Bigler and Eisenman, 1988; Wrutniak et al, 1995); (2) the c-ErbAα2 proteins (Lazar et al, 1988; Mitsuhashi et al, 1988); (3) the various TRβ isoforms (Weinberger et al, 1986; Hodin et al, 1989; Forrest et al, 1990a; Wood et al, 1991; Sjoberg et al, 1992). The size of the protein, in aminoacids, is indicated above each molecule. On the right are shown the species in which the different forms have been described to date and the ability (+) or inability (–) for a given molecule to bind T3. See text for details. DBD = DNA-binding domain.

blasts (Koenig et al, 1989; Lazar et al, 1988; unpublished observations). In chicken, TRα expression has been detected as soon as 4 days of embryonic life in the head, the embryonic yolk and in blood cells; the steady state level of c-erbAα mRNA increases until birth, particularly in the brain and the heart (Hentzen et al, 1987; Forrest et al, 1990a). TRβ mRNA have been detected in most tissues but their expression is prominent versus TRα particularly in brain and liver. TRβ1 mRNA are found mainly in brain, liver and kidneys whereas TRβ2 mRNA are found almost exclusively in pituitary gland. Whereas the level of expression of TRα is similar in the embryo and adult, expression of TRβ strongly increases during development (Forrest et al, 1990a, Forrest et al, 1991; Wood et al, 1991; Thompson et al, 1987; Hodin et al, 1989).

Studies of the influence of T3 upon expression of its receptor indicated that only transcription of the TRβ gene is positively regulated by the hormone (Suzuki et al, 1994; Machuca et al, 1995).

Response elements to receptors of the c-erbA family

All nuclear receptors of the c-erbA family recognise derivatives of the same hexameric DNA core motif 5′-PuGGTCA-3′. Mutations, orientation, duplication and spacing of repeats create the selectivity of receptor recognition. For receptors of the c-erbA family the most potent HRE are direct repeats of the core AGGTCA half-site. Spacing between repeats dictates the preference for specific receptors. However, in addition to spacing, subtle differences in the sequence of the hexad half site and the 5′extension of the HRE can also strongly modulate the response to the receptor (review in Mangelsdorf and Evans, 1995). General structure of the direct repeats and their affinity to receptors are shown in Figure 4.5. Other arrangements of the hexameric half-sites than DR motifs can constitute functional HRE like palindromes (PAL) and inverted palindrome (IP). In all these structures the two halves of the HRE are spaced by a defined number n of bases (DR-n, PAL-n, IP-n). It should be emphasised that in many natural gene promoters, recognition motifs show variations from these idealised sequences and that sometimes several DR, PAL or IP are part of a complex element which mediates hormone response. For an exhaustive review of HRE see Gronemeyer and Laudet (1995).

TR bind preferentially and rather specifically to TRE of the DR-4 structure which are found in many genes regulated by T3. Natural IP TRE are rather rare. Apart from the complex element of the rat GH gene, the synthetic palindromic element without spacer (PAL-0) called TREpal has not yet been found under this perfect structure in any natural gene and is not

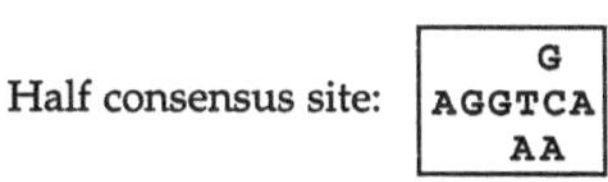

TRE	Sequence	TR Complexes
DR-4	AGGTCAnnnnAGGTCA TCCAGTnnnnTCCAGT	Heterodimers
IP-6	TGACCTnnnnnnAGGTCA ACTGGAnnnnnnTCCAGT	Homodimers
PAL-0	AGGTCATGACCT TCCAGTACTGGA	Heterodimers Homodimers
TRE-spe	(T/C)(A/G)AGGTCA (A/G)(T/C)TCCAGT	Monomers

Figure 4.5 Structure of some T3-response elements (TRE). Shown is the structure of both natural specific (direct repeat: DR-4, inverted palindromic: IP-6 and modified half-site: TRE-spe) and synthetic non-specific (palindromic: PAL-0) response elements. On the right is shown the nature of the bound complexes.

specific to TR since it works efficiently as a retinoic acid responsive element (RARE).

TRα is unique among receptors of the c-erbA family in that it can also bind as a monomer on the half-site motif (T/C)(A/G)AGGTCA (Forman et al, 1992).

Dimerization and binding to DNA of the receptors

In contrast to receptors of class I which work as homodimers, receptors of the c-erbA family work preferentially as heterodimers with RXRs, but in some cases, like TR, as homodimers (Mangelsdorf and Evans, 1995). Depending on the structure of the TRE, TR will bind with greatest affinity as homodimers on an IP-6 and as heterodimers on a DR-4, although binding with low affinity is also observed with the alternative forms on both types of TRE. On the PAL-0 TRE, TR bind either as homodimers or heterodimers with RXR. These observations made in vitro are physiologically relevant since in avian myoblasts, which do not contain any detectable RXR, transfected TR activate transcription only from the PAL-0 and not from the DR-4 TRE. Transfection of an exogenous RXR into these cells restores the activation of DR-4 by TR (Cassar-Malek et al, 1996). RAR bind as heterodimers with RXR on DR-5 and DR-1 motifs.

On all direct repeats except DR-1 the receptors assemble in a head-to-tail arrangement with RXR occupying the upstream hexameric DNA repeat (Figure 4.3). In contrast, the RAR occupies the upstream position when RAR/RXR heterodimers bind on DR-1. It has been proposed that this particular conformation of the heterodimer could inhibit the retinoic acid-induced dissociation of co-repressors from RAR required to alleviate the silencing activity of this receptor (see below; Kurokawa et al, 1995). In the bound receptor complex, the two monomers bind on one side of the DNA double helix, with their respective helix C1 occupying adjacent major grooves. On PAL and IP elements the two receptors are bound symmetrically respectively in head-to-head and tail-to-tail arrangements (Figure 4.3).

Depending on these arrangements, various domains are involved in the dimerization interface. On DR elements, the D box of RXR contacts the tip of the first finger and the T-A box of TR leading to cooperative binding to DNA (Figure 4.3). Dimerization of the ligand binding domains implies that the ligand binding domain of TR rotates 180° relative to its DBD. The long A helix of TR plays a key role in spacer discrimination as it produces sterically unfavourable interfaces on half-site spacing with less than four base pairs (Rastinejad et al, 1995; Figure 4.3 (D)). On an IP element both monomers of TR harbour a 180° rotation of the LBD relative to their DBD to allow dimerization.

These observations demonstrated that the receptors contain two independent dimerization motifs, one within the carboxy-terminal LBD and a second within the conserved DNA binding core. This led to hypothesise a two-step dimerization process. First the receptors would form dimers in solution through their LBD. In a second step the dimeric structure would be stabilised on DNA through interactions between their DBD.

TR can also bind with RAR, VDR and PPAR in solution, but there is no experimental evidence that such heterodimers endogenous to the cells have any physiological function in vivo.

As mentioned above, TRα is unique among receptors of the c-erbA family in that it can also bind as a monomer on some response elements.

Binding of the ligand can affect the pattern of dimerization. Indeed, binding of T3 decreases the affinity of TR homodimers for the IP and DR response elements but does not affect the binding of RXR/TR heterodimers to these elements. Such modulating effects of T3 are not observed for the binding to PAL elements (Ribeiro et al, 1992; Miyamoto et al, 1993). In contrast 9-*cis*-retinoic acid stabilizes RXR homodimers and decreases the affinity of RXR/TR for DR elements but not for IP elements

(Yen et al, 1994b; lehmann et al, 1993). Most of these data come from in vitro experiments and their biological relevance in cells in vivo remains to be demonstrated.

Binding of cofactors to the receptors

Many proteins called TRAP (T3 Receptor Auxiliary Proteins) which bind to receptors of the c-erbA family have been detected in co-Immunoprecipitation assay (Beebe et al, 1991; Lazar et al, 1991). Several of these proteins have been cloned to date but there is some evidence now that RXR were members of the first identified TRAP (Darling et al, 1991; Leid et al, 1992b).

A series of related proteins which interact with TRα, TRβ, RARα and RARβ were cloned by the two hybrid screen. The proteins termed SMRT (Chen and Evans, 1995), N-CoR (Hörlein et al, 1995), TRAC-2 (Sande and Privalsky, 1996) belong to the same family of related proteins. Indeed, SMRT and TRAC-2 are encoded by the same gene which is different from the gene encoding N-CoR. All these proteins interact with a conserved region, Tau 2 (Figure 4.1) forming an amphipatic helix within the D domain of the TR and RAR, previously shown to harbour transcriptional silencing activity (Baniahmad et al, 1995; Casanova et al, 1994; Wagner et al, 1995). Several observations show that these proteins work as co-repressors (Chen and Evans, 1995; Hörlein et al, 1995; Sande and Privalsky, 1996). First, they interact only with unliganded receptors, a context in which these receptors function as repressors. Addition of appropriate ligand alleviates receptor-mediated repression and in parallel releases these proteins from the receptors. Second, their amino termini contain an autonomous repressor domain. Finally, point mutations in the receptor D region which binds these proteins abolish the silencing activity of unliganded receptors and abrogate binding of these proteins.

Several potential positive cofactors of the receptors have also been identified by the two-hybrid assay. Trip1 (Lee et al, 1995; vom Baur et al, 1995) is the vertebrate homologue of the yeast Sug1 factor, a mediator between transactivating factors and the TATA-binding proteins. Trip1 interacts with the AF-2 domain of TR, RAR, VDR and oestrogen receptor in a ligand-dependent fashion, but binds weakly to RXR. TIF1 (Le Douarin et al, 1995) contains a RING finger motif and also interacts with the AF-2 domain of RAR, RXR and oestrogen receptor in a ligand-dependent way. It is suggested that TIF1 is an intermediary factor mediating AF-2 activity. RIP 140 and RIP 160 (also known as ERAP 160) were first identified through their oestrogen-dependent binding to the AF-2 domain of oestrogen receptor. These products also bind to TR, RAR and RXR. They are supposed to function as adapters between the receptors and the transcriptional machinery (Cavaillès et al, 1995; Halachmi et al, 1994). SRC1 (Onate et al, 1995) also binds to the AF-2 domain of all above-mentioned receptors but it can interact also with non receptor-transcription factors like SP1. The CREB-binding protein, CBP, directly binds to the ligand binding domain of nuclear hormone receptors through its amino-terminus. It also binds through its carboxy-terminus to SRC1 (Kamei et al, 1996).

The ligand-dependent interaction of these cofactors with the receptors is allowed by the exposition of the AF-2 domain consecutively to the conformation change induced by binding of the ligand (Renaud et al, 1995).

Most of these factors exhibit a ubiquitous distribution. It is still unclear how so many factors contribute to the function of nuclear hormone receptors in the same tissues.

Interactions of TR with TFIIB, a component of the pre-initiation transcription complex have also been observed and are probably involved in the transcriptional activity of these receptors. Unliganded TRα and TRβ interact with TFIIB through their amino terminus and the carboxy terminus of the E region (Baniahmad et al, 1993; Hadzic et al, 1995; Fondell et al, 1996). As a consequence of sequestration of TFIIB, formation of the preinitiation complex on the promoter linked to the TRE is inhibited leading to the repressive effect of unliganded TRβ (Baniahmad et al, 1993). Data suggesting a similar mechanism for unliganded TRα have also been reported (Fondell et al, 1993). Addition of T3 induces a dissociation of the carboxy terminus region from TFIIB and allows formation of the pre initiation complex thus derepressing transcription (Baniahmad et al, 1993). In addition, abrogation of this carboxy-terminal interaction by T3 could allow the exposition of the AF-2 sequence for the binding of regula-

Figure 4.6 Sequence alignment between the gag-v-ErbA protein (a composite sequence made from v-ErbA (Genbank accession M32090; Damm et al, 1987) and p19 protein (NCBI accession 420624; direct submission) sequences), the chicken c-ErbA protein (THA–CHICK; Swiss-Prot accession P04625; Sap et al, 1986) from which is was derived and, for comparison purposes, the human c-ErbA protein (THA1–HUMAN; Swiss-Prot accession P21205; Pfahl and Benbrook, 1987). This last sequence is two aminoacids longer than the chicken sequence leading to potential numbering confusion throughout the literature. Here, we numbered the v-ErbA protein by a negative numbering of the 252 aminoacids belonging to the gag domain (underlined) and then by a positive numbeling of the 387 erbA-specific aminoacids (top numeration). The human c-ErbA protein is numbered below its sequence. In order to know the position of an aminoacid in the chicken c-ErbA sequence, one has to either add 12 from the v-erbA numbering or to withdraw 2 from the human c-erbA numbering.

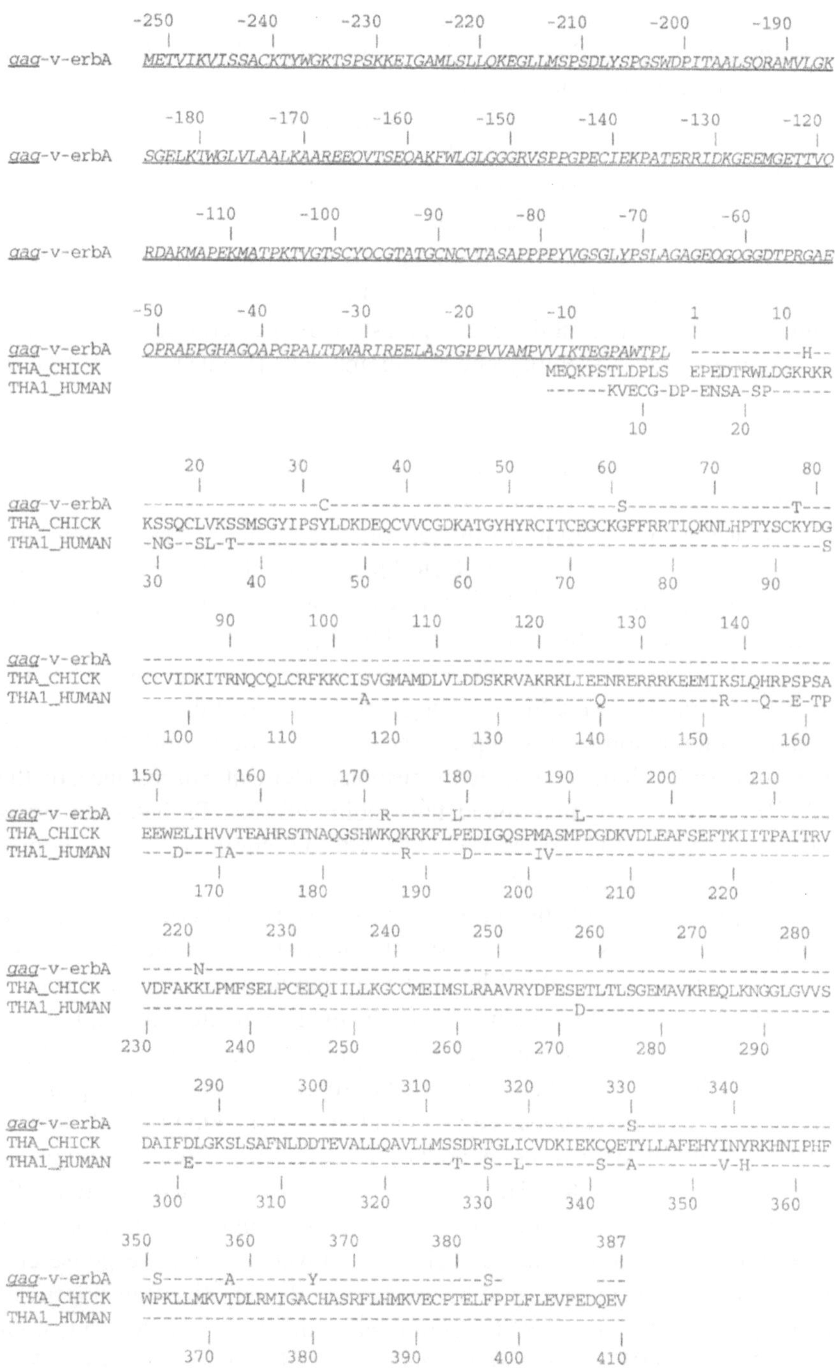
-250 -240 -230 -220 -210 -200 -190
gag-v-erbA METVIKVISSACKTYWGKTSPSKKEIGAMLSLLQKEGLLMSPSDLYSPGSWDPITAALSQRAMVLGK
-180 -170 -160 -150 -140 -130 -120
gag-v-erbA SGELKTWGLVLAALKAAREEQVTSEQAKFWLGLGGGRVSPPGPECIEKPATERRIDKGEEMGETTVQ
-110 -100 -90 -80 -70 -60
gag-v-erbA RDAKMAPEKMATPKTVGTSCYQCGTATGCNCVTASAPPPPYVGSGLYPSLAGAGEQGQGGDTPRGAE
-50 -40 -30 -20 -10 1 10
gag-v-erbA QPRAEPGHAGQAPGPALTDWARIREELASTGPPVVAMPVVIKTEGPAWTPL -----------H--
THA_CHICK MEQKPSTLDPLS EPEDTRWLDGKRKR
THA1_HUMAN ------KVECG-DP-ENSA-SP------
10 20
20 30 40 50 60 70 80
gag-v-erbA -----------------C----------------------------S----------------T---
THA_CHICK KSSQCLVKSSMSGYIPSYLDKDEQCVVCGDKATGYHYRCITCEGCKGFFRRTIQKNLHPTYSCKYDG
THA1_HUMAN -NG--SL-T---S
30 40 50 60 70 80 90
90 100 110 120 130 140
gag-v-erbA ---
THA_CHICK CCVIDKITRNQCQLCRFKKCISVGMAMDLVLDDSKRVAKRKLIEENRERRRKEEMIKSLQHRPSPSA
THA1_HUMAN ---------------------A----------------------Q-----------R---Q--E-TP
100 110 120 130 140 150 160
150 160 170 180 190 200 210
gag-v-erbA ------------------------R------L----------L------------------------
THA_CHICK EEWELIHVVTEAHRSTNAQGSHWKQKRKFLPEDIGQSPMASMPDGDKVDLEAFSEFTKIITPAITRV
THA1_HUMAN ---D---IA----------------R-----D------IV---------------------------
170 180 190 200 210 220
220 230 240 250 260 270 280
gag-v-erbA -----N---
THA_CHICK VDFAKKLPMFSELPCEDQIILLKGCCMEIMSLRAAVRYDPESETLTLSGEMAVKREQLKNGGLGVVS
THA1_HUMAN --D------------------------
230 240 250 260 270 280 290
290 300 310 320 330 340
gag-v-erbA ---S-------------------
THA_CHICK DAIFDLGKSLSAFNLDDTEVALLQAVLLMSSDRTGLICVDKIEKCQETYLLAFEHYINYRKHNIPHF
THA1_HUMAN ----E-------------------------T--S--L-------S--A--------V-H--------
300 310 320 330 340 350 360
350 360 370 380 387
gag-v-erbA -S------A-------Y---------------S- ---
THA_CHICK WPKLLMKVTDLRMIGACHASRFLHMKVECPTELFPPLFLEVFEDQEV
THA1_HUMAN ---
370 380 390 400 410

tory cofactors. We may imagine that in the absence of T3 the binding of co-repressor to the D domain could potentiate or stabilise T3-independent interaction of the receptor with TFIIB.

Calreticulin was also shown to bind to, and to inhibit the activity of nuclear hormone receptors of both class I and class II (Dedhar et al, 1994; Burns et al, 1994). The binding was localised to a six amino-acids sequence (KGFFRR; numbered 74–79 in THA1-HUMAN; Figure 4.6) present in the DNA binding domain of all nuclear hormone receptors. The physiological relevance of that observation was demonstrated by the suppression by calreticulin overexpression of the retinoic acid-induced differentiation of P19 embryonic carcinoma cells (Dedhar et al, 1994).

Transcriptional regulation by the receptors of the c-erbA family

The transcriptional activity of the receptors is modulated by binding of a specific ligand. Indeed, on the RXR/TR complex bound to DR-4, *9-cis*-RA and T3 can individually bind, but when added together, T3 inhibits the binding of *9-cis*-RA. In contrast, on a DR-5 both *9-cis*-RA and *all-trans*-RA can bind together and cooperatively activate transcription (Forman et al, 1995). Binding of the ligand changes the conformation of the receptors and their interaction with co-repressors and co-activators and also changes the nature of the dimers bound to the response element. For instance, on the IP TRE, as previously mentioned, the binding of RXR/TR heterodimers is affected neither by T3 nor by *9-cis*-RA, but the binding of TR/TR homodimers is decreased by T3. On this element, T3 could then induce transcription simply by removing the repressive effect exerted by bound unliganded TR/TR homodimers (Yen et al, 1994b. In contrast, on the DR-4 TRE, *9-cis*-RA dissociates RXR/TR heterodimers and favours homodimerization of RXR in solution (Lehmann et al, 1993) then silencing transcription.

Whereas ligand-activated RAR mainly mediates transcriptional activation, T3-activated TR can either activate or repress transcription depending on the nature of the response element and its context. On the PEPCK promoter TR and RAR can independently activate transcription through respective TRE and RARE. However, TR can repress RAR-mediated activation through binding to the RARE either as monomer, homodimer or heterodimer with RAR (Lucas et al, 1991). Overlapping of response elements have also been shown to be involved in the negative transcriptional activity of TR. For instance, the promoter of the EGF receptor gene contains Sp1 and TRE overlapping sequences, leading to a decrease of Sp1 binding by occupancy of the TRE site by TR, and a decrease in the transcription rate of this gene (Xu et al, 1993).

In addition, some genes are directly negatively regulated by T3, like the genes encoding the β TSH subunit and TRH respectively. The negative regulation by T3-bound TR has also been described in the case of keratin genes (Tomie-Canie et al, 1996). In these genes the TRE contain one or several spaced motifs related to the consensus hemisite AGGTCA which are located either close to the promoters, upstream or downstream of the TATA box or within the first exon (Chatterjee et al, 1989). The transcription from the promoter of the TRH gene is activated by TR in the absence of T3 and repressed in the presence of the hormone and this effect is observed only with TRβ in neuronal cells (Lezoualc'h et al, 1992). In addition, an artificial promoter containing an IP-6 TRE isolated from the LTR of the Rous Sarcoma Virus (RSV) is activated by TRα in the absence of T3 but not by TRβ (Saatcioglu et al, 1993b). The gene encoding chicken lysozyme is also repressed in macrophages by T3 through an IP-6 TRE present in its promoter (Baniahmad et al, 1987). In all these cases, the fashion in which the TRE modulates the response to T3 is highly dependent on the cellular context which leads to assume that specific cellular cofactors are required to dictate the way of the hormone response.

The function of the receptors can also be modulated by the presence of other receptors. For instance transcriptional activation by TR is down-modulated by the TRα2 isoforms which are assumed to bind to TRE competitively with TRα1 or TRβ (Katz and Lazar, 1993). Similarly, natural truncated isoforms to TRα1 devoid of domain C identified in chicken erythrocytic cells can transdominantly inhibit transcription mediated by TRα1 (Bigler and Eisenman, 1988; Bigler and Eisenman, 1988). In early xenopus embryo, the presence of TRα in the absence of endogenous T3 inhibits the transcriptional activation by RAR (Banker and Eisenman, 1993). In mouse embryonic stem cells where the TRα gene has been inactivated by homologous recombination, the activation of the RARβ gene expression by retinoic acid is strongly enhanced (Lee et al, 1994).

In turn, the function of TR can be modulated by RAR. Heterodimers between TR and RAR have been observed in vitro. Negative or positive cooperative effects of TR and RAR have been described on some artificial TRE (Glass et al, 1989). That TR and RAR can functionally interfere is demonstrated by some genes that are regulated by both T3 and RA through common response elements like the genes encoding laminin B1 and Oxytocine (Adan et al, 1993; Williams et al, 1992). Transcriptional activation of the gene encoding ADH3 by RA is inhibited by T3 (Harding and Duester, 1992). Unliganded TR can repress transcriptional activation by ligand-activated RAR of the promoter of the RARβ gene (Barettino et al, 1993). In early xenopus embryo, unliganded TR abrogates the teratogenic effects of RA by repressing RA-induced genes (Banker and Eisenman, 1993).

Other members of the nuclear receptors superfamily could influence TR transcriptional activity. The orphan receptor COUP antagonises transcription mediated by TR on the PAL-0 TRE and on the DR-4 TRE of αMHC (Cooney et al, 1993; Tran et al, 1992). PPAR expression induces formation of TR/PPAR heterodimers, with preferential binding to a DR-2 element (Bogazzi et al, 1994); consequently, expression of this receptor could change the target genes of T3, from genes containing a DR-4 in their promoter to genes containing a DR-2 element.

It is likely that phosphorylation modulates the activity of nuclear hormone receptors. Indeed inhibitors of phosphatases increase transcriptional activity of TRβ (Lin et al, 1992). TRα1 is phosphorylated on two serine residues localised within the A/B region (Ser 12 and Ser 28/29 in chicken TRα) (Goldberg et al, 1988; Glineur et al, 1989). TRα and TRβ are differently phosphorylated on tyrosine and threonine residues (Lin et al, 1992). The phosphorylation could modulate the activity of the receptors by affecting their affinity for T3, their affinity for the TRE or/and their affinity for the dimerization partner. The role of the different phosphorylation sites in the function of the TR isoforms is still unclear.

Functional interactions with AP-1

TR, RAR and RXR can modulate gene expression through an indirect way by inactivating the AP-1 transcription factor in a ligand-dependent fashion. *Vice versa*, high level of expression of the AP-1 factor will abrogate the transcriptional activation by TR and RAR (Desbois et al, 1991 b; Schüle et al, 1991; Zhang et al, 1991c; Nicholson et al, 1990, Salbert et al, 1993). These phenomena are not specific of receptors of the c-erbA family, but were also observed for the glucocorticoid receptor (review in Pfahl, 1993).

The mechanisms of interference between receptors of the c-erbA family and AP-1 are still obscure. The functional interference implies the carboxy terminus and DNA binding domain of TR or RAR and the DNA binding domain of Jun (Yang-Yen et al, 1991; Schüle et al, 1991; Zhang et al, 1991; Saatcioglu et al, 1993 a). For RXR the ninth heptad in the E domain is involved (Salbert et al, 1993). Direct interactions between the glucocorticoid receptor and AP-1 were described, but these interactions do not alter the pattern of binding of AP-1 on its DNA binding site either in vitro or in vivo (Konig et al, 1992). For the c-erbA family direct interactions have been described only between in vitro translated RARβ and purified bacterially produced c-Jun (Yang-Yen et al, 1991). No such direct interactions have yet been detected for the other members of the c-erbA family. Another explanation is that nuclear receptors titre a co-factor of AP-1 which could be CBP. Indeed, CBP binds to nuclear receptors and activates their ligand-dependent transcriptional activity. As CBP is also required for activation of AP-1 and is limiting in the cells, overexpression of the nuclear receptors would indirectly down modulate AP-1 activity and *vice versa* (Kamei et al, 1996).

There is some evidence that AP-1 interferes only with TR/RXR heterodimers and not with TR/TR homodimers. Indeed, in avian myoblasts which do not contain any detectable RXR, no functional negative interference was observed between the AP-1 and T3 dependent transcriptional pathways. However, the expression of an exogenous RXR in these cells restored the AP-1/TR antagonism (Cassar-Malek et al, 1996). In line with these data, coexpression of RXR and TR in CV-1 cells leads to the formation of TR/RXR heterodimers and strongly increases the sensitivity of the receptors to T3 to inhibition by AP-1 (Claret et al, 1996). Functional interference between AP-1 and RAR does not require RXR in myoblasts (Cassar-Malek et al, 1996) which suggests that the mechanisms of interference with AP-1 might be different for the two types of receptors, at least in this cell type. In contrast, in a cell line of ovarian adenocarcinoma resistant to RA, restoration of the inhibition of AP-1 by RA requires expression of exogenous RAR and RXR (Soprano et al, 1996). Since the functional interference between AP-1 and the receptors depends on the nature of the cells, specific cellular factors should be also involved in the interactions.

In agreement with the observation that RXR overexpression strongly potentiates the myogenic effect of T3 (Cassar-Malek et al, 1996), such a functional cross-talk is certainly relevant for the development of tissues. As AP-1 is activated following mitogen stimulation (Angel and Karin, 1991), we may anticipate that hormone regulation of gene expression will be altered in growing cells and alternatively that activation of the hormone response in the cells will slow down their proliferation. Thus TR and RAR are major factors in controlling the balance between the differentiation and proliferation pathways (Figure 4.7).

The v-ErbA Protein

Structure of the protein

As previously mentioned, the v-ErbA oncoprotein is the viral homologue of the c-erbAα proto-oncogene product, TRα. It originates from the transduction of the c-erbAα coding sequence into the viral gag gene during the

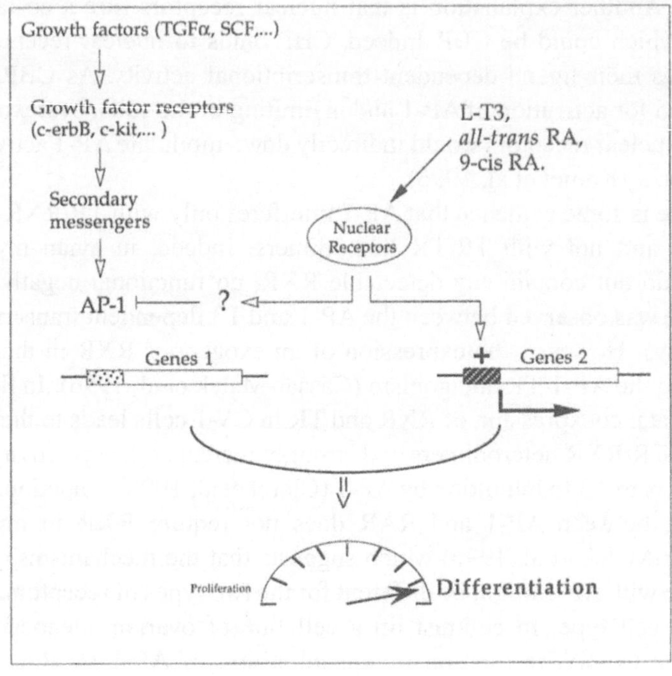
Growth factors (TGFα, SCF,...)
Growth factor receptors
(c-erbB, c-kit,...)
L-T3,
all-trans RA,
9-cis RA.
Secondary
messengers
Nuclear
Receptors
AP-1
?
Genes 1
+
Genes 2
Proliferation
Differentiation

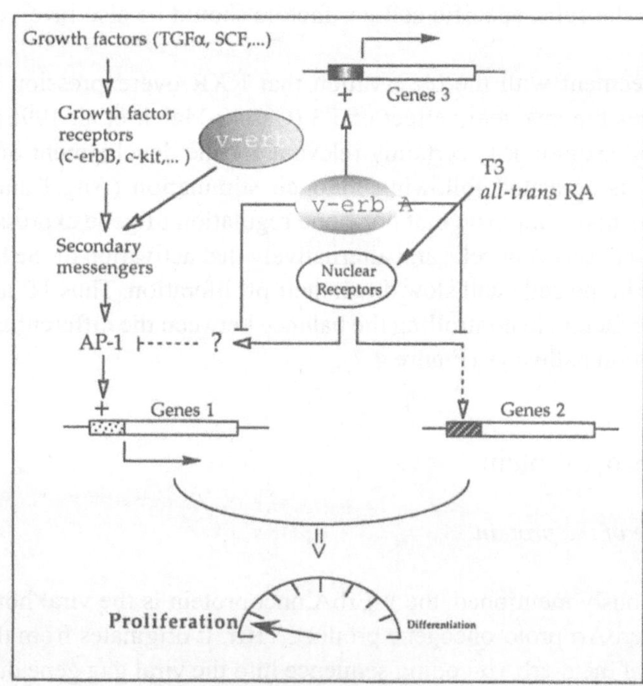
Growth factors (TGFα, SCF,...)
+
Genes 3
Growth factor
receptors
(c-erbB, c-kit,...)
v-erb
T3
all-trans RA
v-erb A
Secondary
messengers
Nuclear
Receptors
AP-1
?
+
Genes 1
Genes 2
Proliferation
Differentiation

natural genesis of AEV. As usual in this kind of process the transduced sequence underwent several rearrangements.

The v-ErbA product is synthesized by AEV as a 75 kilodalton fusion product of gag and v-ErbA sequences (Figure 4.6). The fusion gag-v-ErbA protein will hence be referred to as the v-ErbA protein. As a result of this fusion, the first 12 amino acids of TRα have been replaced by the 255 first residues of gag. The oncoprotein was also deleted of nine amino acids corresponding to the binding site of cofactors within the AF-2 domain. In addition, 13 point mutations appeared, distributed all along the sequence of v-ErbA (Figures 4.1 and 4.6). Two of these mutations are localised in the DNA binding domain. One of these mutations at position 61 changes the P box sequence EGCKG specific of receptors of the c-erbA family into EGCKS, a sequence identical to that of orphan receptors of the COUP family (Figure 4.2). The other mutation is localised in the D box. Two mutations are localised within the residual A/B domain. Nine mutations in the E domain, together with the deletion in AF-2, induce a strong reduction of the affinity of the oncoprotein for T3 (Yen et al, 1994a, Munoz et al, 1988).

The tri-dimensional structure of the v-ErbA oncoprotein is not yet available, but it should reveal interesting changes as compared to TRα. Indeed two proline residues close to each other located between helices E2 and E3 in TRα are mutated into leucines in v-ErbA. These alterations might have important consequences on the folding of the protein. In contrast to TRα which is exclusively nuclear, the v-ErbA protein is localised both in the cytoplasm and the nucleus and pulse chase experiments demonstrated little, if any, exchange between these two compartments (Bigler and Eisenman, 1988; Boucher et al, 1988; Boucher and Privalsky, 1990). The DNA-binding domain, the D region and the beginning of the E region are necessary for the nuclear localisation (Boucher and Privalsky, 1990). Surprisingly, the cytoplasmic form of v-ErbA is associated with hsp90 (Privalsky, 1991) in contrast to all receptors of the c-erbA family. The

Figure 4.7 Control of gene expression by c-erbA, RAR, and AP-1 (upper panel) and its deregulation under the influence of v-erbA and v-erbB oncogenes (lower panel). The —▷ arrows denote an activation and the —| signs repression. The upper panel shows the normal pattern of gene regulation in cells treated with T3 or RA. Genes 1 represent AP-1 dependent genes under the control of TPA-responsive element (TPA-RE: ▒). Genes 2 are hormone-regulated genes under the control of TRE or RARE (▨). The lower panel depicts the pattern of gene expression in v-erbA- and v-erbB-expressing cells. Genes 3 represent hypothetical genes which might be directly activated by v-ErbA through a v-ErbA-responsive element (VRE: ■).

functional relevance of such an association is unclear. The cytoplasmic localisation of the oncoprotein might result from its large size and its fusion to gag since a mutation of residue 230 of gag enhances the migration of the protein into the nucleus (Boucher and Privalsky, 1990). The binding v-ErbA to hsp90 might be a consequence of its accumulation in the cytoplasm.

Dimerization and binding to DNA

The nine heptad motifs of TRα are well conserved in v-ErbA and hence the oncoprotein can dimerize in solution with RXR (Chen and Privalsky, 1993). However, the point mutation in the E domain at position 351 in v-ErbA (corresponding to residue 363 in chicken TRα) decreases the affinity of the oncoprotein for RXR (Barettino et al, 1993). The oncoprotein does not form stable heterodimers with TR in solution (Chen and Privalsky, 1993).

Many conflicting results have been presented on the pattern of binding of v-ErbA to HRE. These discrepancies result mainly from the use of different techniques (in vitro binding vs. transfected cells) and different origins of the proteins (in vitro translated products, baculovirus products or nuclear extracts from different types of cells). From these data it is evident that v-ErbA can bind as heterodimers with RXR on most natural TRE, but with a lower affinity than TR/RXR heterodimers (Hermann et al, 1993, Yen et al, 1994a; Sap et al, 1989; Damm et al, 1989; Bonde and Privalsky, 1990; Barettino et al, 1993). It should be mentioned that, in contrast to TRα/RXR heterodimers, v-ErbA/RXR heterodimers do not significantly bind to an imperfect TRE present in the promoter of the chicken carbonic anhydrase II gene (Rascle et al, 1994) or to the PAL-0 in the context of avian myoblasts (Cassar-Malek et al, 1997). The oncoprotein has been shown by some authors to bind as homodimers and as heterodimers with TRα on some TRE (Yen et al, 1994a, Hermann et al, 1993; Chen and Privalsky, 1993).

v-ErbA/RXR heterodimers binding to response elements for VDR (DR-3) and RAR (DR-5, but not DR-2), but not for RXR (DR-1) has also been reported (Chen and Privalsky, 1993; Hermann et al, 1993). The binding to the DR-5 element depends highly on the fine structure of this element. Indeed, a DR-5 isolated from the mouse RARβ gene and composed of the motifs A/GGTTCA binds weakly v-ErbA either as monomer or heterodimer with RXR, whereas the oncoprotein binds efficiently to the DR-5 motif AGGTCA. Correlatively, v-ErbA inhibits RAR-mediated transcription only from the AGGTCA DR-5 motif (Chen and Privalsky, 1993).

The specificity of binding to DNA of v-ErbA is similar but distinct from that of TRα. By using random DNA pool selection, competition DNA binding assays and functional assays, Subauste and Koening (1995) identified an optimal binding site for v-ErbA as the decamer 5′-T(A/G)AGGT-CACG-3′ which is closely related to a half TRE. Furthermore, the authors showed that all sequences that are T3-responsive are not necessarily responsive to v-ErbA. Moreover, whereas the motif AGGTCA binds both TRα and v-ErbA either as monomers or homodimers, its derivatives degenerated at positions 1, 4, 5 and 6, are bound only by TRα (Chen et al, 1993; Wong and Privalsky, 1995). However heterodimers TRα/RXR and v-ErbA/RXR show a similar efficiency of binding to all these motifs (Judelson and Privalsky, 1996). These observations suggest that depending on whether it works as homodimers or heterodimers with RXR, the v-ErbA oncoprotein could alter the regulation of different sets of TR-target genes.

Binding to other proteins

Despite its dimerization with RXR, the v-ErbA protein was shown not to bind to the TRAPs, most of which are likely cofactors binding to the AF-2 domain (O'Donnell et al, 1991; Darling et al, 1991). In contrast, the oncoprotein binds N-CoR and SMRT constitutively (Hörlein et al, 1995; Chen and Evans, 1995). As an engineered mutation at the residue 144 of v-ErbA abrogates its binding to SMRT and its oncogenicity, this suggests that constitutive binding of co-repressors is a major determinant of the oncogenic properties of v-ErbA (Chen and Evans, 1995; Damm and Evans, 1993).

It is not known whether v-ErbA interacts with TFIIB like unliganded TRα (Fondell et al, 1993, Fondell et al, 1996; Baniahmad et al, 1993). This information would be interesting as it will provide additional information on the silencing effect of the oncoprotein on transcription.

It has not been directly tested whether the v-ErbA protein binds to calreticulin or not. The v-ErbA protein differs from all of the members of the c-ErbA family in the calreticulin-binding region (KSFFRR instead of KGFFRR). Nevertheless, the second residue of the binding sequence is the most variable in the calreticulin-binding consensus sequence (KxFFK/RR; Burns et al, 1994). Direct evidence is therefore needed to solve this question.

Inhibition by v-ErbA of transcription mediated by TR and RAR

The v-ErbA oncoprotein was initially demonstrated to be a constitutive repressor of genes activated by T3 (Sap et al, 1989; Damm et al, 1989; Pain et al, 1990; Zenke et al, 1990; Baniahmad et al, 1992). It was later shown to also inhibit transcription mediated by ligand-activated RARα, RARβ and RARγ (Sharif and Privalsky, 1991; Sande et al, 1993; Chen and Privalsky, 1993). The oncoprotein inhibits transcription by TR on all tested natural and artificial TRE (Hermann et al, 1993). In contrast, the inhibition of RAR seems to depend upon the structure of the RARE and the nature of the host cells. V-erbA inhibits RA-induced transcription from the PAL-0 but not from DR-2 (Hermann et al, 1993). It does not inhibit DR-5 RARE in PC19, COS-7 and MCF-7 cells (Barettino et al, 1993; Sharif and Privalsky, 1991) in contrast to HeLa cells (Rascle et al, 1996). In CV-1 cells the DR-5 RARE composed of the AGGTCA motif is efficiently repressed by v-ErbA in contrast to the imperfect DR-5 made of the AGTTCA and GGTTCA motifs which is in agreement with the binding efficiency of the oncoprotein to these elements (Chen and Privalsky, 1993). In HeLa cells transcription from the imperfect DR-5 is repressed by v-erbA like the whole promoter of RARβ which contains this specific RARE (Rascle et al, 1996).

In addition, it should be mentioned that v-ErbA does not alter the function of glucocorticoid receptor (Sharif and Privalsky, 1991). However, the oncoprotein inhibits transcription by the oestrogen receptor through an ERE in MCF-7 cells but activates it in HeLa cells (Sharif and Privalsky, 1991; Rascle et al, 1996). We do not know whether this difference relies upon the nature of the cells or the structure of the ERE.

The mechanisms underlying the antagonistic influence of v-erbA upon TR and RAR transcriptional activity remain largely unclear as many conflicting data were provided. In many cases the ability of v-ErbA to inhibit transcription from a HRE correlates with its ability to bind to this HRE which suggests that the inhibition results from binding of v-erbA complexes to the promoters in vivo (Chen and Privalsky, 1993; Hermann et al, 1993; Yen et al, 1994). In some cases, no such correlation exists as, for example, with the DR-4 TRE in the LTR of Moloney murine leukemia retrovirus, the synthetic PAL-0 (Barettino et al, 1993; Sap et al, 1989) and the complex TRE of the carbonic anhydrase II gene (Rascle et al, 1994). Also, v-ErbA strongly represses transcription mediated by *9-cis* RA-induced-RXR from the DR-1 response element, whereas it does not bind to this element (Chen and Privalsky, 1993). In those cases we should imagine that v-erbA abrogates transcription by quenching RXR in the cells. In agreement with this hypothesis, several studies bring evidence that transcriptional inhibition of T3-regulated promoters by v-ErbA requires RXR. Indeed, mutants of v-ErbA with deletions in the E domain which do not heterodimerize with RXR cannot inhibit transcription mediated by TR (Hermann et al, 1993). Moreover, in avian myoblasts naturally devoid of RXR, v-ErbA does not inhibit transcription activated by T3 from the PAL-0. Expression of an exogenous RXR in these cells restores the inhibitory effect of the oncoprotein (Cassar-Malek et al, 1997). As PAL-0 can bind both TR homodimers and TR/RXR heterodimers, we conclude that v-ErbA is unable to abrogate transcriptional activation by TR homodimers in myoblasts. This observation is consistent with the fact that v-ErbA poorly heterodimerizes with TR (Chen and Privalsky, 1993). The lack of inhibition of transcription mediated by ligand-activated TR in myoblast suggests that titration of TR by v-ErbA does not even occur in these cells. In contrast, v-erbA does inhibit transcriptional activation by RAR on this same TRE in myoblasts which demonstrates that v-ErbA can directly interfere with RAR homodimers activity. Then the transdominant repression of RAR and TR by v-ErbA in avian myoblasts follows different mechanisms.

In line with these last results, the inability of v-ErbA to inhibit RAR activity in some cells correlates with its poor ability to dissociate preformed RAR/RXR complexes bound to the RARE, due to amino-acid 351 (a serine in v-ErbA and a proline in TRα) in the E domain (Barettino et al, 1993). This natural mutation reduces the affinity of the v-ErbA product for RXR (Selmi and Samuels, 1991). Therefore, in conjunction with data obtained in myoblasts, it appears that the oncoprotein would better interfere with RAR homodimers than RAR/RXR heterodimers in some cell types. Moreover, as the opposite is observed for TR activity, it might be suggested that TR affinity for RXR is probably lower than RAR affinity, leading to an easier dissociation of TR/RXR homodimers by v-erbA (Cassar-Malek et al, 1997).

A TR mutant deleted of its DNA binding domain blocks transcription by TR or RAR, presumably by making crippled heterodimers with the receptors. In contrast, a v-ErbA mutant with the same deletion is unable to inhibit RAR and TR (Bonde et al, 1991). These observations suggest that v-ErbA acts by competing for the binding to the TRE rather than formation of non-functional v-erbA/TRα heterodimers or RXR titration. However, we canot exclude that the DNA binding domain of v-ErbA could bind and titrate cofactors of the nuclear receptors.

In conclusion, v-ErbA seems to inhibit transcription mediated by TR and RAR through several mechanisms depending on cell type and nature of the target gene. Identification of natural target genes of the oncoprotein in erythrocytic cells would definitely help in elucidating the fine molecular mechanisms of action of v-ErbA at the transcriptional level in relation with its oncogenicity.

Whatever the mechanism of action of v-ErbA to inhibit TR and RAR, it is clear that the oncoprotein does not behave simply as an unliganded TR.

Transcriptional activation by v-ErbA

Surprisingly, the v-ErbA protein acts as a constitutive transcriptional activator on a TRE when expressed in yeast. Its transcriptional activity is strongly enhanced by TRIAC (Privalsky et al, 1990), an acetic acid analogue of T3 displaying a very strong affinity for TR. As previously mentioned, the oncoprotein weakly binds T3, but significantly binds TRIAC. TRβ but not TRα behaves similarly under these conditions. This observation suggests that v-erbA has maintained the property to induce transcription in response to ligand binding. Indeed in yeast, the oncoprotein seems to bind TRIAC and to respond to TRIAC with an efficiency 100-fold lower than TRα (Privalsky et al, 1990). From a series of engineered insertion mutants of v-ErbA it could be concluded that integrity of all domains of v-ErbA except the D domain is absolutely required for the transcriptional activation (Smit-McBride and Privalsky, 1993). We may speculate that in yeast the v-ErbA protein does not encounter transcriptional repressors like N-CoR and SMRT and can have a constitutive and TRIAC-upregulated transcriptional activity using its preserved AF-1 transactivating domain. Very interestingly, the domains of v-ErbA required for transactivation in yeast and for oncogenic function in animal cells are closely congruent (Smit-McBride and Privalsky, 1993).

TRα and v-ErbA, but not TRβ, exhibit a constitutive transcriptional activity on an engineered promoter containing part of the LTR of Rous Sarcoma Virus (RSV). In this LTR the erbA proteins recognise a specific response element containing the palindromic motif AAGGCA with the two motifs separated by a 6 bp spacer. With TRα the transcriptional activity on this IP-6 is decreased by T3 (Saatcioglu et al, 1993b). These observations are highly suggestive that v-erbA might be a potent constitutive transcriptional activator on some promoters in animal cells. No natural promoter that might positively respond to v-erbA has yet been identified. However, in avian myoblasts, coexpression of nearly equimolar amounts of v-erbA and TRα induces a T3-independent transcriptional activation through the PAL-0 TRE (Cassar-Malek et al, 1997). A similar activity of v-ErbA was reported upon RARα transcriptional activity through a βRARE (Sharif and Privalsky, 1992). The v-erbA oncogene was shown to constitutively activate keratin gene expression and to block ligand-dependent TR and RAR repression (Tomie-Canie et al, 1996).

These data suggest that in some conditions, v-ErbA could induce a positive response of TR and RAR target genes, in the absence of the respective ligand.

Abrogation of functional interactions between AP-1 and the receptors

In contrast to TRα, v-ErbA is unable to down modulate AP-1 activity either in the presence or absence of T3 (Desbois et al, 1991 b). More interestingly, v-ErbA transdominantly abrogates the ability of TR and RAR to inactivate AP-1 (Desbois et al, 1991 b). The inability of v-ErbA to inhibit AP-1 is due to mutations in the E domain and mainly to the nine amino acids-deletion in the AF-2 domain (Desbois et al, 1991 b; Saatcioglu et al, 1993 a). The v-ErbA oncoprotein was also shown to enhance transcriptional activation by c-Jun from an AP-1-responsive element (Sharif and Privalsky, 1992). Whether v-ErbA directly interacts with c-Jun or titrates repressors of c-jun is not known. Since RXR is required for the antagonism between AP-1 and TR (Cassar-Malek et al, 1996; Claret et al, 1996) titration of RXR by v-ErbA could account for the abrogation by the oncoprotein of the AP-1/TR interferences. It is not yet known whether the v-ErbA product can bind to CBP but because of mutations in the ligand binding domain and because v-ErbA does not inactivate AP-1, we may anticipate that it binds poorly to CBP.

Taken together, all these observations show that v-ErbA should maintain a strong AP-1 activity in the cells even in the presence of T3 or RA. Indeed it was shown that chicken embryo fibroblasts whose growth is normally inhibited by RA, become resistant to this growth inhibition when they express the v-ErbA oncoprotein (Desbois et al, 1991 a). Moreover, Desbois et al, (1991 b) showed that there is a strong correlation between the ability of artificial mutants of v-ErbA to abrogate AP-1 down modulation by RA and to induce resistance to the growth inhibition by RA. These experiments demonstrated then clearly that v-ErbA could contribute to maintain a constitutive growth activity in the cells.

Roles of the mutations in v-ErbA

The inability of v-ErbA to activate transcription in a hormone dependent fashion is due mainly to the nine amino acid deletion in the AF-2 domain (Zenke et al, 1990). Also the mutations in the E domain are responsible for the inability of v-ErbA to down modulate AP-1 (Desbois et al, 1991a; Saatcioglu et al, 1993b). They contribute to the incapacity of v-ErbA to bind T3 (Munoz et al, 1988). As expected, the nine amino acid deletion in the AF-2 domain is responsible for the loss of the transactivating activity of the oncoprotein in mammalian cells (Zenke et al, 1990; Saatcioglu et al, 1993a; Barettino et al, 1994). This same deletion should prevent the binding of the cofactors which normally bind to the AF-2 domain of TR and RAR.

Replacing the E domain of v-ErbA with that of c-ErbAα makes a chimeric protein which shows oncogenic properties only in the absence of T3, suggesting that the inability to induce a T3-dependent transcriptional activation is required for the oncogenic properties of v-ErbA (Pain et al, 1990; Zenke et al, 1990; Forrest et al, 1990b).

The two mutations recorded in the DNA binding domain are essential for the oncogenicity of the protein (Privalsky et al, 1988; Bonde et al, 1991; Hall et al, 1992).

Interestingly, several mutations seem to have coordinated effects. For instance, the mutation at position 61 in the P box of v-ErbA considerably reduces the spectrum of TRE recognized by the oncoprotein in comparison to c-erbAα. However, the mutation at position 32 in the residual A/B domain of v-ErbA restores the spectrum of TRE recognized by the v-ErbA/RXR heterodimers to that of the TRE bound by TRα/RXR heterodimers (Judelson and Privalsky 1996). It then appears that the effects of these two mutations compensate each other as far as the recognition of TRE is considered. As these two mutations were preserved in v-erbA oncogene isolated from the natural virus AEV, we should imagine that they fulfil other biochemical functions than controlling binding to TRE.

The mutation at position 61 is important because it is necessary for v-ErbA to antagonise transactivation by RAR. Reversion of the mutation to the wild type residue of c-erbAα (mutant S61G of v-ErbA) abrogates inhibition of RAR and leukemogenicity (Sharif and Privalsky, 1991). This observation is puzzling because c-erbA in the absence of T3 is a strong antagonist of RAR. We suggest that some of the other mutations outside position 61 abrogate functional interference between v-ErbA and RAR and that this interference is restored by mutating residue 61. Here again, some compensatory effects are observed between the different mutations of the oncoprotein.

We might assume that combined mutations on residue 61 and other residues are useful for v-ErbA to interfere with genes, other than RAR- and TR-target genes in erythrocytic cells. The role of the mutation at position 351 in v-ErbA which changes a proline in c-erbAα into a serine in v-ErbA in unclear. Indeed reversion of this residue to a proline in a selected mutant of AEV, AEVr12, strongly enhanced the inhibitory effect of the v-ErbA mutant protein on RA-induced gene expression in erythroblasts (Barettino et al, 1993). We might then wonder why the wild type AEV has maintained a mutation at this position. However, AEVr12 also showed mutations in the v-erbB oncogene as compared to AEVwt. We might then imagine that the purpose of the serine/proline reversion in v-ErbAr12 was to strengthen some defectiveness of the mutant v-erbB oncogene (Damm et al, 1987).

In summary, it is clear that many of the mutations in the v-erbA oncogene were not fortuitously accumulated, but that they were selected naturally as conferring to the oncoprotein its strongest antagonistic effects against RAR- and TR-mediated transcriptional regulatory pathways.

The v-erbA Oncogene Inhibits Implementation of the Differentiation Program of Erythrocytic Progenitors

v-erbA blocks the commitment of erythrocytic progenitor cells to the final differentiation

The v-ErbA oncogene is one of the most-clear-cut cases of an oncogene that acts by blocking a differentiation program. When this oncogene is expressed in an erythrocytic progenitor cell at the BFU-E stage of differentiation, this cell progresses along the differentiation pathway to the downstream early CFU-E stage where it stops differentiating (Samarut and Gazzolo, 1982). Although this effect was first recognised when acting in conjunction with the v-erbB oncogene (Sealy et al, 1983 a; 1983 b), the use of a selectable retrovirus carrying the v-erbA oncogene as the sole oncogene has directly demonstrated that the v-ErbA gene product induces by itself a blockade of the erythrocytic differentiation program at the CFU-E stage (Gandrillon et al, 1989; Schroeder et al, 1990).

The current view of this phenomenon is that during the BFU-E to CFU-E transition, the expression levels of a number of genes would have to be modulated by ligand-activated endogenous nuclear hormones receptors of the c-erbA family thereby committing the cell from a proliferation to a differentiation program. These genes could either be a small number of master controllers of the erythrocytic differentiation program or an entire group bank of genes, each of which plays its own peculiar and discrete role in the combined phenomenon of erythrocytic differentiation. In any case, the v-ErbA protein would constitutively repress the actions of these endogenous nuclear hormones receptors and thus block the differentiation sequence.

This model predicts that small lipophilic molecules like T3 and retinoic acid acting via their respective nuclear receptors should play an essential, previously unrecognised, role in the final steps of the erythrocytic differentiation sequence. Such a role has indeed been demonstrated in vitro for T3 in chicken (Gandrillon et al, 1994; Schroeder et al, 1992), murine (Dinnen et al, 1994) and human (Dainiak et al, 1978) erythrocytic cells. More conflicting results have been obtained concerning the involvement of *all-trans*-retinoic acid in the control of the erythrocytic differentiation pathway. Some studies concluded that this molecule inhibits erythrocytic differentiation (Labbaye et al, 1994; Rusten et al, 1996), whereas others detected a stimulating effect (Schroeder et al, 1992; Correa and Axelrad, 1992). Part of the discrepancy could originate from different culture conditions, and, notably, different retinoic acid dosages: a recent study has demonstrated that at low doses ($10^{-11}-10^{-12}$ M) retinoic acid elicited an increase in BFU-E formation, whereas at higher doses ($10^{-6}-10^{-7}$ M) a decrease was observed (Zauli et al, 1995). This is in agreement with studies showing that retinoic acid is involved at low concentrations (nM range) in the final commitment process but can induce apoptosis in progenitor cells at higher concentrations in the absence of differentiation-inducing agents (Gandrillon et al, 1994). The v-erbA oncogene expression can protect progenitor cells against both apoptosis and differentiation induced by retinoic acid or T3 (Gandrillon et al, 1994). It was therefore proposed that, during the BFU-E to CFU-E transition, v-erbA blocks a commitment process induced by ligand-activated nuclear hormones receptors that turns a self-renewing cell into a cell facing either apoptosis or terminal differentiation depending on external stimuli (Figure 4.8).

One aspect of this blockade is the ability of v-erbA to abrogate the complex growth requirement of transformed erythroblasts and to enable these cells to proliferate under a wide range of pH and HCO_3^-/Na^+ ion concentration (Damm et al, 1987). This property has been related to the v-erbA-induced inhibition of Band 3 and CAII gene expression (Fuerstenberg et al, 1992; see below).

One open question concerns the physiological relevance of a redundant control of the final erythrocytic differentiation steps by two molecules as different as T3 and retinoic acid. Part of the answer may rely upon the discovery of new natural active metabolites of vitamin A (Achkar et al, 1996; Blumberg et al, 1996) and, above all, on the elucidation of the physiological mechanisms controlling their production rate in living organisms. This might well represent a challenging question for years to come.

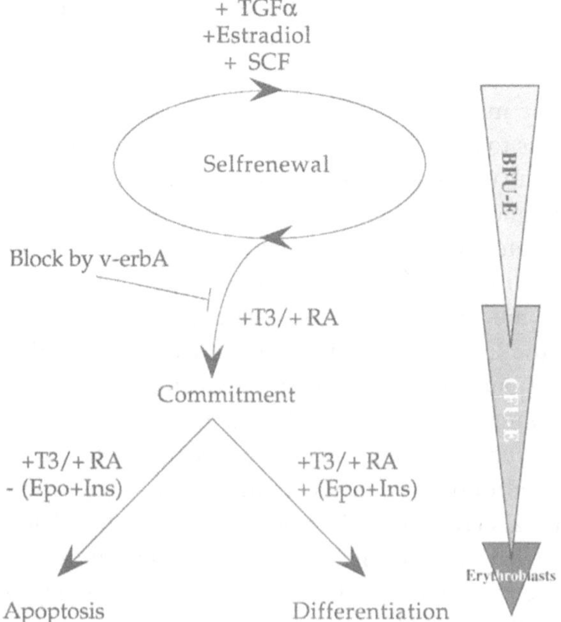

Figure 4.8 Model of the erythrocytic differentiation pathway in the chicken illustrating the respective actions of self-renewal factors (TGF-α, Stem cell factor (SCF) and oestradiol), commitment-inducing agents (RA or T3) and differentiation-inducing factors (erythropoietin plus insulin). This model predicts that the v-erbA oncogene blocks the differentiation sequence by interfering with a very early commitment program that can be elicited either by c-erbA or by RAR. On the right is shown a tentative correspondence with the previously identified cellular compartments of the erythrocytic differentiation pathway; BFU-E: burst-forming unit erythrocytic; CFU-E: colony-forming unit erythrocytic.

v-erbA reverses the pattern of gene expression

The ability of the v-ErbA oncoprotein to behave as a transdominant repressor of TR and RAR and to abrogate the receptor-mediated downmodulation of AP-1 dependent genes (see sections *Inhibition by v-ErbA of transcription mediated by TR and RAR* and *Abrogation of functional interactions between AP-1 and the receptors*) form the basis of the proposed mode of action of v-erbA. The v-erbA oncogene induces a radical change in the gene expression program in transformed cells. As schematised in Figure 4.7, it totally reverses the pattern of regulation of different types of genes. It represses at least a subset of genes that are normally upregulated by T3 or retinoic acid and that are possibly linked to the differentiation program of the cell. In contrast, v-erbA maintains the expression of AP-1-dependent genes, some of which are supposed to be involved in cell proliferation. Moreover, we should consider the possibility that v-erbA activates directly the transcription of a third type of genes that are not normally under the control of RAR or TR.

Whatever the mechanisms of action, one emerging concept is that the oncoprotein has to block both some (but not all) TR-controlled genes and some (but not all) RAR-controlled genes. That both pathways have to be blocked simultaneously is exemplified by the S61G mutation of v-ErbA. This mutant protein is still able to inhibit the TR-controlled response, but has lost the ability to block the RAR-mediated response. The biological consequences are an inability to induce either the protection against apoptosis or the leukemogenicity characteristic of the wild-type v-ErbA protein (Sharif and Privalsky, 1991; Gandrillon et al, 1994; see section *Roles of the mutations in v-ErbA*). Furthermore, this is in good agreement with the observed functional overlap between T3 and *all-trans*-retinoic acid in their control of final differentiation steps (Gandrillon et al, 1994). This concept is also reinforced by re-expression studies. The re-expression of the human RARα gene in chicken erythroblastic cells transformed by AEV induced either no (Sande et al, 1993) or only a partial and clone-dependent (Schroeder et al, 1992) re-expression of differentiation markers.

The evidence that all TR-target genes are not necessary v-ErbA target genes comes mainly from DNA binding specificity studies strengthened by functional assays (Subauste and Koenig, 1995; Judelson and Privalsky, 1996). The evidence that some RAR-target genes are not v-erbA targets has been obtained using different retinoic acid response element and seems to depend upon the cell type tested (see section *Inhibition by v-ErbA of transcription mediated by TR and RAR*). One might therefore envision that the v-erbA oncogene has accumulated mutations that have created a v-erbA-specific subset of target genes that will incorporate some TR targets genes, some RAR targets genes and possibly some target genes for other nuclear hormone receptors. This could lead to the specific repression of a subset of differentiation – and activation of a subset of proliferation – relevant key genes.

v-erbA inhibits transcription of erythrocytic genes

Recent studies using mass culture of normal progenitor cells have shown that during the first 2 days of the differentiation process, the gene expression pattern of the cells is extensively reprogrammed (Dolznig et al, 1995) as exemplified by the silencing of genes characteristic of self-renewing cells (c-myb, c-kit and the oestrogen receptor gene) and the upregulation of erythrocytic-specific transcription factors (GATA-1, GATA-2, SCL and NF-E2), followed later by the activation of late erythrocyte genes (carbonic anhydrase II, β-globin, d-amino-levulinic acid synthase (ALA-S) and Band 3).

By preventing the commitment, the v-erbA oncogene indirectly suppresses the expression of those late erythrocyte genes (Zenke et al, 1988). Furthermore, a specific inhibition was evidenced for the Band 3 and CAII genes (Zenke et al, 1988; Pain et al, 1990). Of these two genes, only the CAII gene has been shown to be transcriptionally upregulated by T3 in erythrocytic chicken cells in a direct fashion that does not requires *de novo* protein synthesis and to be directly repressed by the v-erbA oncogene (Pain et al, 1990; Disela et al, 1991; Rascle et al, 1994).

The relevance of the repression of these genes toward the v-erbA block in differentiation was assessed by re-expressing CAII or Band 3 in erythrocytic progenitor cells blocked by v-erbA expression. Although this re-expression abrogated the v-erbA-induced tolerance to pH variations, no direct effect was seen on the differentiation status of these cells, leading the authors to conclude that a different set of target genes must be directly involved in the block of differentiation (Fuerstenberg et al, 1992). This conclusion is strengthened by recent evidence showing that normal erythropoiesis occurs in cattle fully deficient in Band 3 expression (Inaba et al, 1996), although the mature erythrocytes of the affected animals are defective for their membrane structure and their CO_2 transport ability.

These results therefore stress the absolute necessity for isolating new v-erbA target genes. One attractive candidate is the RARβ gene. The expression level of this gene has been shown to be highly induced in normal non-transformed progenitors by *all-trans*-retinoic acid treatment (Gandrillon et al, 1994). On the other hand, no RARβ gene expression was detected after *all-trans*-retinoic acid treatment in AEV-transformed cell lines (Sande et al, 1993). This gene therefore represents a putative target gene for the v-erbA oncogene. In any case, should the RARβ gene prove to be a v-erbA target gene or not, further isolation of relevant target genes is a challenge still ahead of us.

Conclusion and Future Prospects

Although the involvement of hormones and their receptors in cancer has been assumed for a long time, direct evidence was brought by the identification of the v-erbA oncogene as a mutated form of the gene encoding a nuclear hormone receptor. Nuclear hormone receptors of the c-erbA family clearly appear now as master regulators in the control of cell proliferation and differentiation. It is then not unexpected that alteration of regulatory pathways mediated by these receptors drive cells to neoplastic transformation. Even if the fine mechanisms through which v-erbA transforms cells are not completely understood, it has become obvious that the oncogene has undergone a strong selective pressure in the virus to subvert multiple redundant and/or complementary biochemical pathways in the target cells. All rearrangements and mutations accumulated by the oncogene deserve attention and should help in elucidating the role of the different domains of the oncoprotein and normal nuclear receptors of the c-erbA family. It is puzzling that v-erbA contributes mostly to transformation of erythrocytic cells in chicken and hepatocytes in mouse, whereas receptors of the c-erbA family are present and presumably play determinant functions in many other types of cells. Certainly, interactions of the oncoprotein with erythroblast and hepatocyte specific proteins are to be investigated. Also, identification of target genes of the oncoprotein in these cells will be a major step.

Because of their crucial role in cell differentiation and proliferation, pathways mediated by receptors of the c-erbA family can certainly be impaired by oncoproteins other than v-ErbA. Recently, the fusion oncoprotein Myb-Ets carried by the avian leukemia retrovirus E26 has been shown to work very similarly to v-ErbA in abrogating the function of RAR and TR (Rascle et al, 1996). That two oncogenes as different as v-erbA and myb-ets use the same strategy to transform cells encourages further investigations to clarify if other oncogenes function in the same way.

For all these reasons, the v-erbA oncogene appears as a powerful model in relation to some human neoplasias (review in Gandrillon et al, 1995). Indeed, in the human acute promyelocytic leukemia (APL), the gene encoding RARα is rearranged and fused to the gene encoding the nuclear protein PML (Borrow et al, 1990; De Thé et al, 1990; Kakisuka et al, 1991; see also chapter by Labelle and Delattre). The PML-RARα fusion product is likely to be responsible for leukemogenic transformation (Altabef et al, 1996). In a single case of human hepatocellular carcinoma, the RARβ gene is rearranged as a result of hepatitis B (HBV) virus genome integration (Dejean et al, 1986), and the encoded fusion protein HBV-RARβ exhibits oncogenic properties (Garcia et al, 1993). Loss of response to retinoic acid seems to be a frequent feature in some types of tumors. In human epidermoid lung cancer cells, transfection of an exogenous RARβ gene dramatically reduces tumourigenicity (Gebert et al, 1991; Houle et al, 1993). The investigation of the molecular basis for the loss of RARβ response in lung cancer cells led to the discovery that those cells were deficient into an AF-2 transactivating function that also affected the TRα function (Moghal and Neel, 1995). This strengthens the importance of alterations in the hormone-receptor-mediated transcription mechanisms during the genesis of tumour cells.

Due to their crucial role, nuclear receptors of the c-erbA family may be privileged targets for impairment by carcinogenic agents. The inactivation of these genes or their products might then represent a critical step in multistage carcinogenesis and there is no doubt that further studies on the v-erbA model will help in elucidating these general processes of oncogenic transformation.

Acknowledgments

Works in our laboratories have been supported by ARC, the Ligue Nationale contre le Cancer, and AFM. A.R. was a fellowship recipient of the Comité de l'Yonne de la Ligue contre le Cancer. We thank O. Chassande for critical reading of the manuscript.

References

Achkar CC, Derguini F, Blumberg B, Langston A, Levin AA, Speck J, Evans RM, Bolado J Jr., Nakanishi K, Buck J, Gudas LJ (1996): 4-Oxoretinol, a new natural ligand and transactivator of the retinoic acid receptors. *Proc Natl Acad Sci USA* 93: 4879–4884

Adan RAH, Cox JJ, Beischlag TY, Burbach JPH (1993): A composite hormone response element mediates the transactivation of the rat oxytocin gene by different classes of nuclear hormone receptors. *Mol Endocrinol* 7: 47–57

Altabef M, Garcia M, Lavau C, Bae S-C, Dejean A, Samarut J (1996): A retrovirus carrying the promyelocyte-retinoic acid receptor PML-RARα fusion gene transforms haematopoietic progenitors in vitro and induces acute leukemia. *The EMBO J* 15: 2707–2716

Angel P, Karin M (1991): The role of jun, fos, and AP-1 complex in cell-proliferation and transformation. *Biochim Biophys Acta* 1072: 129–157

Baniahmad A, Muller M, Steiner C, Renkawitz R (1987): Activity of two different silencer elements of the chicken lysozyme gene can be compensated by enhancer elements. *The EMBO J* 6: 2297–2303

Baniahmad A, Kohne AC, Renkawitz R (1992): A transferable silencing domain is present in the thyroid hormone receptor, in the v-erbA oncogene product and in the retinoic acid receptor. *The EMBO J* 11: 1015–1023

Baniahmad A, Ha I, Reinberg D, Tsai S, Tsai MJ, O'Malley BW (1993): Interaction of human thyroid hormone receptor beta with transcription factor TFIIB may mediate target gene derepression and activation by thyroid hormone. *Proc Natl Acad Sci USA* 90: 8832–8836

Banker DE, Eisenman RN (1993): Thyroid hormone receptor can modulate retinoic acid-mediated axis formation in frog embryogenesis. *Mol Cell Biol* 13: 7540–7552

Barettino D, Bugge TH, Bartunek P, Vivanco Ruiz MD, Sonntag-Buck V, Beug H, Zenke M, Stunnenberg HG (1993): Unliganded T3R, but not its oncogenic variant, v-erbA, suppresses RAR-dependent transactivation by titrating out RXR. *EMBO J* 12: 1343–1354

Barettino D, Ruiz M, Stunnenberg HG (1994): Characterization of the ligand-dependent transactivation domain of thyroid hormone receptor. *The EMBO J* 13: 3039–3049

Barlow C, Meister B, Lardelli M, Lendahl U, Vennström B (1994): Thyroid abnormalities and hepatocellular carcinoma in mice transgenic for v-erbA. *The EMBO J* 13: 4241–4250

Beebe JS, Darling DS, Chin WW (1991): 3,5,3′-triiodothyronine receptor auxiliary protein (TRAP) enhances receptor binding by interactions within the thyroid hormone response element. *Mol Endocrinol* 5: 85–93

Beug H, Palmieri S, Freudenstein C, Zentgraf H, Graf T (1982): Hormone-dependent terminal differentiation in vitro of chicken erythroleukemia cells transformed by ts mutants of avian erythroblastosis virus. *Cell* 28: 907–919

Bigler J, Eisenman RN (1988): c-erbA encodes multiple proteins in chicken erythroid cells. *Mol Cell Biol* 8: 4155–4161

Bigler J, Hokanson W, Eisenman RN (1992): Thyroid hormone receptor transcriptional activity is potentially autoregulated by truncated forms of the receptor. *Mol Cell Biol* 12: 2406–2417

Blumberg B, Bolado J Jr., Derguini F, Craig AG, Moreno TA, Chakravarti D, Heyman RA, Buck J, Evans RM (1996): Novel retinoic acid receptor ligands in *Xenopus* embryos. *Proc Natl Acad Sci USA* 93: 4873–4878

Bogazzi F, Hudson LD, Nikodem VM (1994): A novel heterodimerization partner for thyroid hormone receptor. *J Biol Chem* 269: 11683–11686

Bonde BG, Privalsky ML (1990): Sequence-specific DNA binding by the v-erbA oncogene protein of avian erythroblastosis virus. *J Virol* 64: 1314–1320

Bonde BG, Sharif M, Privalsky ML (1991): Ontogeny of the v-erbA oncoprotein from the thyroid hormone receptor: an alteration in the DNA binding domain plays a role crucial for c-erbA function. *J Virol* 65: 2037–2046

Borrow J, Goddard A, Sheer D, Salomon E (1990): Molecular analysis of acute promyelocytic leukemia breakpoint cluster region on chromosome 17. *Science* 249: 1577–1580

Boucher P, Privalsky ML (1990): Mapping of functional domains with the v-erbA oncogene protein: the remnants of the hormone binding domain play multiple, vital roles in protein action. *Oncogene* 5: 1303–1311

Boucher P, Koning A, Privalski ML (1988): The avian erythroblastosis virus erb A oncogene encodes a DNA – binding protein exhibiting distinct nuclear and cytoplasmic subcellular localizations. *J Virol* 62: 534–544

Bourguet W, Ruff M, Chambon P, Gronemeyer H, Moras D (1995): Crystal structure of the ligand-binding domain of the human nuclear receptor RXR-α. *Nature* 375: 377–382

Burns K, Duggan B, Atkinson EA, Famulski KS, Nemer M, Bleackley RC, Michalak M (1994): Modulation of gene expression by calreticulin binding to the glucocorticoid receptor. *Nature* 367: 476–480

Casanova J, Helmer E, Selmi-Ruby S, Qi J-S, Au-Fliegner M, Desai-Yajnik V, Koudinova N, Yarm F, Raaka F, Samuels HH (1994): Functional evidence for ligand-dependent dissociation of thyroid hormone and retinoic acid receptors from an inhibitory cellular factor. *Mol Cell Biol* 14: 5756–5765

Casini T, Graf T (1995): Bicistronic retroviral vector reveals capacity of v-erbA to induce erythroleukemia and to co-operate with v-myb. *Oncogene* 11: 1019–1026

Cassar-Malek I, Marchal S, Altabef M, Wrutniak C, Samarut J, Cabello G (1994): Stimulation of quail myoblast terminal differentiation by the v-erbA oncogene. *Oncogene* 9: 2197–2206

Cassar-Malek I, Marchal S, Rochard P, Casas F, Wrutniak C, Samarut J, Cabello G (1996): Induction of c-ErbA/AP-1 interactions and c-ErbA transcriptional activity in myoblasts by RXR – Consequences for muscle differentiation. *J Biol Chem* 271: 11392–11399

Cassar-Malek I, Marchal S, Rochard P, Wrutniak C, Samarut J, Cabello G (1997): Molecular basis of the cell-specific activity of v-erbA in quail myoblasts. *Oncogene*, 14: (in press)

Cavaillès V, Dauvois S, L'Horset F, Lopez G, Hoare S, Kushner PJ, Parker MG (1995): Nuclear factor RIP140 modulates transcriptional activation by the estrogen receptor. *EMBO J* 14: 3741–3751

Chatterjee VKK, Lee J-K, Rentoumis A, Jameson JL (1989): Negative regulation of the thyroid-stimulating hormone α gene by thyroid hormone: receptor interaction adjacent to the TATA boy. *Proc Natl Acad Sci USA* 86: 9114–9118

Chen HW, Privalsky ML (1993): The erbA oncogene represses the actions of both retinoid X and retinoid A receptors but does so by distinct mechanisms. *Mol Cell Biol* 13: 5970–5980

Chen HW, Smit-McBride Z, Lewis S, Sharif M, Privalsky ML (1993): Nuclear hormone receptors involved in neoplasia: erb-A exhibits a novel DNA sequence specificity determined by amino acids outside of the zinc-finger domain. *Mol Cell Biol* 13: 2366–2376

Chen JD, Evans RM (1995): A transcriptional co-repressor that interacts with nuclear hormone receptors. *Nature* 377: 454–457

Claret F-X, Antakly T, Karin M, Saatcioglu F (1996): a shift in the ligand responsiveness of thyroid hormone receptor α induced by heterodimerization with retinoid X receptor α. *Mol Cell Biol* 16: 219–227

Cooney AJ, Leng XH, Tsai SY, Omalley BW, Tsai MJ (1993): Multiple mechanisms of chicken ovalbumin upstream promoter transcription factor-dependent repression of transactivation by the vitamin-D, thyroid hormone, and retinoic acid receptors. *J Biol Chem* 268: 4152–4160

Correa PN, Axelrad AA (1992): Retinyl acetate and *all-trans*-retinoic acid enhance erythroid colony formation in vitro by circulating human progenitors in an improved serum-free medium. *Int J Cell Cloning* 10: 286–291

Dainiak N, Hoffman R, Maffei LA, Forget BG (1978): Potentiation of human erythropoiesis in vitro by thyroid hormone. *Nature* 272: 260–262

Damm K, Evans RM (1993): Identification of a domain require for oncogenic activity and transcriptional suppression by v-erbA and thyroid-hormone receptor alpha. *Proc Natl Acad Sci USA* 90: 10668–10672

Damm K, Beug H, Graf T, Vennstrom B (1987): A single mutation in erbA restores the erythroid transforming potential of a mutant avian erythroblastosis virus (AEV) defective in both erbA and erbB oncogenes. *EMBO J* 6: 375–382

Damm K, Thompson CC, Evans RM (1989): Protein encoded by v-erbA functions as a thyroid-hormone receptor antagonist. *Nature* 339: 593–597

Darling DS, Beebe JS, Burnside J, Winslow ER, Chin WW (1991): 3,5,3′-triiodothyronine (T3) receptor-auxiliary protein (TRAP) binds DNA and forms heterodimers with the T3 receptor. *Mol Endocrinol* 5: 73–84

De Thé H, Chomienne C, Lanotte M, Degos L, Dejean A (1990): The t(15;17) translocation of acute promyelocytic leukemia fuses the retinoic acid receptor α gene to a novel transcribed locus. *Nature* 347: 558–561

Dedhar S, Rennie PS, Shago M, Leung HCY, Yang H, Filmus J, Hawley RG, Bruchovsky N, Cheng H, Matusik RJ, Giguère V (1994): Inhibition of nuclear hormone receptor activity by calreticulin. *Nature* 367: 480–483

Dejean A, Bouguelerey L, Grzeschik KH, Tiollais P (1986): Hepatitis B virus DNA integration in a sequence homologous to v-erbA and steroid receptor genes in a hepatocellular carcinoma. *Nature* 322: 70–72

Desbois C, Pain B, Guilhot C, Benchaibi M, French M, Ghysdael J, Madjar JJ, Samarut J (1991 a): v-erbA oncogene abrogates growth inhibition of chicken embryo fibroblasts induced by retinoic acid. *Oncogene* 6: 2129–2135

Desbois C, Aubert D, Legrand C, Pain B, Samarut J (1991b): A novel mechanism of action for the v-erbA oncogene: abrogation of the inactivation of AP-1 transcription factor by retinoic acid receptor and thyroid hormone receptor. *Cell* 67: 731–740

Dinnen RD, White SR, Elsayed S, Yeh Y-I, Ebisuzaki K (1994): An endogenous signal triggering erythroid differentiation: identification as thyroid hormone. *Cell Growth Differ* 5: 855–861

Disela C, Glineur C, Bugge T, Sap J, Stengl G, Dodgson J, Stunnenberg H, Beug H, Zenke M (1991): v-erbA overexpression is required to extinguish c-erbA function in erythroid cell differentiation and regulation of the erbA target CAII. *Genes Dev* 5: 2033–2047

Dolznig H, Bartunek P, Nasmyth K, Müllner EW, Beug H (1995): Terminal differentiation of normal erythroid progenitors: shortening of G1 correlates with loss of D-cyclin/cdk4 expression and altered cell size control. *Cell Growth Differ* 6: 1341–1352

Evans RM (1988): The steroid and thyroid hormone receptor superfamily: transcriptional regulators of development and physiology. *Science* 240: 889–895

Fondell JD, Roy AL, Roeder RG (1993): Unliganded thyroid hormone receptor inhibits formation of a functional preinitiation complex: implications for active repression. *Genes Dev* 7: 1400–1410

Fondell JD, Brunel F, Hisatake K, Roeder RG (1996): Unliganded thyroid hormone receptor alpha can target TATA-Binding protein for transcriptional repression. *Mol Cell Biol* 16: 281–287

Forman BM, Samuels HH (1990): Dimerization among nuclear hormone receptors. *New Biol* 2: 587–594

Forman BM, Casanova J, Raaka BM, Ghysdael J, Samuels HH (1992): Half-site spacing and orientation determines whether thyroid hormone and retinoic acid receptors and related factors bind to DNA response elements as monomers, homodimers, or heterodimers. *Mol Endocrinol* 6: 429–442

Forman BM, Umesono K, Chen J, Evans RM (1995): Unique response pathways are established by allosteric interactions among nuclear hormone receptors. *Cell* 81: 541–550

Forrest D, Sjöberg M, Vennström B (1990 a): Contrasting developmental and tissue-specific expression of α and β thyroid hormone receptor genes. *EMBO J*: 1519–1528

Forrest D, Munoz A, Raynoschek C, Vennström B, Beug H (1990 b): Requirement for the C-terminal domain of the v-erbA oncogene protein for biological function and transcriptional repression. *Oncogene* 5: 309–316

Forrest D, Hallböök F, Persson H, Vennström B (1991): Distinct functions for thyroid hormone receptors α and β in brain development indicated by differential expression of receptor genes. *EMBO J* 10: 269–275

Frykberg L, Palmieri S, Beug H, Graf T, Hayman MJ, Vennström B (1983): Transforming capacities of avian erythroblastosis virus mutants deleted in the erbA or erbB oncogenes. *Cell* 32: 227–238

Fuerstenberg S, Leitner I, Schroeder C, Schwarz H, Vennstrom B, Beug H (1992): Transcriptional repression of band 3 and CAII in v-erbA transformed erythroblasts accounts for an important part of the leukaemic phenotype. *The EMBO J* 11: 3355–3365

Gandrillon O, Jurdic P, Benchaibi M, Xiao JH, Ghysdael J, Samarut J (1987): Expression of the v-erbA oncogene in chicken embryo fibroblasts stimulates their proliferation in vitro and enhances tumor growth in vivo. *Cell* 49: 687–697

Gandrillon O, Jurdic P, Pain B, Desbois C, Madjar J-J, Moscovici MG, Moscovici C, Samarut J (1989): Expression of the v-erbA product, an altered nuclear hormone receptor, is sufficient to transform erythrocytic cells in vitro but not to induce acute erythroleukemia *in vivo*. *Cell* 58: 115–121

Gandrillon O, Ferrand N, Michaille J-J, Roze L, Zile MH, Samarut J (1994): c-erbAα/T3R and RARs control commitment of hematopoietic self-renewing progenitor cells to apoptosis or differentiation and are antagonized by the v-erbA oncogene. *Oncogene* 9: 749–758

Gandrillon O, Rascle A, Samarut J (1995): The v-erbA oncogene: a superb tool for dissecting the involvement of nuclear hormone receptors in differentiation and neoplasia. *Int J Oncol* 6: 215–231

Garcia M, Dethe H, Tiollais P, Samarut J, Dejean A (1993): A hepatitis-B virus pre-S-retinoic acid receptor-beta chimera transforms erythrocytic progenitor cells in vitro. *Proc Natl Acad Sci USA* 90: 89–93

Garrido C, Li RP, Samarut J, Gospodarowicz D, Saule S (1993): v-erbA cooperates with bFGF in neuroretina cell transformation. *Virology* 192: 578–586

Gebert JF, Moghal N, Frangioni JV, Sugarbaker DJ, Neel BG (1991): High frequency of retinoic acid receptor beta abnormalities in human lung cancer [published erratum appears in Oncogene 1992 Apr; 7(4): 821]. *Oncogene* 6: 1859–1868

Glass CK, Lipkin SM, Devary OV, Rosenfeld MG (1989): Positive and negative regulation of gene transcription by a retinoic acid-thyroid hormone receptor heterodimer. *Cell* 59: 697–708

Glineur C, Bailly M, Ghysdael J (1989): The c-erbAα-encoded thyroid hormone receptor is phosphorylated in its amino terminal domain by casein kinase II. *Oncogene* 4: 1247–1254

Goldberg Y, Glineur C, Gesquiere JC, Ricouart A, Sap J, Vennstrom B, Ghysdael J (1988): Activation of protein kinase C or cAMP-dependent protein kinase increases phosphorylation of the c-erbA-encoded thyroid hormone receptor and of the v-erbA-encoded protein. *EMBO J* 7: 2425–2433

Graf T, Beug H (1983): Role of the v-erbA and v-erbB oncogenes of avian erythroblastosis virus in erythroid cell transformation. *Cell* 34: 7–9

Gronemeyer H, Laudet V (1995): Transcription factors 3: nuclear receptors. *Protein Profile* 2: 1173–1308

Hadzic E, Desai-Yajnik V, Helmer E, Guo S, Wu S, Koudinova N, Casanova J, Raaka BM, Samuels HH (1995): A 10-amino-acid sequence in the N-terminal A/B domain of thyroid hormone receptor alpha is essential for transcriptional activation and interaction with the general transcription factor TFIIB. *Mol Cell Biol* 15: 4507–4517

Halachmi S, Marden E, Martin G, MacKay H, Abbondanza C, Brown M (1994): Estrogen receptor-associated proteins: possible mediators of hormone-induced transcription. *Science* 264: 1455–1458

Hall BL, Bonde VG, Judelson C, Privalsky ML (1992): Functional interaction between the two zinc finger domains of the v-erbA oncoprotein. *Cell Growth Differ* 3: 207–216

Harding PP, Duester G (1992): Retinoic acid activation and thyroid hormone repression of the human alcohol dehydrogenase gene ADH3. *J Biol Chem* 267: 14145–14150

Hentzen D, Renucci A, le Guellec D, Benchaibi M, Jurdic P, Gandrillon O, Samarut J (1987): The chicken c-erbA proto-oncogene is preferentially expressed in erythrocytic cells during late stages of differentiation. *Mol Cell Biol* 7: 2416–2424

Hermann T, Hoffmann B, Piedrafita FJ, Zhang K-K, Pfahl M (1993): v-erbA requires auxiliary proteins for dominant negative activity. *Oncogene* 8: 55–65

Hodin RA, Lazar MA, Wintman BI, Darling DS, Koening RJ, Larsen PR, Moore DD, Chin WW (1989): Identification of a thyroid hormone receptor that is pituitary-specific. *Science* 244: 76–79

Hörlein AJ, Näär AM, Heinzel T, Torchia J, Gloss B, Kurokawa R, Ryan A, Kamei Y, Söderström M, Glass CK, Rosenfeld MG (1995): Ligand-independent repression by the thyroid hormone receptor mediated by a nuclear hormone co-repressor. *Nature* 377: 397–404

Houle B, Rochette-Egli C, Bradley WEC (1993): Tumor-suppressive effect of the retinoic acid receptor β in human epidermoid lung cancer cells. *Proc Natl Acad Sci USA* 90: 985–989

Iglesias T, Llanos S, López-Barahona M, Seliger B, Rodríguez-Pena A, Bernal J, Muñoz A (1995): Induction of platelet-derived growth factor B/c-sis by the v-erbA oncogene in glial cells. *Oncogene* 10: 1103–1110

Inaba M, Yawata A, Koshino I, Sato K, Takeuchi M, Takakuwa Y, Manno S, Yawata Y, Kanzaki A, Sakai J-I, Ban A, Ono K-I, Maede Y (1996): Defective anion transport and marked spherocytosis with membrane instability caused by hereditary total deficiency of red cell Band 3 in cattle due to a nonsense mutation. *J Clin Invest* 97: 1804–1817

Izumo S, Mahdavi V (1988): Thyroid hormone receptor α isoforms generated by alternative splicing differentially activate myosin HC gene transcription. *Nature* 334: 539–542

Judelson C, Privalsky ML (1996): DNA recognition by normal and oncogenic thyroid hormone receptors – Enexpected diversity in half-site specificity controlled by non-zinc-finger determinants. *J Biol Chem* 271: 10800–10805

Kakisuka A, Miller WH, Umesono RP, Warrel S, Frankel VVVS, Marty E, Dmitrowsky E, Evans RM (1991): Chromosomal translocation t(15;17) in human acute promyelocytic leukemia fuses RARα with a novel putative transcription factor. *Cell* 66: 663–674

Kamei Y, Xu L, Heinzel T, Torchia J, Kurokawa R, Gloss B, Lin S-C, Heyman RA, Rose DW, Glass CK, Rosenfeld MG (1996): A CBP integrator complex mediates transcriptional activation and AP-1 inhibition by nuclear receptor. *Cell* 85: 403–414

Katz D, Lazar MA (1993): Dominant negative activity of an endogenous thyroid hormone receptor variant (alpha 2) is due to competition for binding sites on target genes. *J Biol Chem* 268: 20904–20910

Khazaie K, Panayotou G, Aguzzi A, Samarut J, Gazzolo L, Jurdic P (1991): EGF promotes in vivo tumorigenic growth of primary chicken embryo fibroblasts expressing v-myc and enhances in vitro transformation by the v-erbA oncogene. *Oncogene* 6: 21–28

Knegtel RM, Katahira M, Schilthuis JG, Bonvin AM, Boelens R, Eib D, van der Saag PT, Kaptein R (1993): The solution structure of the human retinoic acid receptor-beta DNA-binding domain. *J Biomol NMR* 3: 1–17

Koenig RJ, Lazar MA, Hodin RA, Brent GA, Larsen PR, Chin WW, Moore DD (1989): Inhibition of thyroid hormone action by a non-hormone binding c-erbA protein generated by alternative mRNA splicing. *Nature* 337: 659–661

Konig H, Ponta H, Rahmsdorf HJ, Herrlich P (1992): Interference between pathway-specific transcription factors: glucocorticoids antagonize phorbol ester-induced AP-1 activity without altering AP-1 site occupation in vivo. *EMBO J* 11: 2241–2246

Kurokawa R, Söderström M, Hörlein AJ, Halachmi S, Brown M, Rosenfeld MG, Glass CK (1995): Polarity-specific activities of retinoic acid receptors determined by a co-repressor. *Nature* 377: 451–454

Labbaye C, Valtieri M, Testa U, Giampaolo A, Meccia E, Sterpetti P, Parolini I, Pelosi E, Bulgarini D, Cayre YE, Peschle C (1994): Retinoic acid downmodulates erythroid differentiation and GATA1 expression in purified adult-progenitor culture. *Blood* 83: 651–656

Lazar MA, Hodin RA, Darling DS, Chin WW (1988): Identification of a rat c-erbAα-related protein which binds deoxyribonucleic acid but does not bind thyroid hormone. *Mol Endocrinol* 2: 893–901

Lazar MA, Berrodin TJ, Harding HP (1991): Differential DNA binding by monomeric, homodimeric, and potentially heterodimeric forms of the thyroid hormone receptor. *Mol Cell Biol* 11: 5005–5015

Le Douarin B, Zechel C, Garnier JM, Lutz Y, Tora L, Pierrat B, Heery D, Gronemeyer H, Chambon P, Losson R (1995): The N-terminal part of TIF1, a putative mediator of the ligand-dependent activation function (AF-2) of nuclear receptors, is fused to B-raf in the oncogenic protein T18. *EMBO J* 14: 2020–2033

Lee JW, Ryan F, Swaffield JC, Johnston SA, Moore DD (1995): Interaction of thyroid-hormone receptor with a conserved transcriptional mediator. *Nature* 374: 91–94

Lee LR, Mortensen RM, Larson CA, Brent GA (1994): Thyroid hormone receptor-alpha inhibits retinoic acid-responsive gene expression and modulates retinoic acid-stimulated neural differentiation in mouse embryonic stem cells. *Mol Endocrinol* 8: 746–756

Lee Y, Mahdavi V (1993): The D-domain of the thyroid hormone receptor-alpha1 specifies positive and negative transcriptional regulation functions. *J Biol Chem* 268: 2021–2028

Lehmann JM, Zhang X, Graupner G, Lee MO, Hermann T, Hoffmann B, Pfahl M (1993): Formation of retinoid X receptor homodimers leads to repression of T3 response: Hormonal cross talk by ligand-induced squelching. *Mol Cell Biol* 13: 7698–7707

Leid M, Kastner P, Chambon P (1992a): Multiplicity generates diversity in the retinoic acid signalling pathway. *TIBS* 17: 427–433

Leid M, Kastner P, Lyons R, Nakshatri H, Saunders M, Zacharewski T, Chen JY, Staub A, Garnier JM, Mader S, Chambon P (1992b): Purification, cloning, and RXR identity of the HeLa cell factor with which RAR or TR heterodimerizes to bind target sequences efficiently [published erratum appears in Cell 1992 Nov 27; 71(5): following 886]. *Cell* 68: 377–395

Lezoualc'h F, Hassan AHS, Giraud P, Loeffler J-P, Lee SL, Demeneix BA (1992): Assignment of the β-thyroid hormone receptor to 3,5,3′-triiodothyronine-dependent inhibition of transcription from the thyrotropin-releasing hormone promoter in chick hypothalamic neurons. *Mol Endocrinol* 6: 1797–1804

Lin KH, Ashizawa K, Cheng SY (1992): Phosphorylation stimulates the transcriptional activity of the human beta 1 thyroid hormone nuclear receptor. *Proc Natl Acad Sci USA* 89: 7737–7741

Llanos S, Iglesias T, Riese HH, Garrido T, Caelles C, Munoz A (1996): V-erbA oncogene induces invasiveness and anchorage-independent growth in cultured glial cells by mechanisms involving platelet-derived growth factor. *Cell Growth Differ* 7: 373–382

Lucas PC, Forman BM, Samuels HH, Granner DK (1991): Specificity of a retinoic acid response element in the phosphoenolpyruvate carboxykinase gene promoter: consequences of both retinoic acid and thyroid hormone receptor binding [published erratum appears in *Mol Cell Biol* 1991 Dec; 11(12): 6343]. *Mol Cell Biol* 11: 5164–5170

Machuca I, Esslemont G, Fairclough L, Tata JR (1995): Analysis of structure and expression of the Xenopus thyroid hormone receptor-β gene to explain its autoinduction. *Mol Endocrinol* 9: 96–107

Mangelsdorf DJ, Evans RM (1995): The RXR heterodimers and orphan receptor. *Cell* 83: 841–850

Mangelsdorf DJ, Thummel C, Beato M, Herrlich P, Schutz G, Umesono K, Blumberg B, Kastner P, Mark M, Chambon P, Evans RM (1995): The nuclear receptor superfamily: the second decade. *Cell* 83: 835–839

Mitsuhashi T, Tennyson GE, Nikodem V (1988): Alternative splicing generates messages encoding rat c-erbA proteins that do not bind thyroid hormone. *Proc Natl Acad Sci USA* 85: 5804–5808

Miyamoto T, Suzuki S, DeGroot LJ (1993): High affinity and specificity of dimeric binding of thyroid hormone receptors to DNA and their ligand-dependent dissociation. *Mol Endocrinol* 7: 224–231

Moghal N, Neel BG (1995): Evidence for impaired retinoic acid receptor-thyroid hormone receptor AF-2 cofactor activity in human lung cancer. *Mol Cell Biol* 15: 3945–3959

Munoz A, Zenke A, Gehring U, Sap J, Beug H, Vennstrom B (1988): Characterization of the hormone-binding domain of the chicken c-erbA/thyroid hormone receptor protein. *EMBO J* 7: 155–159

Munoz A, Wrighton C, Seliger B, Bernal J, Beug H (1993): Thyroid hormone receptor/c-erbA: control of commitment and differentiation in the neuronal/chromaffin progenitor line PC12. *J Cell Biol* 121: 423–438

Nicholson RC, Mader S, Nagpal S, Leid M, Rochette-Egly C, Chambon P (1990): Negative regulation of the rat stromelysin gene promoter by retinoic acid is mediated by an AP-1 binding site. *EMBO J* 9: 4443–4454

O'Donnell AL, Rosen ED, Darling DS, Koenig RJ (1991): Thyroid hormone receptor mutations that interfere with transcriptional activation also interfere with receptor interaction with a nuclear protein. *Mol Endocrinol* 5: 94–99

Onate SA, Tsaï M, O'Malley B (1995): Sequence and characterization of a coactivator for the steroid hormone receptor superfamily. *Science* 270: 1354–1356

Pain B, Melet F, Jurdic P, Samarut J (1990): The carbonic anhydrase II gene, a gene regulated by thyroid hormone and erythropoietin, is repressed by the v-erbA oncogene in erythrocytic cells. *New Biol* 2: 284–294

Pain B, Woods CM, Saez J, Flickinger T, Raines M, Kung HJ, Peyrol S, Moscovici C, Moscovici G, Jurdic P, Lazarides E, Samarut J (1991): EGF-R as a hemopoietic growth factor receptor: The c-erbB product is present in normal chicken erythrocytic progenitor cells and controls their self-renewal. *Cell* 65: 37–46

Pfahl M (1993): Nuclear receptor/AP-1 interaction. *Endocr Rev* 14: 651–658

Pfahl M, Benbrook D (1987): Nucleotide sequence of cDNA encoding a novel human thyroid hormone receptor. *Nucleic Acids Res* 15: 9613

Privalsky ML (1991): A subpopulation of the v-erb A oncogene protein, a derivative of a thyroid hormone receptor, associates with heat shock protein 90. *J Biol Chem* 266: 1456–1462

Privalsky ML, Boucher P, Koning A, Judelson C (1988): Genetic dissection of functional domains within the avian erythroblastosis virus v-erbA oncogene. *Mol Cell Biol* 8: 4510–4517

Privalsky ML, Sharif M, Yamamoto KR (1990): The viral erbA oncogene protein, a constitutive repressor in animal cells, is a hormone-regulated activator in yeast. *Cell* 63: 1277–1286

Rascle A, Ghysdael J, Samarut J (1994): c-ErbA, but not v-ErbA, competes with a putative erythroid repressor for binding to the carbonic anhydrase II promoter. *Oncogene* 9: 2853–2867

Rascle A, Ferrand N, Gandrillon O, Samarut J (1996): Myb-Ets fusion oncoprotein inhibits T3R/c-erbA and RAR functions: a novel mechanism of action for leukemogenic transformation by E26 avian retrovirus. *Mol Cell Biol* 16: 6338–6351

Rastinejad F, Perlmann T, Evans RM, Sigler PB (1995): Structural determinants of nuclear receptor assembly on DNA direct repeats. *Nature* 375: 203–211

Renaud JP, Rochel N, Ruff M, Vivat V, Chambon P, Gronemeyer H, Moras D (1995): Crystal structure of the RAR-gamma ligand-binding domain bound to *all-trans* retinoic acid. *Nature* 378: 681–689

Ribeiro RC, Kushner PJ, Apriletti JW, West BL, Baxter JD (1992): Thyroid hormone alters in vitro DNA binding of monomers and dimers of thyroid hormone receptors. *Mol Endocrinol* 6: 1142–1152

Rusten LS, Dybedal I, Blomhoff HK, Blomhoff R, Smeland EB, Jacobsen SE (1996): The RAR-RXR as well as the RXR-RXR pathway is involved in signalling growth inhibition of human CD34+ erythroid progenitor cells. *Blood* 87: 1728–1736

Saatcioglu F, Bartunek P, Deng T, Zenke M, Karin M (1993 a): A conserved C-terminal sequence that is deleted in v-ErbA is essential for the biological activities of c-ErbA (the thyroid hormone receptor). *Mol Cell Biol* 13: 3675–3685

Saatcioglu F, Deng T, Karin M (1993b): A novel cis element mediating ligand-independent activation by c-ErbA: implications for hormonal regulation. *Cell* 75: 1095–1105

Salbert G, Fanjul A, Piedrafita FJ, Lu XP, Kim SJ, Tran P, Pfahl M (1993): Retinoic acid receptors and retinoid X receptor-alpha down-regulate the transforming growth factor-beta 1 promoter by antagonizing AP-1 activity. *Mol Endocrinol* 7: 1347–1356

Samarut J, Gazzolo L (1982): Target cells infected by avian erythroblastosis virus differentiate and become transformed. *Cell* 28: 921–929

Sande S, Privalsky ML (1996): Identification of TRACs, a family of co-factors that associate with, and modulate the activity of nuclear hormone receptors. *Mol Endocrinol* 10: 813–825

Sande S, Sharif M, Chen H, Privalsky M (1993): v-erbA acts on retinoic acid receptors in immature avian erythroid cells. *J Virol* 67: 1067–1074

Sap J, Munoz A, Damm K, Goldberg Y, Ghysdael J, Leutz A, Beug H, Vennstrom B (1986): The c-erb-A protein is a high-affinity receptor for thyroid hormone. *Nature* 324: 635–640

Sap J, Munoz A, Schmitt J, Stunnenberg H, Vennström B (1989): Repression of transcription mediated at thyroid hormone response element by the v-erbA oncogene product. *Nature* 340: 242–244

Schroeder C, Raynoschek C, Fuhrmann U, Damm K, Vennström B, Beug H (1990): The v-erbA oncogene causes repression of erythrocyte-specific genes and an immature, aberrant differentiation phenotype in normal erythroid progenitors. *Oncogene* 5: 1445–1453

Schroeder C, Gibson L, Zenke M, Beug H (1992): Modulation of normal erythroid differentiation by the endogenous thyroid hormone and retinoic and receptors: a possible target for v-erbA oncogene action. *Oncogene* 7: 217–227

Schüle R, Rangarajan P, Yang N, Kliewer S, Ransone LJ, Bolado J, Verma IM, Evans RM (1991): Retinoic acid is a negative regulator of AP-1-responsive genes. *Proc Natl Acad Sci USA* 88: 6092–6096

Sealy L, Privalsky ML, Moscovici G, Moscovici C, Bishop JM (1983 a): Site-specific mutagenesis of avian erythroblastosis virus: erb B is required for oncogenecity. *Virology* 130: 155–173

Sealy L, Privalsky ML, Moscovici G, Moscovici C, Bishop JM (1983 b): Site-specific mutagenesis of avian erythroblastosis virus: v-erb A is not required for transformation of fibroblasts. *Virology* 130: 179–194

Selmi S, Samuels HH (1991): Thyroid hormone receptor/and v-erbA. A single amino acid difference in the C-terminal region influences dominant negative activity and receptor dimer formation. *J Biol Chem* 266: 11589–11593

Sharif M, Privalsky ML (1991): v-erbA oncogene function in neoplasia correlates with its ability to repress retinoic acid receptor action. *Cell* 66: 885–893

Sharif M, Privalsky ML (1992): v-erbA and c-erbA proteins enhance transcriptional activation by c-jun. *Oncogene* 7: 953–960

Sjöberg M, Vennström B, Forrest D (1992): Thyroid hormone receptors in chick retinal development: differential expression of mRNAs for alpha and N-terminal variant beta receptors. *Development* 114: 39–47

Smit-McBride Z, Privalsky ML (1993): Functional domains of the v-erbA protein necessary for oncogenesis are required for transcriptional activation in Saccharomyces cerevisiae. *Oncogene* 8: 1465–1475

Soprano DR, Chen LX, Wu S, Donigan AM, Borghaei RC, Soprano KJ (1996): Overexpression of both RAR and RXR restores AP-1 repression in ovarian adenocarcinoma cells resistant to retinoic acid-dependent growth inhibition. *Oncogene* 12: 577–584

Subauste JS, Koenig RJ (1995): Comparison of the DNA binding specificity and function of v-ErbA and thyroid hormone receptor alpha1. *J Biol Chem* 270: 7957–7962

Suzuki S, Miyamoto T, Opsahl A, Sakurai A, DeGroot LJ (1994): Two thyroid hormone response elements are present in the promoter of human thyroid hormone receptor beta1. *Mol Endocrinol* 8: 305–314

Thompson CC, Weinberger C, Lebo R, Evans RM (1987): Identification of a novel thyroid hormone receptor expressed in the mammalian central nervous system. *Science* 237: 1610–1614

Tomie-Canie M, Day D, Samuels HH, Freedberg IM, Blumenberg M (1996): Novel regulation of keratin gene expression by thyroid hormone and retinoid receptors. *J Biol Chem* 271: 1416–1423

Tran P, Zhang XK, Salbert G, Hermann T, Lehmann JM, Pfahl M (1992): COUP orphan receptors are negative regulators of retinoic acid response pathways. *Mol Cell Biol* 12: 4666–4676

Trapasso F, Martelli ML, Battaglia C, Angotti E, Mele E, Stella A, Samarut J, Avvedimento VE, Fusco A (1996): The v-erbA oncogene selectively inhibits iodide uptake in rat thyroid cells. *Oncogene* 12: 1879–1888

Wagner RL, Apriletti JW, McGrath ME, West BL, Baxter JD, Fletterick RJ (1995): A structural role for hormone in the thyroid hormone receptor. *Nature* 378: 690–697

Weinberger C, Thompson CC, Ong ES, Lebo R, Gruol DJ, Evans RM (1986): The c-erb-A gene encodes a thyroid hormone receptor. *Nature* 324: 641–646

Williams GR, Harney JW, Moore DD, Larsen PR, Brent GA (1992): Differential capacity of wild type promoter elements for binding and trans-activation by retinoic acid and thyroid hormone receptors. *Mol Endocrinol* 6: 1527–1537

Wong CW, Privalsky ML (1995): Role of the N terminus in DNA recognition by the v-erbA protein, an oncogenic derivative of a thyroid hormone receptor. *Mol Endocrinol* 9: 551–562

Wood WM, Ocran KW, Gordon DF, Ridgway EC (1991): Isolation and characterization of mouse complementary DNAs encoding α and β thyroid hormone receptors from thyrotrope cells: the mouse pituitary-specific β2 isoform differs at the amino terminus from the corresponding species from rat pituitary tumor cells. *Mol Endo* 1049–1061

Wrutniak C, Cassar-Malek I, Marchal S, Rascle A, Heusser S, Keller JM, Fléchon J, Dauça M, Samarut J, Ghysdael J, Cabello G (1995): A 43-kDA protein related to c-ErbA alpha1 is located in the mitochondrial matrix of rat liver. *J Biol Chem* 270: 16347–16354

Xu J, Thompson KL, Shepard LB, Hudson L, Gill GN (1993): T3 receptor suppression of Sp1-dependent transcription from the epidermal growth factor receptor promoter via overlapping DNA-binding sites. *J Biol Chem* 268: 16065–16073

Yang-Yen HF, Zhang XK, Graupner G, Tzukerman M, Sakamoto B, Karin M, Pfahl M (1991): Antagonism between retinoic acid receptors and AP-1 – Implications for tumor promotion and inflammation. *New Biol* 3: 1206–1219

Yen PM, Ikeda M, Brubaker JH, Forgione M, Sugawara A, Chin WW (1994a): Roles of v-erbA homodimers and heterodimers in mediating dominant negative activity by v-erbA. *J Biol Chem* 269: 903–909

Yen PM, Brubaker JH, Apriletti JW, Baxter JD, Chin WW (1994b): Roles of 3,5,3′-triiodothyronine and deoxyribonucleic acid binding on thyroid hormone receptor complex formation. *Endocrinology* 134: 1075–1081

Zauli G, Visani G, Vitale M, Gibellini D, Bertolaso L, Capitani S (1995): All-trans retinoic acid shows multiple effects on the survival, proliferation and differentiation of human fetal CD34+ haemopoietic progenitor cells. *Br J Haematol* 90: 274–282

Zenke M, Kahn P, Disela C, Vennstrom B, Leutz A, Keegan K, Hayman MJ, Choi HR, Yew N, Engel JD, Beug H (1988): v-erbA specifically suppresses transcription of the avian erythrocyte anion transporter (band 3) gene. *Cell* 52: 107–119

Zenke M, Munoz A, Sap J, Vennstrom B, Beug H (1990): v-erbA oncogene activation entails the loss of hormone-dependent regulator activity of c-erbA. *Cell* 61: 1035–1049

Zhang XK, Wills KN, Husmann M, Hermann T, Pfahl M (1991): Novel pathway for thyroid hormone receptor action through interaction with jun and fos oncogene activities. *Mol Cell Biol* 11: 6016–6025

5

Oncogenes as Transcriptional Regulators
Vol. 1: Retroviral Oncogenes
ed. by M. Yaniv and J. Ghysdael

Rel Proteins and Their Inhibitors: A Balancing Act

MARY LEE MACKICHAN AND ALAIN ISRAËL

Introduction

Cell differentiation and proliferation are influenced by the extracellular environment and internal cell state. Transcription factors play a central role in integrating intra- and extracellular cues and assuring appropriate gene expression. The rel family of transcription factors, initially described as a DNA binding activity called NF-κB, mediate responses to a large number of stimuli in many cell types. These transcription factors are important regulators of immune system function and of HIV transcription (recent reviews include Baeuerle and Henkel, 1994; Gilmore, 1995; Siebenlist et al, 1994; Verma, 1995). Certain rel proteins and their regulatory subunits also play a role in lymphoid and myeloid proliferation and differentiation and others have been implicated in oncogenesis.

The activity of transcription factors is highly regulated at multiple levels. A novel mechanism of regulation via specifically associated inhibitory subunits has been described for rel transcription factors. These inhibitors, or IκBs, retain dimers of rel proteins such as NF-κB in the cytoplasm until an appropriate signal induces release of the rel dimer to the nucleus. The release of rel dimers from inactive cytoplasmic complexes with at least one such inhibitor, IκBα, involves the phosphorylation, ubiquitination and proteolysis of the IκB subunit. This mechanism allows rapid induction of rel-dependent transactivation, via specific DNA sequences called κB sites, without the need for intervening protein synthesis. As described here, evidence from many systems suggests that the balance between the IκB inhibitors and rel activators is a critical determinant of nuclear κB binding activity and transcription of target genes.

The rel Proteins

The rel homology domain (RHD)

The study of the 5′ regulatory regions of the major histocompatibility complex (MHC) Class I H-2K^b promoter and the immunoglobulin (Ig) kappa light chain enhancer led to the identification of two DNA binding activities, termed KBF1 and NF-κB, with similar sequence specificity (Israël et al, 1989a; Lenardo et al, 1987; Sen and Baltimore, 1986a, b). The isolation of cDNAs encoding the p50 subunits of NF-κB and KBF1 demonstrated the two activities share a common subunit (Ghosh et al, 1990; Kieran et al, 1990). KBF1 is a homodimer of p50, while NF-κB is a heterodimer of p50 and a second rel family protein called p65 (or relA). Surprisingly, the cDNA encoding p50 has an open reading frame predicting a 105-kd protein with p50 encoded in the N-terminal half of the molecule. The sequence of the p105/p50 gene revealed similarity in the N-terminal p50 half to a previously described proto-oncogene, c-*rel* and a *Drosophila* maternal morphogen, *Dorsal*. These results established the existence of a multi-gene family of transcription factors known as rel proteins.

Five genes containing the 300-amino acid rel homology domain (RHD) have been cloned in mammalian cells (Figure 5.1). The RHD contains DNA binding and dimerization domains, as well as a nuclear localization signal (NLS) (Blank et al, 1991; Henkel et al, 1992). All rel proteins, with the exception of relB, can freely form homo- and heterodimers with other family members. The crystal structure of a DNA-bound p50 homodimer reveals a structural similarity to the immunoglobulin superfamily (Ghosh et al, 1995; Muller et al, 1995a). The crystal structure also confirmed earlier work localizing the dimerization domain to the more C-terminal portion of the RHD, with important DNA contacts concentrated in the N-terminal portion of the RHD, but also present in the dimerization domain (Kieran et al, 1990; Kumar et al, 1992; Logeat et al, 1991).

Outside the RHD, rel proteins exhibit little conservation of sequence. However, they can be divided into two classes, the transactivators and the precursors. The transactivators are p65, c-rel and relB; the precursors are p105 and p100, which are proteolytically processed to the DNA-binding subunits p50 and p52, respectively. The C-termini of the two precursors are homologous and consist of multiple ankyrin repeats also found in the IκBs. In their full-length form, the precursors do not bind DNA and are generally cytoplasmic. The corresponding DNA-binding subunits, produced by proteolysis of the C-terminal portion of the precursors, consist essentially of a RHD, with little transactivating potential of its own.

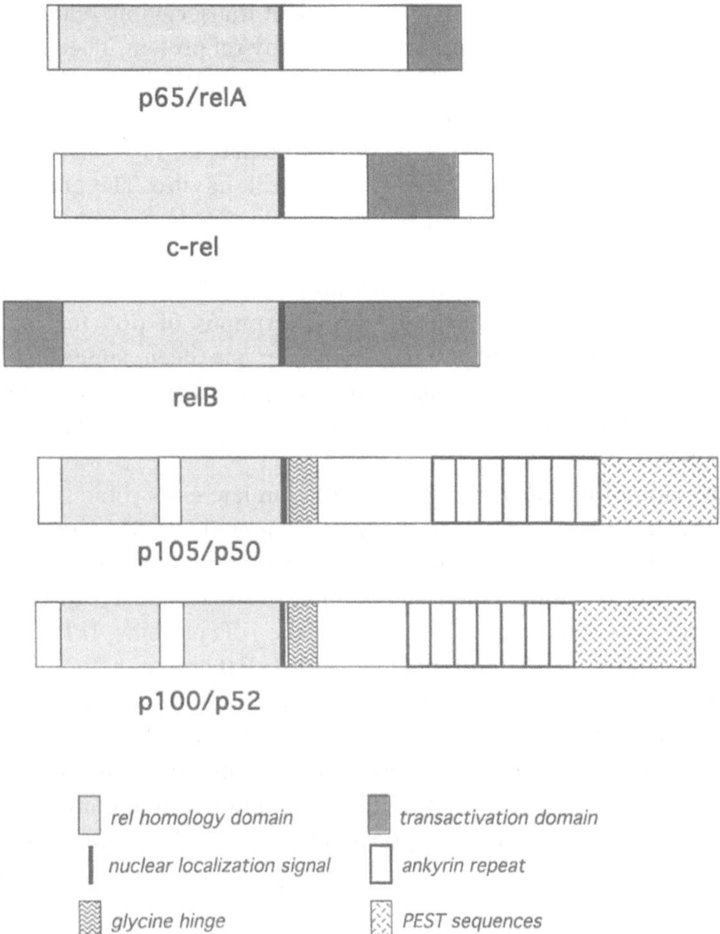

Figure 5.1 The five known members of the rel family in vertebrates. The various functional domains discussed in the text are indicated. The p105 and p100 precursors give rise to the p50 and p52 DNA binding subunits by processing.

The transactivators

Although p65, c-rel, and relB all effectively activate transcription from κB sites, usually in heterodimers with p50 or p52, each has a specific role in the regulation of κB-dependent transcription, as reflected by their differential expression and the diverse phenotypes observed in mice with targeted disruption of each of the corresponding genes.

Nuclear κB binding activity is usually induced within minutes of treatment of cells with rel activators (Baeuerle and Henkel, 1994; Siebenlist et al, 1994). In most assays of κB-dependent transcription, cotransfected p65 induces the greatest transactivation of any rel protein. The cloning of the gene encoding p65 showed it is composed of a RHD and a 250-amino acid C-terminal extension, and is able to bind κB sites as a homodimer in the absence of p50 (Nolan et al, 1991; Ruben et al, 1991; Schmid et al, 1991), although such binding is relatively weak in vitro. The greater transcription activating potential of p65 relative to p50 was shown to depend on the C-terminal sequences (Fujita et al, 1992). Further structure-function analysis has identified two (Schmitz et al, 1994) or three (Blair et al, 1994) related acidic motifs in the C-terminus of p65 that activate transcription synergistically. DNA bending has been suggested as a mechanism of transcription enhancement by p65 (Schreck et al, 1990), as has direct binding to TFIIB and TATA binding protein (TBP) (Blair et al, 1994).

Disruption of the p65/relA gene is lethal in late embryonic stages, and is accompanied by apoptosis in the liver (Beg et al, 1995b), suggesting a possible role for p65 in hepatocyte proliferation or survival. Embryonic fibroblasts from homozygous p65 mutants are less responsive to TNF than their wildtype counterparts. The levels of the IκB proteins, IκBα and -β, were reduced in the p65 null mutants, with IκBβ protein being undetectable (Beg et al, 1995b). Conversely, in transgenic mice overexpressing p65/relA in the thymus, a compensating increase in IκBα was observed, and no effect on thymocyte development or lymphocyte responses was seen (Perez et al, 1995). The difficulty encountered in attempting to experimentally alter the balance of p65 and IκBα is likely due to two characteristics of their interaction. First, free IκBα is rapidly degraded in unstimulated cells and its half-life is greatly increased by interaction with p65 (Rice and Ernst, 1993). Second, IκBα transcription is regulated in large part by rel activity, as discussed below.

Several studies using p65 antisense RNA have documented functional effects of reducing p65 protein levels, notably altered cell adhesion properties and loss of tumorigenicity (Higgins et al, 1993; Narayanan et al, 1993; Perez et al, 1994; Sokoloski et al, 1993). Antisense RNA to p65 caused the regression of fibroblastic tumors induced by ectopic expression of the *tax* gene of human T-cell leukemia virus (HTLV-1), which induces nuclear NF-κB (Kitajima et al, 1992). Similarly, the tumorigenicity of a fibrosarcoma cell line was abolished by p65 antisense expression (Perez et al, 1994). The target genes affected by the reduction in p65 levels were not identified in all studies, but fibrosarcoma cell adhesion was recovered in the presence of p65 antisense upon addition of exogenous transforming growth factor β1 (TGFβ1) (Perez et al, 1994). A connection between rel activity and regulation of cell adhesion is also suggested by the finding that p50 antisense inhibits cell adhesion in differentiating embryonic stem (ES) cells (Narayanan et al, 1993), and phorbol ester-induced adhesion of HL-60 cells is inhibited by excess κB oligo (Eck et al, 1993). A splice variant of p65 (Δp65) reportedly transforms rat embryo fibroblasts (Narayanan et al, 1992, but see Grimm and Baeuerle, 1994).

Another transactivator, c-rel, is the cellular homolog of v-rel, the transforming protein of Rev-T, an avian retrovirus (Stephens et al, 1983; Wilhelmsen et al, 1984) that likely transforms a common progenitor of dendritic cells and neutrophils in chicken (Boehmelt et al, 1992). Although expression of neither v- nor c-rel has been shown to transforms human cells in culture, c-rel has been mapped to a common site of chromosome rearrangement in non-Hodgkin's lymphomas and found to be altered or amplified at a significant frequency in several other lymphomas (Houldsworth et al, 1996; Lu et al, 1991).

The gene encoding c-rel was cloned more than a decade ago (Brownell et al, 1985; Wilhelmsen et al, 1984), but it was not until the cloning of the p105/p50 gene that c-rel was identified as a transcription factor specifically binding κB sites (Kabrun et al, 1991; Kieran et al, 1990). The C-terminal domain of c-rel, like that of p65, contains a potent acidic transactivation domain (Bull et al, 1990), and c-rel may also interact directly with TBP (Kerr et al, 1993). Interestingly, v-rel is truncated in the middle of the transactivating domain, leading to the suggestion that transformation by v-rel might depend on repression of transcription by binding of v-rel to κB sites (Ballard et al, 1990). However, others have found that transformation by v-rel requires the remaining transactivating sequences (Sarkar and Gilmore, 1993). Thus other mechanisms, including altered interaction with IκBs and DNA, likely also contribute to the transforming properties of v-rel (see Siebenlist et al, 1994 for review, and below).

The expression of c-rel is highest in lymphoid tissues in mammals (Brownell et al, 1987; Grumont and Gerondakis, 1990a), in contrast to that of p65, which is ubiquitous (Nolan et al, 1991). High levels of c-rel expression in the developing chick embryo are reported to be localized to cells undergoing programmed cell death, and although overexpression of c-rel transforms avian fibroblasts, similar overexpression in bone marrow cells induces apoptosis (Abbadie et al, 1993). Conversely, v-rel has been shown to block apoptosis induced by several types of stimuli in transformed cells, which may be another mechanism of v-rel oncogenesis (Nieman et al, 1991; White and Gilmore, 1993). However, inactivation of murine c-rel by homologous recombination did not interfere detectably with apoptosis, and development including that of hematopoietic cells, is normal in these mice (Kontgen et al, 1995). The proliferation of murine B- and T-cells in response to mitogens (LPS and ConA, respectively) is defective in c-rel$^{-/-}$ mice. Deficient T-cell proliferation could be corrected by addition of exogenous interleukin-2 (IL-2), suggesting that the critical requirement in this cell type is for c-rel's contribution to IL-2 transcription (Kontgen et al, 1995).

The search for other genes encoding a RHD of the rel family led to the cloning of relB (also called I-rel) (Ruben et al, 1992; Ryseck et al, 1992). RelB has a unique 100-amino acid domain N-terminal to the RHD, as well as unconserved C-terminal sequences, both of which contribute to its trans-activing potential (Dobrzanski et al, 1993). RelB heterodimerizes only with p50 and p52 and is unable to form homodimers (Ryseck et al, 1992).

RelB is highly expressed in lymphoid cells and interdigitating dendritic cells (Carrasco et al, 1993; Lernbecher et al, 1993). Several studies find constitutively nuclear relB in lymphoid tissues, which may reflect the low affinity of relB-containing dimers for IκBα (Dobrzanski et al, 1994; Lernbecher et al, 1994; Weih et al, 1994). The phenotype observed upon targeted disruption of relB supports the idea that its expression, although dispensible for lymphocyte development, is required for the development of dendritic cells and related cells in the thymic medulla and for T-cell-mediated immunity (Burkly, 1995; Weih et al, 1995). RelB$^{-/-}$ mice have a deficit in antigen presenting cells (APCs) and an excess of myeloid cells, perhaps indicating a role for relB in cell-fate determination of certain APCs (Burkly, 1995). The organs of relB$^{-/-}$ mice, notably liver and lung, are inflamed and expression of inflammatory cytokines is elevated (Weih et al, 1995), which may imply relB represses these genes in the wildtype immune system.

The precursors and their products

Following the RHD of the rel precursor proteins, p105 and p100, is a glycine-rich hinge region that links it to a C-terminal domain with homology to the IκBs. p50 and p52 are derived from their respective precursors by proteolytic degradation of the C-terminal half of the protein (Chang et al, 1994; Fan and Maniatis, 1991). The C-termini of the precursors are composed mainly of multiple ankyrin repeats, followed by a "PEST" domain. The ankyrin repeat is a 33-amino acid motif whose conserved residues are thought to confer a structure that facilitates protein-protein interactions, while the variable residues determine interaction specificity (Michaely and Bennett, 1992). Although ankyrin repeats are present in many proteins with diverse functions, those of the IκBs and rel precursors are closely related and constitute a subfamily (Hatada et al, 1992). (See

Blank et al, 1992; Michaely and Bennett, 1992 for reviews). PEST sequences are rich in P, E, D, S, and T residues, and their presence has been correlated with rapid turnover (Rechsteiner, 1990).

The p105/p50 subunit is ubiquitously expressed and, in contrast to most other rel activities, some p50/p50 homodimer is constitutively nuclear in nearly all cells. Within the N-terminal portion of the RHD of p50 is a 30-amino acid insert; a similar insert is present in the RHD of p52. The function of this domain is unknown, but crystal-structure results suggest it could be involved in DNA binding in association with high mobility group (HMG) proteins (Ghosh et al, 1995; Muller et al, 1995a). Although the capacity of p50 homodimers to enhance transcription has been a point of controversy, it appears p50 alone is at best a weak activator and may even repress transcription directly or indirectly (see Baeuerle and Henkel, 1994 for discussion).

Given the ubiquitous expression of p50 and its presence in NF-κB, the most commonly induced rel dimer, it is somewhat surprising that disruption of the p105/p50 gene causes no developmental abnormalities and lymphocyte numbers are normal (Sha et al, 1995). The constitutively nuclear NF-κB present in normal B-cells is absent, and there does not appear to be a compensating constitutive binding activity by other rel proteins, as assayed in vitro. In B-cells, the levels of several other rel proteins are reduced rather than increased in the absence of p50, and Ig κ light chain and MHC Class I are expressed normally. However, the p50$^{-/-}$ mice are defective in some immune system functions, including B- and T-cell proliferation in response to mitogens, clearance of *S. pneumoniae*, and antibody production. In contrast to these deficits in immune response, p50 null mice were more resistant to a viral agent than wildtype, possibly due to increased interferon-β (IFNβ) production (Sha et al, 1995).

Structure-function studies demonstrated that the C-terminal domain of p105 is necessary for its cytoplasmic localization, probably because it masks the NLS in the RHD (Blank et al, 1991; Henkel et al, 1992). Thus, the C-terminus of p105 can be considered a covalently attached inhibitor for p50. The exact site of processing of the p105 precursor has not been determined, but this domain can confer susceptibility to proteolytic cleavage, though not degradation, on an unrelated protein (Lin and Ghosh, 1996). This finding suggests that an endoprotease for p105 exists that is distinct from the proteolytic activity that simultaneously degrades the C-terminus (probably the proteasome; see below).

In addition to serving as a precursor, p105 has a second function analogous to that of the IκBs as an inhibitor of other rel proteins. p105 forms complexes with other rel subunits, including p65, c-rel, and p50, in lymphoid and nonlymphoid cells (Mercurio et al, 1993; Naumann et al,

1993; Rice et al, 1992). p105 associates with a single rel subunit via its ankyrin repeats and the dimerization domains of the two RHDs (Rice et al, 1992). The cytoplasmic p105-containing complexes give rise to nuclear activity as p105 is processed to p50 (Rice et al, 1992). Although some processing of p105 occurs constitutively, this process also appears to be regulated by mechanisms similar to those regulating other IκBs.

p100 resembles p105 both in structure and function, serving as a precursor for p52 (Chang et al, 1994; Mercurio et al, 1992) as well as forming inactive cytoplasmic complexes with other rel subunits (Mercurio et al, 1993). The DNA binding specificity of p52 resembles that of p50 (Duckett et al, 1993; Perkins et al, 1992), and like p50, p52 is a poor transactivator in homodimer form, but is an effective one when coupled to p65 (Chang et al, 1994; Duckett et al, 1993). The expression of p100/p52 is more restricted than that of p105/p50, however, it is present at low levels in immature lymphoid cell lines and at a higher level in mature B- and T-lines (Chang et al, 1994). The constitutive rate of processing of p100 is reportedly slower than that of p105 in the same cell line, and p100 processing appears less responsive to stimuli (Mercurio et al, 1993; Sun et al, 1994b; Watanabe et al, 1994).

p100 was cloned from a chromosomal translocation associated with B-cell lymphoma (Neri et al, 1991). The oncogenic form of the gene initally described, *lyt-10*, lacks multiple C-terminal ankyrin repeats, and similar truncations of the p100 gene have been found in a variety of lymphoid tumors (Migliazza et al, 1994; Neri et al, 1991). One study found 2% of lymphoid malignancies had a similar alteration of p100 (Fracchiolla et al, 1993). Although the mechanism of transformation is not well-understood, the truncated proteins are nuclear (Migliazza et al, 1994) and may activate transcription inappropriately.

The IκBs

Initial purification of rel inhibitor subunits led to the isolation of two proteins with the ability to inhibit DNA binding of NF-κB in vitro, termed IκBα and -β (Baeuerle and Baltimore, 1988a, b; Zabel and Baeuerle, 1990). The gene for IκBα was rapidly cloned and has been studied extensively. Less is known about IκBβ, which was cloned more recently, or IκBγ, an inhibitor with limited tissue expression that is encoded by the p105 gene. There are reports, both published and unpublished, of the existence of other IκB molecules. Bcl-3 is an additional regulator of rel proteins and a putative human proto-oncogene. Although bcl-3 shares structural similarity to the IκBs it differs fundamentally in function.

IκBα

The gene encoding human IκBα, also called MAD3, was cloned as an immediate early gene induced upon adherence of monocytes (Haskill et al, 1991). IκBα is a 37-kd protein containing five central ankyrin repeats and short N- and C-terminal sequences (Figure 5.2). A short stretch of acidic amino acids separates the last two ankyrin repeats, a feature also found in the precursors, and the IκBα C-terminus beyond the repeats is rich in PEST residues. IκBα has been suggested to retain associated rel dimers in the cytoplasm by masking their NLS (Beg et al, 1992; Ganchi et al, 1992; Zabel et al, 1993). More recently, the solution of the p50 homodimer crystal structure reveals a groove created by the interface of the dimer subunits, including their NLS, where an IκB molecule might bind (Ghosh et al, 1995; Muller et al, 1995a). A stoichiometry of one IκBα per rel dimer has been deduced (Ernst et al, 1995). IκBα binds preferentially to p50/p65 or p50/c-rel heterodimers, and although it can associate with p50 or p52 homodimers, it does not inhibit binding of the latter to a κB site in vitro (see Siebenlist et al, 1994 and references therein).

IκBα has a short half-life in most cells, in comparison to either precursor (Rice and Ernst, 1993). Somewhat surprisingly, the half-life of IκBα in unstimulated cells is also shorter than that of cytoplasmic NF-κB, indicating that a rel dimer interacts sequentially with multiple IκBα molecules without undergoing detectable nuclear translocation. Following stimulation not all p65 is translocated to the nucleus but is found associated with newly synthesized IκBα (Rice and Ernst, 1993). IκBα is stabilized and protected from degradation by its association with p65 (Scott et al, 1993; Sun et al, 1993), and nearly all IκBα in unstimulated cells is complexed either to that protein or c-rel, suggesting that free IκBα is rapidly degraded (Rice and Ernst, 1993).

Recently, the inhibition by IκBα of binding to DNA by rel dimers has been shown to depend on the presence of the IκBα C-terminal domain following the ankyrin repeats (Jaffray et al, 1995; Sachdev et al, 1995). The crystal structure of p50/p50 and previous data suggest IκBα binds the side of a rel dimer opposite the DNA binding contacts. Thus it is not obvious how interaction with IκB subunits alters rel DNA binding activity, nor is it clear that this effect is physiologically relevant, as complexes with IκBα are almost entirely cytoplasmic.

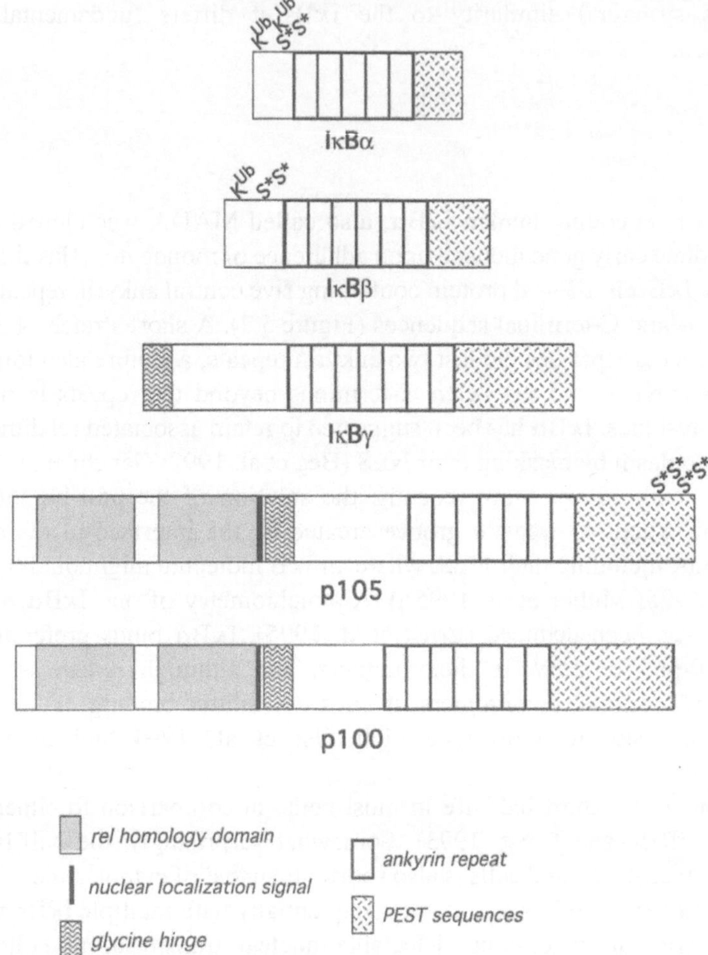

Figure 5.2 A schematic representation of the mammalian IκB proteins thus far isolated. The various functional domains discussed in the text are indicated. The Serine residues which are phosphorylated in response to signalling and are necessary for the inducible degradation of IκBα and IκBβ and for the regulated processing of p105 are represented by S*S*. The Lysines which represent the major sites of ubiquitination of IκBα and IκBβ in response to signalling are represented by K^{ub}.

However, overexpressed IκBα has been observed in the nuclear fraction (Cressman and Taub, 1993; Zabel et al, 1993) and in vitro IκBα has been shown to dissociate rel proteins from κB sites (Zabel and Baeuerle, 1990). On the basis of these observations, it has been proposed that IκBα may also strip rel proteins from κB sites in vivo (Zabel et al, 1993). A recent report has shown some endogenous IκBα can be detected in the nuclear fraction when it is resynthesized following stimulus-induced degradation (Arenzana-Seisdedos et al, 1995).

Targeted disruption of the murine IκBα gene reveals its importance in limiting the response to activation of κB binding activity (Beg et al, 1995a; Klement et al, 1996). Nuclear rel activity is observed constitutively in the hematopoietic tissues of IκB$\alpha^{-/-}$ mice (Beg et al, 1995a). Fibroblasts contain only cytoplasmic rel activity, however, probably associated with other IκBs, and these rel dimers are rapidly released to the nucleus upon stimulation. In the wildtype situation nuclear rel activity disappears within hours of activation, probably due to rel-dependent resynthesis of IκBα. The consequences of removing this negative feedback loop in the IκB$\alpha^{-/-}$ mouse is prolonged activation of nuclear rel, which may explain the elevated tumor necrosis factor alpha (TNF) expression in the skin and dermatitis of these mice (Klement et al, 1996). Myelopoiesis appears to be disregulated in the IκB$\alpha^{-/-}$ mice, perhaps due to elevated constitutive cytokine production, notably of granulocyte colony stimulating factor (G-CSF). Despite a normal appearance at birth, IκB$\alpha^{-/-}$ mice fail to grow and die within a week. Mice lacking both IκBα and p50 developed similar problems, although the onset is delayed (Beg et al, 1995a).

IκBβ

IκBβ, like IκBα, is ubiquitously expressed and associates with p65 and c-rel-containing dimers rather than p50 or p52 homodimers (Thompson et al, 1995). It is also degraded in response to stimuli that induce IκBα degradation. However the responses of the two inhibitors are not identical, as the kinetics of IκBβ degradation are slower, and IκBβ is not immediately resynthesized (Thompson et al, 1995). Thompson et al reported that phorbol ester and TNF stimulation did not affect IκBβ, but induced rapid degradation of IκBα. The absence of degradation of IκBβ was suggested to account for the transience of activation of NF-κB by these stimuli, compared to the long-lasting effects of LPS or interleukin-1 (IL-1) treatment on nuclear rel activity, both of which induce IκBα and IκBβ degradation (Thompson et al, 1995). However, other workers have observed degradation of IκBβ in cells treated with TNF or PMA (Didonato et al, 1996; R. Weil and AI, unpublished results).

Other IκBs

IκBγ is synthesized from an mRNA transcribed from the p105/p50 gene and is a 70-kd protein identical to the C-terminal 607 amino acids of p105 (Inoue et al, 1992; Liou et al, 1992). Expression of IκBγ is limited to certain murine lymphoid and erythroid cell types (Inoue et al, 1992); it has not been observed in human cells. IκBγ is cytoplasmic and, unlike IκBα or -β, associates with p50 homodimers as well as NF-κB in vitro (Hatada et al, 1992; Inoue et al, 1992; Leveillard and Verma, 1993; Liou et al, 1992). Alternately spliced forms of IκBγ have been identified; one is entirely nuclear and may be specific for p50 homodimers (Grumont and Gerondakis, 1994).

It has been suggested that the p100/p52 gene could give rise to a C-terminal inhibitor similar to IκBγ (Dobrzanski et al, 1995). A bacterially expressed protein containing the p100 ankyrin repeats, called IκBδ, is the only IκB able to inhibit relB-containing dimers in vitro (Dobrzanski et al, 1995). However, there is currently no evidence for the existence of IκBδ in vivo.

A recent report announces the isolation of an IκB-like molecule, IκBR (for IκB-related), which is specifically expressed in human lung epithelium, skeletal muscle and heart (Ray et al, 1995). IκBR is a 52-kd protein containing just three ankyrin repeats (Ray et al, 1995). IκBR is able to inhibit DNA binding of p50 homodimers and p50/p65 in vitro and inhibits κB-driven reporter gene expression in transient transfection assays, suggesting the structural homology to IκBs may reflect a similar function in vivo, which remains to be elucidated.

Another novel IκB-like cDNA encoding two to three ankyrin repeats has been cloned (Albertella and Campbell, 1994). The corresponding mRNA is expressed in lymphoid and myeloid cells and hepatocytes. Yet another cDNA with homology to the IκBs, IκBε, has been isolated in our laboratory from a yeast two-hybrid screen using p100/p52 as bait and represents the major c-rel and p65 associated IκB species in several cell types (S. Whiteside, J. C. Epinat and A. Israël, unpublished data).

Bcl-3

Bcl-3 was cloned from a common translocation break point in chronic lymphocytic leukemia (CLL) (Ohno et al, 1990). It contains seven ankyrin repeats, which resemble those of p105 and IκBα more closely than those of non-IκB ankyrin repeat-containing proteins (Hatada et al, 1992), and short serine- and proline-rich N- and C-terminal sequences (Ohno et al, 1990). In the CLL translocation, the bcl-3 coding sequences are intact, but its expression is increased by translocation into the IgH chain locus, which may contribute to oncogenesis (Ohno et al, 1990). Bcl-3 associated with the E4TF1/GABP transcription factor complex has recently been reported to transactivate the retinoblastoma (Rb) gene, and antisense bcl-3 RNA suppressed induction of Rb and muscle cell differentiation (Shiio et al, 1996). This recent result suggests bcl-3′s effects on cell proliferation and differentiation may be independent of rel proteins.

Bcl-3 mRNA is present in a wide range of tissues and is most highly expressed in spleen and liver (Nolan et al, 1993). In vitro experiments suggested its function might be similar to that of IκBs. Bcl-3 can inhibit DNA binding of rel proteins, particularly p50 or p52 homodimers (Franzoso et al, 1993; Hatada et al, 1992; Kerr et al, 1992; Nolan et al, 1993; Wulczyn et al, 1992), and bcl-3 association with homodimers of p50 or p52 in the absence of DNA was shown using in vitro translated proteins (Wulczyn et al, 1992) and in transiently transfected cells (Franzoso et al, 1993; Nolan et al, 1993). However, bcl-3 is a nuclear protein (Franzoso et al, 1993; Nolan et al, 1993), and thus does not fit the model of IκBs as cytoplasmic inhibitors of rel translocation to the nucleus.

The exact role of bcl-3 in regulation of rel activity is somewhat unclear, and may be dual. In some transient transfection assays, bcl-3 reduces transactivation of a cotransfected κB-dependent reporter gene (Kerr et al, 1992); more often, it increases such activity (Bours et al, 1993; Franzoso et al, 1992, 1993; Fujita et al, 1993), perhaps as a function of the amount of bcl-3 expressed or the cotransfected rel proteins. Two mechanisms have been proposed to account for increased transactivation by bcl-3. In the first scenario, bcl-3 removes constitutively nuclear p50 homodimers from κB sites, thus "derepressing" transcription of target genes by allowing more transcriptionally active dimers such as NF-κB to bind (Franzoso et al, 1992, 1993; Leveillard and Verma, 1993). In the second, bcl-3 provides a transactivation domain via association with the p50 or p52 homodimers bound to DNA (Bours et al, 1993; Fujita et al, 1993).

These apparently contradictory hypotheses may be reconcilable. Transactivation of a κB reporter gene by bcl-3 with cotransfected p52 depends on the unique N- and C-terminal domains of bcl-3, while the association of bcl-3 and rel proteins depends on the ankyrin repeats (Bours et al, 1993), suggesting these sequences provide a surrogate transactivation domain. A construct consisting only of the ankyrin repeats of bcl-3 is able to reverse the inhibitory effect of p50 on NF-κB dependent transcription (Franzoso et al, 1993), lending support to the model in which bcl-3 acts by dissociating p50/p50 from DNA. In vitro, bcl-3 dissociates bound p50 homodimers from DNA more rapidly than p52 homodimers, leading to the suggestion that direct transactivation by bound bcl-3 may occur specifically with p52 (Bours et al, 1993). In addtion, bcl-3 is highly phosphorylated (Nolan et al, 1993), and variable phosphorylation may account for inconsistent experimental results obtained with p50 homodimers.

Rel Target Genes

Rel proteins have been shown to contribute to the transcriptional activation of a large number of genes (see Table 1 adapted and appended from reviews by Baeuerle and Henkel, 1994; Grilli et al, 1993; Kopp and Ghosh, 1995; Siebenlist et al, 1994). The cellular proteins encoded by these genes can be grouped into several classes: receptors involved in determining immune specificity, cytokines, acute phase response proteins, cell adhesion molecules, and transcription factors. Many viruses, notably HIV, also harness NF-κB to activate their own transcription.

Immunoreceptors

Rel proteins contribute to activation of transcription of many of the proteins involved in a specific immune response to antigen. The genes encoding antigen receptors on both T- and B-lymphocytes are positively regulated through κB sites (Jamieson et al, 1989; Sen and Baltimore, 1986a, b). Rel-dependent transcription also contributes to the cytokine-induced expression of Class I molecules of the major histocompatibility complex (MHC) and their associated subunit, β_2-microglobulin, and to that of the invariant chain of MHC Class II (Benoist and Mathis, 1990; David-Watine et al, 1990; Zhu and Jones, 1990).

Expression of interleukin 2 receptor α chain (IL2Rα) is also inducibly up-regulated by rel DNA binding activity, along with other transcription factors (Ballard et al, 1988; Leung and Nabel, 1988; Lowenthal et al, 1989; Pierce et al, 1995). This upregulation appears to be an important aspect of the transformation of T-cells by HTLV-1.

Cytokines

The transcription of the inflammatory cytokines TNF and interleukins -1, -6, and -8 are regulated by NF-κB (see Kopp and Ghosh, 1995 and references therein). A recent report adds IL-12 p40, a factor important for induction of T helper type 1 differentiation, to this list (Murphy et al, 1995). Several chemokines may also depend on rel proteins for their transcription, including IL8, Gro-α, -β, and -γ, macrophage inflammatory protein 1 alpha (MIP1α), MCP-1/JE, and Rantes (Siebenlist et al, 1994). Rel proteins also regulate factors promoting the growth and differentiation of hematopoietic cell lines, such as granulocyte/macrophage colony stimulating factor (GM-CSF), G-CSF, and M-CSF (Kopp and Ghosh,

Table 1. Rel target genes

Immunoreceptors	T-cell receptor (TCR) immunoglobulin κ light chain major histocompatibility complex (MHC) Class I β_2-microglobulin MHC Class II in variant chain interleukin 2 receptor α chain (IL2 Rα) tissue factor-I
Cytokines and growth factors	Interleukin-2(IL-2) IL-6 IL-8 IL-12 p40 IL-1 tumor necrosis factor α (TNF) lymphotoxin (LT) β-interferon (β-IFN) granulocyte colony-stimulating factor (G-CSF) granulocyte/macrophage-CSF (GM-CSF) macrophage-CSF (M-CSF) Gro-α, -β, -γ (MIP1α (MCP-1/JE) (RANTES) transforming growth factor β
Acute phase proteins	serum amyloid A precursor α1 acid glycoprotein angiotensinogen complement factors B, C3, and C4
Adhesion molecules	endothelial-leukocyte adhesion molecule (ELAM-1) vascular cell adhesion molecule-1 (VCAM-1) intracellular cell adhesion molecule-1 (ICAM-1)
Viruses	human immunodeficiency virus 1 (HIV-1) cytomegalo virus (CMV) herpes simplex virus 1 (HSV-1) adenovirus simian virus 40 (SV40)
Transcription factors and their regulators	Myc interferon regulatory factor I (IRF-I) A-20 c-rel IκBα p105 p100
Other	vimentin inducible NO synthase ferritin H chain

1995; Siebenlist et al, 1994). The activation of transcription of the antiviral cytokine β-interferon by dsRNA also involves κB-binding proteins.

Transcriptional activation of IL-2 expression, like effective T-cell activation, requires two signals, one from increased Ca^{2+} and the other mimicked by phorbol esters. The 300 bp IL-2 enhancer is similarly responsive and contains binding sites for multiple transcription factors, including NF-κB and NF-AT (Ullman et al, 1990), and a CD28 response element (CD28RE), which also binds rel family proteins (Bryan et al, 1994; Fraser et al, 1991; Ghosh et al, 1993). While NF-AT is the only binding activity that requires both signals for induction of nuclear DNA binding activity in cell lines, κB-dependent transcription in activated primary T-cells also requires both signals (Kang et al, 1992; Mattila et al, 1990). As mentioned above, the finding that stimulated T-cell proliferation in c-rel-deficient mice can be restored by addition of exogenous IL-2 suggests c-rel may be required for IL-2 transcription in vivo (Kontgen et al, 1995).

Acute phase response elements

The acute phase response is a systemic reaction to tissue injury or infection, and in contrast to adaptive immunity, this innate immune response does not require specific recognition of antigen by lymphocyte receptors (see Baumann, 1994 and Kopp and Ghosh, 1995 for reviews). The acute phase response involves the release of cytokines and chemotactins by activated macrophages, the production by hepatocytes of acute phase proteins, including complement components, and changes in vasodilation mediated by secreted factors. Changes in endothelial cell adhesion are also part of the response and allow migration of monocytes and lymphocytes into the site of infection or injury. Rel proteins contribute to the activation of genes in each of these categories, including that of the cytokines already listed above (Kopp and Ghosh, 1995).

The IL6 promoter, a critical mediator of the acute phase response, provides an example of synergistic transactivation of a promoter by rel proteins associated with other transcription factors. In reporter gene assays, transcription from the interleukin-6 promoter requires the κB site and is activated synergistically by NF-κB and C/EBPβ (also called NF-IL6), a transcription factor of the bZIP family proteins (Betts et al, 1993), which contain a basic DNA binding domain and a leucine-zipper motif. This synergy likely depends on the direct interaction of the RHD and bZIP domain, which can occur in the absence of DNA (LeClair et al, 1992; Matsusaka et al, 1993; Stein et al, 1993). A recent report suggests NF-κB and C/EBPβ mediate the repression of the IL-6 promoter by estrogen receptor (ER), although the promoter has no ER binding site, suggesting rel or bZIP proteins also interact directly with ER (Stein and Yang, 1995). (For a comprehensive review of interactions of rel proteins with other transcription factors, see Gilmore, 1995).

Cell-adhesion molecules

The cytokine-induced transcription of three cell adhesion molecules, E-selectin or ELAM-1 (endothelial-leukocyte adhesion molecule-1), VCAM-1 (vascular cell adhesion molecule-1), and ICAM-1 (intracellular adhesion molecule-1), requires κB sites in each promoter (reviewed in Collins et al, 1995). The ELAM-1 promoter resembles that of the IFN-β promoter, both are synergistically activated by rel and ATF-2, a bZIP protein, via interaction of each with HMGI(Y) (Du et al, 1993; Whitley et al, 1994). The association between RHD proteins and HMG factors is highly conserved, as Dorsal also interacts with a *Drosophila* homolog of HMG, Dsp (dorsal switch protein), which is required for dorsal-mediated repression of transcription (Lehming et al, 1994).

Viruses

Human immunodeficiency virus 1 (HIV-1) contains two κB sites, which activate transcription from the viral LTR (Nabel and Baltimore, 1987). These sites mediate responsiveness of viral transcription to most stimuli that activate T-cells, including cytokines, antigen, and infection by other viruses (Gimble et al, 1988; Israël et al, 1989b; Osborn et al, 1989; Tong-Starksen et al, 1989). The HIV-2 LTR, which has a single κB site, is less responsive to T-cell activation signals than HIV-1 (Tong-Starksen et al, 1990) and clinical progression may be slower.

Although NF-κB is a critical element for HIV transcription, other factors contribute as well, including SP1, which has three binding sites in the LTR, and the viral gene product Tat (Liu et al, 1992; Perkins et al, 1993; Zimmermann et al, 1991). Interaction of p65 with SP1, a ubiquitously expressed zinc-finger protein, synergistically activates the HIV-1 enhancer (Perkins et al, 1993). This interaction is direct, depending on the RHD and a zinc finger of SP1, and specific, as the RHD domain does not interact with zinc fingers of MyoD, E12, or Kox15 (Perkins et al, 1994).

Two other target genes regulated by NF-κB participate in regulating cell redox state, which appears to be a critical factor for rel activation. Nitric oxide (NO) synthase II (iNOS) contributes to the antioxidant capacity of LPS-activated macrophages (Xie et al, 1994), and recent results suggest NO can inhibit activation of NF-κB by stabilizing IκBα (Peng et al, 1995). TNF-induced expression of ferritin H chain, which helps maintain iron homeostasis, has also been shown to depend on κB sites (Kwak et al, 1995).

Transcription factors and IκBs

Rel activity or κB sites have been shown to participate in the expression of the transcription factors c-Myc, interferon response factor 1 (IRF1), and A20 (Duyao et al, 1990; Krikos et al, 1992). The transcription of each of these is upregulated by T-cell activating stimuli, or by TNF in the case of A20, and their protein products in turn regulate proliferation (c-Myc), resistance to apoptosis (A-20), and the immune response (IRF1).

The promoters of p105, p100, and IκBα all contain κB sites and are regulated by nuclear rel activity, as is that of c-rel (DeMartin et al, 1993; Hannink and Temin, 1990; LeBail et al, 1993; Ten et al, 1992). In contrast, p65 expression is unaltered by κB-binding activity. Newly synthesized IκBα replenishes the pool of inhibitor within 1–2 h following stimulus-induced degradation (Brown et al, 1993, DeMartin et al, 1993; Sun et al, 1993). The importance of this negative feedback loop, and the critical role of IκBα is underlined by the effects of IκBα gene disruption, as described above.

Expression of the p105 and p100 precursors has also been shown to be upregulated in response to various rel-activating stimuli via κB sites in the promoters of each gene (Cogswell et al, 1993; Liptay et al, 1994, Ten et al, 1992). In cells stimulated with IL-2, a threefold increase in p105 mRNA was observed within 4 h (Arima et al, 1992). p105 mRNA levels remained elevated at least 12–24 h in this model and in an LPS-treated monocytic cell line (Cordle et al, 1993). Expression of p100 mRNA has been shown to be upregulated by transiently transfected p65 (Sun et al, 1994b) and by treatment with PMA or TNF (Liptay et al, 1994), both of which activate the endogenous NF-κB. These regulatory loops may also be important to maintaining a proper ratio of inhibitors and rel subunits.

Rel Activating Stimuli

The activators of nuclear translocation of rel proteins and κB-dependent transcription are numerous, as detailed below and in Table 2 (adapted and appended from Baeuerle and Henkel, 1994; Grilli et al, 1993; Kopp and

Table 2. Rel activating stimuli

Mitogens	
B-cell	bacterial lipopolysaccharides (LPS) and other bacterial products antigen, anti-IgM CD40 ligand
T-cell	lectins (PHA, Con A) antigen, anti-TCR, -CD3, -CD2 B7.1, anti-CD28 calcium ionophores (ionomycin, A2837)
Other cell types	serum phorbol ester (PMA, TPA) diacyl glycerol (DAG) platelet-derived growth factor (PDGF) insulin
Cytokines and mediators of inflammation	tumor necrosis factor α (TNF) nerve growth factor (NGF) lymphotoxin (LT) interleukin 1 (IL-1) IL-2 leucocyte inhibitory factor (LIF) leukotriene B4 (LB4) prostaglandin E2 (PGE2) platelet activating factor (PAF)
Viruses/viral products	human immunodeficiency virus 1 (HIV-1)/gp160, (HIV protease?) dsRNA Epstein-Barr virus (EBV)/LMP Human T-cell leukemia virus I (HTLV-1)/Tax hepatitis B virus (HBV)/pX, MHBs cytomegalo virus (CMV) herpes simplex virus 1 (HSV-1) human herpes virus 6 (HHV-6)/(B701?) Sendai virus (SV) Newcastle disease virus
Stress	UV light γ-, X-irradiation oxidative stress (peroxides) hypoxia
Parasite	*Theileria parva*
Chemical agents	phosphatase 1 and 2A inhibitors (okadaic acid, calyculin A) protein synthesis inhibitors (e.g. cycloheximide, anisomycin) inhibitors of transport from the endoplasmic reticulum (e.g. tunicamycin, brefeldin A)

Ghosh, 1995; Siebenlist et al, 1994). They include mitogens, cytokines, viruses, as well as various environment stresses, such as exposure to UV light or gamma irradiation, and certain chemical agents. Many of these stimuli are associated specifically with activation or regulation of lymphoid and myeloid cells, reflecting the importance of rel family proteins in the immune system. However, some rel-activating stimuli are common to most cells, and the relative abundance of examples of rel activators in lymphoid cells may in part reflect the bias of study in this field.

Mitogens and differentiation

Many lines of evidence suggest that rel proteins specifically regulate the proliferation and differentiation of T- and B-lymphocytes. Results from relB$^{-/-}$ mice suggest it is required for the normal differentiation of antigen presenting cells and myeloid cells (Burkly et al, 1995; Weih et al, 1995). The lethality of the p65 knock-out (Beg et al, 1995b), and the compensating increase in endogenous IκBα when p65 is overexpressed (Perez et al, 1995) prevented conclusions concerning the contribution of p65 to proliferation and differentiation of most cell types.

B-cells

NF-κB was initially identified by virtue of the correlation between its nuclear DNA binding activity and expression of Ig kappa light chain in B-cell lines and not pre-B lines (Sen and Baltimore, 1986a, b), suggesting a role for rel proteins in the maturation of B-cells. In fact, mature B and some monocytic cell lines are the only cell types with constitutively nuclear NF-κB. However, recent data from gene disruption experiments show that at least the p50 subunit of NF-κB is not in fact required for kappa light chain expression or normal B-cell development (Sha et al, 1995). However, both p50 and c-rel are required for effective humoral immunity and B lymphocyte proliferation in response to bacterial products (Kontgen et al, 1995; Sha et al, 1995).

Comparative studies of pre-B, B, and plasmacytoma cell lines show that rel expression patterns change over the course of differentiation (Liou et al, 1994; Miyamoto et al, 1994a). Complexes in pre-B-cells consist primarily of p50/p65, which is exclusively cytoplasmic unless nuclear translocation is induced by treating cells with lipopolysaccharide (LPS) a bacterial cell wall component, or cytokines like IL-1 or TNF (Osborn et al, 1989; Sen and Baltimore, 1986b). LPS treatment of the 70Z/3 pre-B line activates nuclear rel and induces differentation to an IgM^{+} phenotype

(Paige et al, 1978), Pre-B-cell lines transformed by v-*abl*, which do not rearrange light chain genes or differentiate into B-cells, also lack constitutive nuclear rel activity, suggesting rel proteins might be needed for these steps (Kerr, 1994; Klug et al, 1994).

The constitutive κB binding activity observed in mature B-cells is composed principally of p50/c-rel dimers, not NF-κB. Additional nuclear rel activity, principally NF-κB, is induced in these cells by antigen or anti-IgM antibody treatment (Liu et al, 1991), as well as by the stimuli mentioned above. In plasmacytoma cell lines or 70Z/3 cells induced to differentiate by a three-day treatment with LPS, constitutively nuclear binding of p52 and relB is detected (Liou et al, 1994), again suggesting a role for rel proteins in the induction or maintenance of the terminally differentiated state.

The mechanism(s) accounting for constitutive nuclear rel activity in mature B-cells is not well defined. IκBα is expressed in B-cells and associates normally with rel proteins (except with relB, as mentioned above). In fact, expression of IκBα and p105 is higher in B-cell lines than in pre-B lines (Liou et al, 1994; Miyamoto et al, 1994a). However, the half-life of IκBα is significantly shorter in the B-cell lines, and this difference has been proposed to account for the constitutive nuclear p50/c-rel activity, which is estimated to amount to only 10% of the total rel activity in the cell (Liou et al, 1994; Miyamoto et al, 1994a). In support of this hypothesis, treatment of B-cells with inhibitors of IκBα degradation and precursor processing abolishes the constitutive nuclear κB binding activity (Miyamoto et al, 1994a). An alternate proposal, not necessarily in conflict with the first, is that B-cells activate rel via autocrine and paracrine stimulation by TNF and IL-1 (Baeuerle and Henkel, 1994; Siebenlist et al, 1994).

T-cells

Activation of resting T-cells induces proliferation as well as a complex differentiation program, ultimately involving the activation of transcription of over a hundred genes (Crabtree, 1989). The rel proteins appear to play an important role in the initiation and execution of this response. Both the signals provided by the TCR and an APC, or agents that mimic them (calcium ionophores and phorbol esters), induce rel DNA binding and transactivation to varying degrees. In most T-cell lines, phorbol esters are more effective inducers of rel activity than agents that increase intracellular calcium (Emmel et al, 1989; Tong-Starksen et al, 1989). However, in an antigen-specific T-cell clone, phorbol ester alone did not induce any new rel binding activity, while ionomycin alone did (Kang et al, 1992). In primary human T-cells, neither antibodies to CD3 (of the TCR) nor anti-CD28 (an APC signal) alone induced any rel DNA binding activity, but the two together were highly activating (Los et al, 1995).

Results obtained in mice with targeted disruption of rel subunits provide good evidence for their importance in activation of peripheral T-lymphocytes. Abnormalities in T-cell activation are apparent in mice lacking p50 (Sha et al, 1995), c-rel (Kontgen et al, 1995), or relB (Weih et al, 1995).

Other cell types

Treatment of most nonlymphoid cells with phorbol esters induces NF-κB, and some stimuli that increase proliferation of many cell types also activate rel, including serum (Bull et al, 1989; Grumont and Gerondakis, 1990b), insulin-like growth factor I (IGF-I) (Bertrand et al, 1995), and platelet-derived growth factor (PDGF) (Olashaw et al, 1992). Nerve growth factor, a cytokine with homology to TNF, was recently shown to activate NF-κB via the p75 neurotrophin receptor (Carter et al, 1996). The relevant target genes in nerve cells have not been identified, but might include extra-cellular matrix proteins. NGF may also link immune and nervous systems in an inflammatory response. Intriguingly, an autocrine loop involving NGF production was proposed to account for neurofibromas seen in transgenic mice expressing the Tax protein of human T-cell leukemia virus I (HTLV-I) (Green, 1991), an activator of NF-κB. More recently, memory B-cells were shown to depend on autocrine NGF for their survival (Torcia et al, 1996).

Cytokines

Among the best studied cytokines that induce rel activity are TNF and IL-1, which are both synthesized during an immune response. They have pleiotropic effecs that depend in part on their ability to induce expression of other cytokines (reviewed in Vassilli, 1992), largely mediated by stimulation of nuclear rel activity (Osborn et al, 1989). Because TNF is both an inducer of rel activity and a target gene, a feed-forward mechanism can amplify TNF production at a site of inflammation. A similar autocrine loop involving IL-2 (a rel activator) (Arima et al, 1992) and IL2Rα (a target gene), may accompany activation of rel by HTLV-I in T-cells (Yodoi and Uchiyama, 1992).

Other mediators of inflammation, prostaglandin E2 (PGE2), leukotrience B4 (LTB4), and platelet activating factor (PAF) are also reported to activate NF-κB (Brach et al, 1992; Kravchenko et al, 1995). However, PGE2 and LTB4 likely require the presence of an additional signal for their rel-inducing activity (Los et al, 1995; Muroi and Suzuki 1993). Interestingly, interleukin 10, which inhibits synthesis of multiple inflammatory cytokines in response to LPS, also inhibits LPS-induced activation of NF-κB (Wang et al, 1995).

Viruses and viral products

An early model for viral induction of κB-dependent transcription was the induction of the interferon-β (IFN-β) promoter by some negative-strand RNA viruses (Lenardo et al, 1989). Infection by many DNA and RNA viruses has since been found to activate rel-dependent transcription (see Table 2), perhaps by activating signal transduction pathways. These viruses are often lymphotrophic, and some can transform cells or are associated with increased cancer risk, notably HTLV-1, Epstein-Barr virus (EBV), and hepatitis B (HBV). Viral activation of rel may increase virus propagation directly by transactivating the viral genes, or indirectly by inducing the infected cell to proliferate.

HIV is a rel activator as well as a target of rel-dependent transactivation. In monocytic cell lines or macrophages chronically infected with HIV, NF-κB is constitutively nuclear (Bachelerie et al, 1991; Roulston et al, 1993). The nuclear rel activity in infected cells is probably the result of increased degradation of IκBα and increased proteolytic processing of p105, according to a recent report (McElhinny et al, 1995).

Several HIV gene products interact with rel/IκB proteins. The HIV-1 protease, for example, can proteolytically cleave p105 in vitro, though at a different site than the endogenous proteolytic activity (Riviere et al, 1991). HIV-1 can also stimulate nuclear rel activity via the interaction of the viral protein gp160 with cellular CD4 (Chirmule et al, 1994). The Tat protein of HIV enhances activation of NF-κB by TNF, probably by suppressing expression of Mn-dependent superoxide dismutase (Mn-SOD) and creating pro-oxidative conditions that may favor rel activation (Westendorp, 1995).

The Tax protein of HTLV-1 is required for immortalization of T-cells by the virus (Yodoi and Uchiyama, 1992). Although it does not bind DNA, Tax alters gene expression through its effects on rel and other transcription factors. In cells infected with HTLV-1 or expressing Tax, there is constitutive nuclear rel activity, and deregulated expression of IL-2 and IL2Rα in T-cells (Yodoi and Uchiyama, 1992) and of NGF in fibroblasts (Green, 1991). Tax protein likely activates rel through more than one mechanism;

one important effect is stimulation of phosphorylation and degradation of IκBα (Brockman et al, 1995; Kanno et al, 1995; Lacoste et al, 1995; Maggirwar et al, 1995; Sun et al, 1994a). Tax is also reported to interact directly with the ankyrin repeats of both precursor proteins and IκBγ (Beraud et al, 1994; Hirai et al, 1992, 1994; Munoz et al, 1994; Watanabe et al, 1993). This interaction is proposed to dissociate the inhibitors from their complexed rel proteins, allowing nuclear translocation (Hirai et al, 1994; Munoz et al, 1994; Watanabe et al, 1993). However, more recently Tax has been shown to associate in a trimolecular complex with p105 and a subunit of the proteasome and to increase p105 processing, suggesting Tax acts by targeting the precursor to the proteasome (Rousset et al, 1996).

Latent membrane protein 1 (LMP1) of Epstein-Barr virus is a transmembrane protein required for EBV transformation of B-lymphoblasts that induces nuclear rel activity (Hammarskjold and Simurda, 1992; Laherty et al, 1992). Activation of NF-κB by LMP1 appears to involve increased proteolytic turnover of IκBα and probably increased processing of p105 and p100, as reflected in elevated p50 and p52 levels (Herrero et al, 1995; Paine et al, 1995). The increased proteolysis of IκBs likely involves the assembly of a TNF receptor (TNFR)-like signaling complex, as LMP1 associates with signaling molecules that have homology to TRAFs, or TNFR associated factors (Mosialos et al, 1995).

pX, a protein product of hepatitis B virus, activates rel dependent transcription as well as the Ras/Raf/MAP kinase pathway (Doria et al, 1995). These two effects of pX are likely separate functions of the protein, as coexpressed dominant-negative Raf blocked pX activation of c-jun but not of nuclear rel (Chirillo et al, 1996). The mechanism of pX-induced rel activation may involve altering IκB levels, as expression of pX leads to decreased IκBα in HeLa cells (Chirillo et al, 1996). This down-regulation of IκBα protein has been suggested to involve interaction of pX with cellular proteases.

Stress stimuli and chemical agents

The infection of an organism by a bacterium or virus is a physiological challenge to homeostasis, and can be considered a stress stimulus. In keeping with the role of rel proteins in coordinating a response to such challenges, other less physiological stresses have also been shown to activate rel. These stress stimuli include agents that damage DNA, like UV light (Stein et al, 1989), γ-irradiation, and chemically induced oxidative DNA damage (Legrand-Poels et al, 1995). Extremes in environmental oxygen concentration can also activate NF-κB, whether induced by hypoxia or by H_2O_2 treatment (Koong et al, 1994; Schreck et al, 1991).

Treatment of cells with various chemicals that interrupt normal cell processes can induce nuclear translocation of rel transcription factors. A recent report suggests that protein accumulation in the endoplasmic reticulum (ER), as a result of treatment with inhibitors of glycosylation or of protein export or from overexpression of Igμ heavy chain, can activate rel translocation (Pahl and Baeuerle, 1995). The relatively slow kinetics of this activation suggest the stimulation of nuclear rel activity may be indirect, and the authors propose that generation of oxidative stress is involved (Pahl and Baeuerle, 1995).

Inhibitors of serine/threonine phosphatases (PP) types 1 and 2A, such as okadaic acid and calyculin A, are strong activators of rel (Thevenin et al, 1990). They induce phosphorylation and proteolysis of IκBα (Finco et al, 1994; Traenckner et al, 1994) and processing of p105 (MacKichan et al, 1996). These compounds may block the activity of IκB phosphatases, or their effect may be indirect. Treatment of cells with inhibitors of protein synthesis, such as cycloheximide (CHX), also activates nuclear κB-binding. This activation is relatively slow and is likely due to progressive loss of IκBα as it is rapidly turned over.

Regulation of rel and IκB Activities

Cytoplasmic rel inhibitors, both IκB subunits and precursor proteins, are subject to constitutive and/or stimulus-induced proteolysis. The likely site of the stimulus-induced proteolysis of IκBs is the 26S proteasome, which is responsible for regulated proteolysis of many cellular proteins and for the generation of antigenic peptides (reviewed in Deshaies, 1995; Jentsch 1995 and summarized in Figure 5.3). Proteins are targeted to the proteasome by a series of enzymatic reactions that covalently attach ubiquitin (Ub), a 76-amino acid peptide, to lysine residues of a protein substrate, and to Ub itself, tagging the protein with covalently bound multiubiquitin chains. There are many isoforms of the ubiquitin conjugating enzyme, E2, and a variety of primary protein sequence motifs have been described that target proteins for ubiquitination. In particular, PEST sequences appear to be necessary for ubiquitination of several proteins, including the yeast transcription factor Matα2p. Specific phosphorylation (or dephosphorylation) is required for ubiquitination of several proteasome substrates, e.g. p40^{SIC1} and c-mos (Deshaies, 1995; Peters, 1994). Stimulus-induced degradation of IκBs in the proteasome also requires phosphorylation, ubiquitination, and PEST-rich sequences, as discussed below.

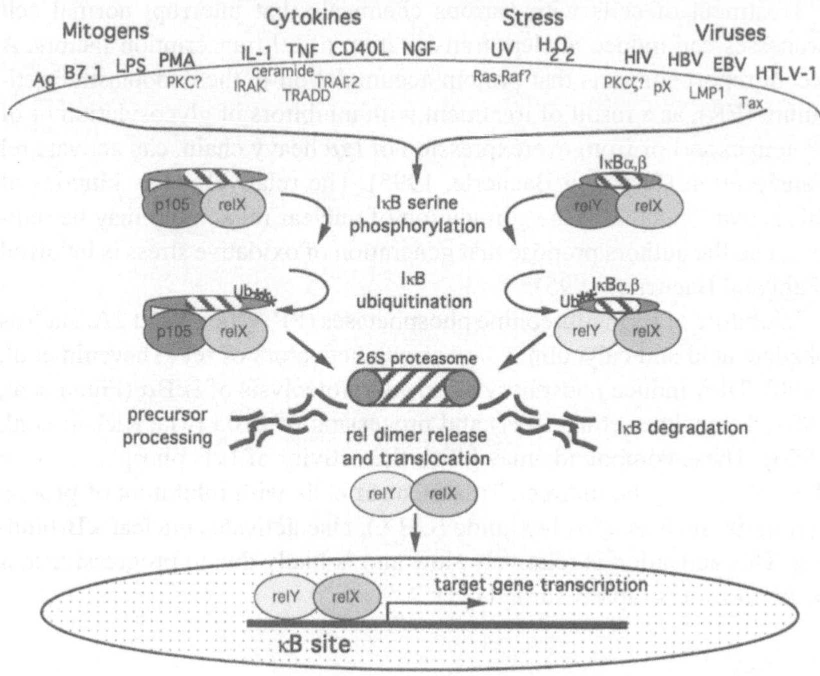

Figure 5.3 Pathways of NF-κB activation. Treatment of cells with NF-κB-activating stimuli (some of them are indicated at the top of the figure, with some known intermediates shown under the plasma membrane) induces a signal transduction cascade which is shown in parallel for the complexes containing the classical IκBα or -β inhibitors (associated with relX/relY dimers), or the p105/p100 IκB's (associated with a relX monomer). This cascade results in the induced phosphorylation of IκBα upon the two Serine residues situated in its N-terminus, and of p105 (and probably p100) on Serine residues located in the C-terminal PEST region. This is followed by ubiquitination and degradation in the case of IκBα and IκBβ, and by increased processing and degradation of the C-terminal region in the case of p105 and p100. Released rel dimers (relX/relY in the case of IκBα or -β, p50/relX or p52/relX in the case of p105 or p100, respectively), are then free to translocate to the nucleus to activate transcription of their target genes.

IκBα regulation

After much experimental effort, the regulation of stimulus-induced degradation of IκBα in the proteasome by phosphorylation and ubiquitination is now fairly well understood. A possible role for phosphorylation in the regulation of IκBα activity was suggested initially by experiments showing that in vitro phosphorylation of an IκB/NF-κB complex caused release of κB-binding activity, and that phosphorylation impinged on the function of IκBα and not NF-κB (Ghosh and Baltimore, 1990). In this model, phosphorylation was proposed to cause dissociation of IκBα from the cytoplasmic complex with NF-κB, allowing translocation of the transcription factors to the nucleus. The cloning of IκBα and generation of specific antibodies allowed the observation of the rapid degradation of IκBα following treatment of cells with various rel-activating stimuli (Beg et al, 1993; Brown et al, 1993, Cordle et al, 1993; Henkel et al, 1993; Sun et al, 1993). The importance of IκBα proteolysis in activation of NF-κB was underlined by the finding that antioxidants and certain serine protease inhibitors (notably *N*-toxyl-L-phenylalanine chloromethyl ketone, or TPCK) blocked both IκBα degradation and the inducible nuclear κB-binding (Beg et al, 1993; Henkel et al, 1993; Mellits et al, 1993). However, the possible role of phosphorylation in the regulation of stimulus-induced IκBα degradation remained obscure as TPCK blocks IκBα phosphorylation as well as proteolysis (Finco et al, 1994).

The role of phosphorylation was clarified following the description of protease inhibitors that specifically block proteolysis of IκBα and p105 processing, but not phosphorylation (Palombella et al, 1994). These peptide-aldehyde inhibitors block both proteasome and calpain protease activities, but as other inhibitors of calpain do not affect IκBs, the proteasome is likely to be the relevant target (Palombella et al, 1994). In stimulated cells pretreated with such proteasome inhibitors, a newly phosphorylated form of IκBα is readily detected and has been shown to remain bound to NF-κB (Alkalay et al, 1994, Didonato et al, 1994; Lin et al, 1995; Miyamoto et al, 1994b; Traenckner et al, 1994).

Two serines in the IκBα N-terminus, residues 32 and 36, are required for stimulus-induced phosphorylation and degradation of the protein and have been shown to be phosphorylated in vivo (Brockman et al, 1995; Brown et al, 1995; Didonato et al, 1996; Traenckner et al, 1995; Whiteside et al, 1995). These serines are conserved across IκBα proteins of several species and in IκBβ as well. Serines 32 and 36 are required for the inducible degradation of IκBα in response to all rel activating stimuli, including Tax.

Interestingly, substitution of serine 32 of IκBα by threonine also blocks stimulus-induced degradation, suggesting the IκBα kinase is not a serine/threonine kinase, but is highly serine-specific (Didonato et al, 1996). The kinase responsible for the phosphorylation of these residues has not been positively identified, but in vitro results suggest that a ubiquitin-dependent kinase activity may be responsible (Chen et al, 1996). Using purified fractions from unstimulated HeLa cells, Chen et al, have characterized an IκBα kinase activity that specifically phosphorylates the relevant serines and does not induce dissociation of IκBα from p65. As purified, the kinase activity is composed of multiple subunits in a large (700 kDa) complex. The kinase activity of the complex depends on the availability of ubiquitin and specific ubiquitin-activating (E1) and carrier proteins (E2) but not the proteasome (Chen et al, 1996). This is the first description of a regulatory function for ubiquitination other than targeting of proteins to the proteasome. The ubiquitin-dependent IκBα kinase activity has not been shown to be upregulated by stimuli that induce IκBα phosphorylation, and the data are equally consistent with the inactivation of a constitutive IκBα phosphatase in the purified enzyme complex.

IκBα is a substrate for ubiquitination in vitro and in vivo in response to rel activating stimuli (Chen et al, 1995). Mutation of serines 32 and 36 to alanine abolished ubiquitination in vitro, whereas mutation to glutamic acid did not, suggesting that the negative charge provided by phosphoryl groups is critical for ubiquitin conjugation (Baldi et al, 1996). Mutation of lysines 21 and 22, but not other lysines, has been shown to greatly reduce the stimulus-induced ubiquitination and proteolysis of IκBα, although the mutant protein is phosphorylated normally (Baldi et al, 1996; Didonato et al, 1996), suggesting these are probable sites of ubiquitin attachment in vivo.

As might be predicted from results with proteasome inhibitors, ubiquitinated IκBα remains bound to NF-κB and is degraded from the complex (Chen et al, 1995; Didonato et al, 1996). Interestingly, treatment of unstimulated HeLa cell extracts with okadaic acid in vitro induced phosphorylation and multiubiquitination of IκBα (Chen et al, 1995), suggesting that phosphorylation and dephosphorylation of the IκBα N-terminus may occur constitutively.

While IκBα PEST sequences are not the target of stimulus-induced phosphorylation (Aoki et al, 1996), they have been reported to be required for the ensuing proteolysis (Brown et al, 1995; Rodriguez et al, 1995; Whiteside et al, 1995). The C-terminal PEST domain of IκBα is constitutively phosphorylated (Ernst et al, 1995). Casein kinase II (CKII) is likely responsible for this constitutive S/T phosphorylation, on the basis of similar peptide maps generated in vivo and with CKII in vitro, and physical association of CKII with IκBα in vitro (Barroga et al, 1995) and in vivo (McElhinny et al, 1996). The role of PEST sequences in stimulus-induced proteolysis is somewhat controversial, however, as another study found they were dispensable for stimulus-induced degradation of IκBα, while the ankyrin repeats were required, possibly because they confer association with a rel dimer (Aoki et al, 1996). The lack of stimulus response observed by other groups in the absence of PEST sequences may reflect weaker affinity of human IκBα constructs for murine rel proteins.

Regulation of other IκBs

The inducible degradation of IκBβ appears to the regulated by mechanisms nearly identical to those controlling IκBα, although the kinetics of proteolysis differ. Both serine residues and one of the lysines required for stimulus-induced degradation of IκBα are conserved in the N-terminus of IκBβ, and mutation of these serines (at positions 19 and 23) in IκBβ blocks stimulus-induced degradation (Didonato et al, 1996; McKinsey et al, 1996).

The regulation of the precursor IκBs has been less studied than that of IκBα, but the available evidence suggests similar mechanisms of induced phosphorylation and selective proteolysis of ankyrin-repeat domains in the proteasome are involved. The signal-induced upregulation of p105 and p100 precursor processing is less dramatic than the rapid and total degradation of IκBα or -β and must be measured against a background of basal processing, making their study more difficult. Nonetheless, upregulated processing of p105 has been observed in response to LPS (Cordle et al, 1993; Donald et al, 1995; Liou et al, 1992) (but see Rice and Ernst, 1993). TNF (Mellits et al, 1993; Naumann and Scheidereit, 1994), T-cell activation (Bryan et al, 1994, MacKichan et al, 1996; Mellits et al, 1993; Mercurio et al, 1993), dsRNA (MacKichan et al, 1996; Mellits et al, 1993). HIV infection or expression of LMP1 of EBV or Tax of HTLV-1 have also been reported to increase the rate of p105 processing (McElhinny et al, 1995; Paine et al, 1995; Rousset et al, 1996).

The rate of both induced and constitutive processing may be cell-type dependent. LPS treatment reduced p105 levels in monocytes (Cordle et al, 1993, Donald et al, 1995) but not in 70Z/3 pre-B and WEHI231 B-cells (Rice and Ernst, 1993). Evidence suggests B-cells may already have upregulated processing of p105 (Naumann et al, 1993) and be somewhat refractory to further stimulation. In contrast, both p100 and p105 precursors are reported to be processed at lower rates in Jurkat T-cells than in HeLa cells (Mercurio et al, 1993).

The possibility that stimulus-induced upregulation of p105 processing to p50 may be regulated by phosphorylation was suggested by observations of increased phosphorylation of p105 in response to several stimuli. TNF treatment and T-cell activation each induce rapid phosphorylation of p105 and increase processing within 30 min (Li et al, 1994b; MacKichan et al, 1996; Mellits et al, 1993; Naumann and Scheidereit, 1994). Like IκBα, phosphorylated p105 remains associated with bound rel subunits (Li et al, 1994b; Mellits et al, 1993; Naumann and Scheidereit, 1994). Treatment of cells with phorbol ester, or phorbol ester and PHA, or hydrogen peroxide (H_2O_2) also increase phospholabeling of p105 (Bryan et al, 1994; Li et al, 1994b; Mellits et al, 1993; Naumann and Scheidereit, 1994; Neumann et al, 1992).

Despite many observations of changes in p105 phosphorylation and processing, a link between the two phenomena has only recently been established. Two recent reports show that serine phosphorylation of p105 increases the susceptibility of the protein to specific proteolysis of the C-terminal domain. Phosphorylation of p105 in response to T-cell activating stimuli has been localized to multiple serines in the PEST domain (MacKichan et al, 1996). Both the inducible phosphorylation and increased processing of p105 to p50 were abolished by a 80-amino acid C-terminal truncation, suggesting that the stimulus-induced phosphorylation of serines in this region is required for up-regulated processing (MacKichan et al, 1996). The inducibly phosphorylated serines in the C-terminus were not identified; of the 70 relevant amino acids, ten are serine residues.

Another study reports that Ser-Pro sites in the p105 PEST domain are phosphorylated in vitro by cyclin-dependent kinase purified from mitotic HeLa cells (Fujimoto et al, 1995). Mutation of the serines identified in vitro, residues 894 and 908, increased the half-life of p105 expressed in COS7 cells both in unstimulated cells and in those exposed to UV irradiation or treated with cycloheximide (Fujimoto et al, 1995). However, it is unclear whether the greater stability of the mutant p105 reflects a block in processing or in protein turnover, as p50 levels were not examined directly.

p105 is a substrate for ubiquitination in vivo. Recently, some of the biochemistry involved has been elaborated in vitro. A specific E2 and a novel ubiquitin-ligase (E3) were found to be required for ubiquitination and processing of p105 in a cell-free system (Orian et al, 1995).

Less information is available concerning the regulation of the p100 precursor. The rate of p100 processing is generally reported to be lower than that of p105. The rate of p100 processing was 30% that of p105 in PMA-stimulated HeLa cells and PMA/ConA-stimulated Jurkat cells (Mercurio et al, 1993). However, TNF treatment of HeLa cells induced the phosphorylation and processing of p100 within 10 min, i.e. with kinetics identical to those of p105 (Naumann and Scheidereit, 1994). This response is apparently cell-type dependent, as no effect on p100/p65 complexes was observed within 2 h of stimulation of Jurkat cells with TNF (Sun et al, 1994b). p100 processing, like that of p105, may be up-regulated by LMP1 of EBV, as a fivefold increase in p52 (fourfold for p50) is observed in LMP1-expressing cells relative to controls (Paine et al, 1995).

Shifting the balance of IκBα and rel

Evidence from many sources suggests that altering the balance of κBs and rel proteins is an effective way to alter κB-dependent transcription and cell state. Such changes appear to be important for B-cell differentiation, for replication or maintenance of some viruses, and are implicated in oncogenesis, as discussed above. The profound effects of altering IκB/rel homeostasis suggest that drugs affecting this balance might be useful in treating numerous pathological conditions, and recent work shows that some widely used drugs alter levels of IκBα.

In a study of NIH3T3 cells, overexpression of IκBα decreased rel activity and cell proliferation. Overexpression of IκBα antisense mRNA to reduce IκBα protein, resulted in increased rel DNA binding and transactivation (Beauparlant et al, 1994), effects also seen in IκB$\alpha^{-/-}$ mice. Furthermore, IκBα antisense induced transformation, as marked by anchorage-independent growth and tumorigenicity in nude mice, suggesting IκBα functions in some cells as a tumor suppressor (Beauparlant et al, 1994). IκBγ antisense had no effect in this system. In combination with p65 antisense experiments, these results suggest a role for nuclear p65 activity in neoplastic growth, which is counterbalanced by IκBα and other IκBs.

Altered IκBα regulation in pathology may be involved in transformation of chicken cells by v-rel. In v-rel transformed cells, most IκBα is associated with v-rel, suggesting a particularly stable association of the two proteins, perhaps allowing more c-rel to translocate to the nucleus (Hrdlickova et al, 1995). Two temperature-sensitive v-rel mutants have been shown to stabilize IκBα and block CHX-induced apoptosis (White and Gilmore, 1995). When cells expressing these mutants are shifted to the nonpermissive, nontransforming temperature, IκBα is degraded following CHX treatment, presumably due to a change in v-rel conformation that alters the stability of associated IκBα, and the cells apoptose (White and Gilmore, 1995). These results suggest that the stability of the v-rel/IκBα interaction may be important for transformation.

Ataxia-telangiectasia (AT) is a human autosomal recessive disorder that causes multiple defects, including immunodeficiency, neurological abnormalities and radiation sensitivity. A screen for cDNAs able to complement the radiation sensitivity in an AT fibroblast cell line yielded two N-terminally-truncated IκBα clones lacking serines 32 and 36 (Jung et al, 1995). In the parental cell line, NF-κB is constitutively activated and no increase in κB-binding activity is seen upon irradiation. In the line complemented by N-terminally truncated IκBα, the constitutive κB binding is lower and increases in response to irradiation (Jung et al, 1995). These results suggest some AT-cells may have a defect that results in constitutive stimulation of phosphorylation and proteolysis of IκBα. Interestingly, the primary defect in AT was recently identified by linkage analysis, and is probably a loss-of-function mutation in a gene with similarity to that encoding phosphatidylinositol (PtdIns) 3-kinase (Lehmann and Carr, 1995), which is activated by many growth factors.

Although research on rel proteins and their inhibitors has not yet led to the discovery of new therapeutic drugs, several recent studies underline the importance of IκBα in inflammation. Two widely used anti-inflammatory agents, salicylates and glucocorticoids, inhibit the activation of NF-κB (Auphan et al, 1995; Kopp and Ghosh, 1994; Scheinman et al, 1995). Certain salicylates, including aspirin, block IκBα degradation, inhibiting the synthesis of the many inflammatory cytokines and chemokines shown to be activated by rel proteins. The immunosuppressive effects of glucocorticoids (GC) may also depend in large part on inhibition of NF-κB.

Two mechanisms have been proposed to account for the inhibition of induced κB-binding activity by glucocorticoids. First, direct interaction of the glucocorticoid receptor and NF-κB is suggested by experimental results with overexpressed proteins (Gilmore, 1995). More recently, however, reduction of nuclear rel activity in the presence of GC has been shown to depend on ongoing protein synthesis, and specifically, on increased IκBα mRNA and protein (Auphan et al, 1995; Scheinman et al, 1995). While the GC-induced IκBα is sensitive to degradation following activitating stimuli, because of the higher total amount of inhibitor a substantial amount of IκBα remains associated with p65, drastically reducing the amount of NF-κB translocated to the nucleus (Auphan et al, 1995; Scheinman et al, 1995).

Rel phosphorylation

Dorsal, a RHD protein in *Drosophila*, is phosphorylated in a signal-dependent manner, and this phosphorylation is important for nuclear import and transactivating function (Whalen and Steward, 1993; Gillespie and Wasserman, 1994), suggesting mammalian rel factors may also be subject to phosphoregulation. Phosphorylation of mammalian rel proteins, particularly the transactivators p65 and c-rel, has been observed in multiple cell types. Although the regulation of rel phosphorylation is not understood, it may have functional effects on subcellular localization, DNA binding affinity, and possibly transactivation.

A consensus site for phosphorylation by PKA is highly conserved in the RHD near the NLS. In the context of c-rel, the mutation of the serine residue at this site to Asp or Glu, to mimic the effect of phosphorylation, disregulated subcellular localization of c-rel, rendering the protein partly nuclear, perhaps due to a change in its affinity for IκBs or its ability to dimerize (Mosialos et al, 1991). Although this suggests a potential regulatory role for phosphorylation of the PKA site, it was not found to the phosphorylated in vivo (Mosialos et al, 1991). However, further study is needed to clarify this point.

c-rel is phosphorylated on serine in activated and resting Jurkat T-cells (Li et al, 1994b); although another study in activated Jurkat cells detected phosphotyrosine in c-rel (Neumann et al, 1992). In resting purified T-cells c-rel is not detectably phosphorylated, but phospholabeled c-rel is detected when cells are activated by PMA and PHA or anti-CD28-treatment (Bryan et al, 1994). The increase in c-rel phosphorylation induced by anti-CD28 costimulation is correlated with more rapid translocation of c-rel to the nucleus than in PMA-stimulated cells (Bryan et al, 1994).

p65 phosphorylation is reportedly induced by TNF in HeLa and osteosarcoma cells (Mellits et al, 1993; Naumann and Scheidereit, 1994), and phosphorylation in vitro by PKA has been shown to increase p65 DNA binding (Naumann and Scheidereit, 1994). Similarly, p50 is reported to bind DNA with higher affinity after phosphorylation in vitro by PKC, and to be inducibly phosphorylated in PMA/PHA-treated Jurkat cells (Li et al, 1994a, b). In this connection, a 43-kd serine kinase able to associate with NF-κB and phosphorylate both subunits in vitro has been described (Hayashi et al, 1993), but cloning of the corresponding cDNA has not been reported.

Signaling pathways

Among the wide variety of stimuli that activate proteolysis of IκBs and translocation of rel dimers to the nucleus, it is difficult to discern a common element that would explain their shared effects on rel-dependent transcription. IκBα is degraded in response to all rel-activating stimuli, and most stimuli have been shown to induce phosphorylation of the same two N-terminal serines. The IκBα N-terminal kinase and/or phosphatase activities involved have not yet been identified with any certainty, although some evidence for IκBα phosphorylation by known kinases is discussed below. The signaling elements upstream of the enzymes responsible for modifying IκBα have not been definitely determined either. There is evidence for the possible involvement of the Ras/Raf-MAP kinase pathway, sphingomyelinase-ceramide pathway, and protein kinase C (PKC), both classical PKCs and the Ca^{2+}- and DAG-independent PKC zeta (PKC-ζ). Serine-threonine protein phosphatases, and more recently tyrosine phosphatases, have also been implicated in the inactivation of IκBα.

The ability of phorbol esters to induce nuclear translocation of NF-κB led to the suggestion early on that classical PKCs might be a common upstream component in the activation pathway (Baeuerle et al, 1988). Purified PKC was shown to phosphorylate IκBα and release NF-κB activity in vitro (Ghosh and Baltimore, 1990; Shirakawa and Mizel, 1989).

However, other activators of NF-κB, such as TNF, are not thought to stimulate the activity of classical DAG-sensitive PKCs, and staurosporin, an inhibitor of PKC, blocks rel activation by PMA but not TNF (Hohmann et al, 1992). Similarly, LPS-induced phosphorylation of IκBα is still apparent in cells made refractory to activation of classical PKCs by long-term treatment with PMA (Cordle et al, 1993).

Another kinase reported to phosphorylate IκBα is the ds-RNA-activated protein kinase (PKR) (Kumar et al, 1994). A GST-PKR fusion protein phosphorylates IκBα in vitro, and expression of a kinase-defective PKR blocked κB-dependent activation of a reporter gene by dsRNA (Kumar et al, 1994). Similarly, cells treated with antisense to PKR were unresponsive to activation of nuclear NF-κB by the dsRNA (Maran et al, 1994). In PKR-deficient mice activation of NF-κB by dsRNA, but not TNF, was impaired; however, pretreatment of the PKR$^{-/-}$ cells with interferon restored some rel response (Yang et al, 1995). These results suggest that although signaling through PKR can activate NF-κB, PKR is not required for rel activation, even by dsRNA.

Ras and Raf proto-oncogenes are required for signal transduction from many receptors and have been implicated in rel activation by a variety of stimuli. Dominant negative Raf and Ras have been shown to block activation of a κB-dependent reporter gene in NIH3T3 cells in response to UV light, PMA, TNF, and serum (Devary et al, 1993; Finco and Baldwin, 1993; Li and Sedivy, 1993). Constitutively, active Ras or Raf activated a κB-dependent reporter gene in NIH3T3 cells, and this activation was inhibited by cotransfection with a plasmid encoding IκBα (Finco and Baldwin, 1993; Li and Sedivy, 1993). Activated Ras p21 when microinjected into *Xenopus* oocytes stimulated nuclear translocation of NF-κB and activation of a reporter gene driven by the HIV LTR (Dominguez, 1993). However, ras-independent activation of NF-κB can occur, e.g. in response to TNF, and in several studies constitutively active Raf did not activate NF-κB (Hambleton et al, 1995; Tobin et al, 1996; G. Courtois and A. Israël, unpublished data).

The Raf kinase domain and IκBα were shown to interact in a yeast two-hybrid system, and incubation of HeLa extracts with a GST-Raf fusion protein and ATP caused the release of κB binding activity (Li and Sedivy, 1993). However, IκBα is not released from rel dimers by the endogenous serine kinase, so Raf seems unlikely to be the stimulus-activated IκBα kinase acting in vivo. IκBα is not a substrate in vitro for kinases downstream of Raf, MEK and MAPK (Diaz-Meco et al, 1994).

The MAP kinases of the JNK/SAPK and p38/RK families are activated by many of the stress stimuli that induce nuclear rel translocation (Cano and Mahadevan, 1995), and it is plausible that one of the kinases, or a related kinase, is an upstream element in the pathway leading to IκBα phosphorylation. This question has not been addressed directly, but evidence from studies of TNF receptor signaling suggest that the stimulation of JNK activity and activation of NF-κB are not dependent on the same upstream elements and may be unrelated events (Belka et al, 1995; Westwick et al, 1995). A specific chemical inhibitor of p38 MAP kinase activity, SB203580, was found to block transactivation of a κB-dependent reporter gene in response to TNF treatment (Beyaert et al, 1996). However, the p38 inhibitor had no effect on IκB proteolysis nor on the phosphorylation, nuclear translocation, and DNA binding of rel proteins (Beyaert et al, 1996).

Although the signaling pathway(s) activated by TNF through its two receptors have not been entirely worked out, one important effect of TNFR I signaling may be the generation of ceramide, a lipid second messenger, by a sphingomyelinase (see Kolesnick and Golde, 1994 for review). Treatment of cells with TNF or IL-1 has been reported to increase ceramide within minutes (Yang et al, 1993; Machleidt et al, 1994, Mathias et al, 1993; Schutze et al, 1992; Wiegmann, 1992). Addition of either sphingomyelinase or cell-permeable analogs of ceramide to intact or permeabilized cells induced nuclear rel activity and IκBα degradation in some studies (Yang et al, 1993; Schutze et al, 1992), though not all (Betts et al, 1994; Westwick et al, 1995). A recent report suggests that LPS may induce ceramide-activated kinase activity by structural mimicry of ceramide by the lipid moiety of LPS (Joseph et al, 1994).

The downstream target of ceramide in the rel activation signaling cascade is not known, but a 97-kd ceramide-activated kinase, as well as raf- and PKC-ζ are candidates. Sphingomyelinase and ceramide are reported to activate PKC-ζ (Lozano et al, 1994; Muller et al, 1995b), however, lipids other than ceramide are also reported to activate PKC-ζ (Nakanishi et al, 1993). Other experiments unrelated to the question of ceramide involvement in rel activation also suggest a role for PKC-ζ in rel activation. A study in *Xenopus* oocytes showed that microinjection of an activated form of phosphatidylcholine-phospholipase C (PC-PLC) induced nuclear κB-binding activity, and this effect was blocked by a peptide inhibitor of PKC-ζ (Dominguez et al, 1993). Further work by this group showed a PKC-ζ fusion protein bound a 50-kd kinase from *Xenopus* oocytes that was able to phosphorylate IκBα in vitro (Diaz-Meco et al, 1994). NIH3T3 cells transformed by expression of v-*ras* or PC-PLC reportedly have constitutive κB-dependent transactivating activity, and this activity is returned to basal levels of nontransformed cells by overexpression of a kinase-defective PKC-ζ, thought to act as a dominant negative PKC-ζ (Bjorkoy et al, 1995).

Transactivation of a κB-dependent reporter gene in HIV-infected monocytes was also blocked by dominant negative PKC-ζ or by antisense to PKC-ζ (Folgueira et al, 1996). However, in these experiments dominant negative PKC-ζ had no effect on rel activation induced by LPS or PMA (Folgueira et al, 1996), and another study found dominant negative PKC-ζ did not alter translocation of NF-κB in response to TNF (Montaner et al, 1995). Given the indirect nature of the evidence supporting a role for PKC-ζ in rel activation and other negative results, a place for PKC-ζ in a rel signaling pathway remains to be established.

Progress in the understanding of signal transduction from receptors for TNF and IL-1 has brought to light other molecules involved in rel signal transduction. cDNAs encoding proteins interacting with TNFRs have been isolated recently, some of which can activate NF-κB (see Tewari and Dixit, 1996 for review). Overexpression of TRADD, which interacts with TNFR I and may serve an adapter function, induces both apoptosis and activation of NF-κB (Hsu et al, 1995). CrmA, a viral protein that blocks activity of cysteine proteases involved in programmed cell death, blocked TRADD-induced apoptosis, but not activation of NF-κB (Hsu et al, 1995), demonstrating that TRADD activation of NF-κB is not a secondary effect of the cell-death program induced by TNF. Overexpression of TRAF2, which interacts with TNFR II and CD40, also activates NF-κB, and a truncated TRAF2 construct blocked rel activation by TNF or CD40 ligand (Hsu et al, 1996; Rothe et al, 1995).

The intracellular domain of the type I receptor for IL-1 (IL1R) shares homology with a *Drosophila* receptor, Toll, that is required for activation of Dorsal during development (Hashimoto et al, 1988) and during response to injury (Peterson et al, 1995). A serine/threonine kinase, Pelle, and a putative adaptor protein, Tube, act downstream of Toll (Galindo et al, 1995; Grosshans et al, 1994; Shelton and Wasserman, 1993). IRAK, a recently cloned kinase inducibly associated with human IL1R, is required for activation of rel by IL-1 and has homology to Pelle (Cao et al, 1996; Croston et al, 1995).

In contrast to the kinases discussed above, which may help activate rel proteins, protein kinase A (PKA) is thought to inhibit rel activation. In T-cells, the addition of forskolin, which increases PKA activity, inhibited the PMA/ionomycin-induced activation of an IL-2 enhancer reporter gene, but not that of a similar construct mutated in the κB-binding site (Neumann et al, 1995). In the forskolin-treated T-cells the degradation of IκBα and the translocation of p65 to the nucleus in response to T-cell activation were reduced (Neumann et al, 1995). The inhibitory effects of PKA on rel activity may in part explain the suppression of IL-2 synthesis and T-cell proliferation by agents that activate this kinase.

Experimental evidence also supports a role for a Ca^{2+}-calmodulin-dependent serine/threonine phosphatase, calcineurin, in induction of nuclear rel by T-cell activation. Calcineurin is thought to be the cellular target of the immunosuppressants cyclosporin A (CsA) and FK-506 (see Kunz and Hall, 1993). These compounds block IL-2 transcription in response to T-cell activation by blocking activation of NF-AT and blocking the Ca^{2+}-dependent activation of rel. Interestingly, the need for a Ca^{2+} costimulus for κB-dependent transactivation in T-cells can be bypassed by expression of a constitutively active calcineurin construct (Frantz et al, 1994). This result suggests that calcineurin, or a similar activity, may mediate the Ca^{2+}-dependent activation of NF-κB, at least in T-cells.

As mentioned above, treatment of cells with okadaic acid or calyculin A, inhibitors of serine/threonine phosphatases of classes 1 and 2A, activates nuclear rel activity (Suzuki et al, 1994; Thevenin et al, 1990), accompanied by phosphorylation and degradation of IκBα (Sun et al, 1995). These effects may reveal the existence of constitutive IκBα serine kinase and phosphatase activities that keep most of the inhibitor in the unphosphorylated state in unstimulated cells. In contrast to the phosphorylation and degradation of IκBα induced by TNF, the effects of calyculin A are not blocked by antioxidant treatment, suggesting calyculin A acts downstream of the antioxidant-sensitive step in IκBα phosphorylation (Sun et al, 1995).

Overexpression of the protein-tyrosine phosphatase HPTPa is reported to induce constitutive NF-κB-like binding activity (Menon et al, 1995), and inhibitors of protein tyrosine phosphatases, such as phenylarsine oxide (PAO) and pervanadate, can block or reduce the activation of NF-κB by TNF, PMA, IL1, and a ceramide analog (Menon et al, 1995; Singh and Aggarwal, 1995). However, in one report the effect of PAO was shown to be reversed by addition of reducing agents (Singh and Aggarwal, 1995). This finding suggests the inhibitors affect DNA binding, not tyrosine phosphatase-dependent signaling, as other agents that modify free sulfhydryls are known to inhibit DNA binding of rel proteins in vitro (Toledano and Leonard, 1991).

The role of redox

The diversity of rel activating signals and signal-transducing elements is evident from the preceding discussion. Few of these signaling molecules are likely to be involved in the response to all stimuli that activate IκBα phosphorylation and degradation and the nuclear translocation of rel proteins. Therefore it is particularly interesting that nearly all rel activating stimuli are largely inhibited by pretreatment of cells with antioxidants like N-acetyl-L-cysteine (NAC), a scavenger of ROIs, or pyrrolidine-dithiocarbamate (PDTC), a precursor to GSH, a major reservoir of intracellular thiol (Schreck and Baeuerle, 1991; Schreck et al, 1991).

NAC inhibits PMA- and TNF-induced κB-dependent transactivation (Staal et al, 1990). Nuclear rel activation by LPS, IL-1, dsRNA, calcium ionophore, and cycloheximide are also inhibited by NAC or PDTC treatment (Schreck et al, 1991). The ability of these and other antioxidants to block activation of NF-κB and phosphorylation of IκBα suggests that the redox state of the cell affects a common final pathway leading to rel activation, and that an oxidizing environment is required for transmission of the signal. At least two models are possible, in the first, an excess of reducing agent interferes with the signaling apparatus, possibly by altering disulfide bonds or inhibiting lipid peroxidation; in the second, oxygen radicals themselves are second messengers that directly mediate rel activation by all stimuli.

The effect of hydrogen peroxide (H_2O_2), which is reported to directly increase intracellular ROI concentration, is relevant to distinguishing between these two mechanisms of action of antioxidants. A study using a T-cell line demonstrated that H_2O_2 could induce nuclear rel DNA-binding activity, suggesting increased intracellular ROI concentration is sufficient to activate NF-κB (Schreck et al, 1991). However, this effect of H_2O_2 was not observed in other T-cell lines (Anderson et al, 1994; Israël et al, 1992). In addition, the activation of NF-κB in the H_2O_2-sensitive cell line could be blocked by tyrosine kinase inhibitors (erbstatin and herbimycin A) that also blocked activation by PMA or TNF, suggesting the effect of H_2O_2 was not mediated directly by increased ROI concentration but via tyrosine kinase-dependent signaling (Anderson et al, 1994).

Although most rel activating stimuli are thought to cause oxidative stress, i.e. increased ROI concentration, few studies directly measure changes in the redox status of the cell induced by the rel activating stimulus. In one report, changes in GSH levels, which represents 30% of the thiol pool in lymphocytes (Messina and Lawrence, 1989), were measured in cells stimulated with PMA, TNF, or both, with or without NAC (Staal et al, 1990). While a similar decrease in GSH was detected in cells stimulated with TNF or PMA alone or with both, the transactivation of a κB-dependent reporter gene was synergistically activated by the combination of the two stimuli (Staal et al, 1990). This result appears inconsistent with a direct role of ROIs in rel-dependent signal transduction, as the greater activation of rel activity by PMA and TNF together should presumably be the result of greater ROI production and be reflected by a corresponding decrease in GSH levels.

Although the effects of antioxidants on rel activity could be multiple and involve different enzymes depending on the activating stimulus, it seems more likely that the thiol/redox-sensitive element in the signal transduction pathway leading to IκBα phosphorylation is shared by all rel activators. As antioxidants appear to act upstream of IκBα phosphorylation, the relevant target may emerge as the enzyme(s) directly responsible for inhibitor phosphorylation are identified.

Conclusion

The rel family of transcription factors has emerged as a important component of the cellular machinery with profound influences on development, cell growth, viral pathogenesis and immunity. In each of these areas, the balance between the IκBs and rel subunits determines the κB-dependent transactivation of target genes. A better understanding of the regulation of this balance may provide insight into a variety of normal and pathological responses and ultimately permit therapeutic manipulation of the system.

References

Abbadie C, Kabrun N, Bouali F, Smardova J, Stehelin D, Vandenbunder B, Enrietto PJ (1993): High levels of c-rel expression are associated with programmed cell death in the developing avian embryo and in bone marrow cells in vitro. *Cell* 75: 899–912

Albertella MR, Campbell RD (1994): Characterization of a novel gene in the human major histocompatibility complex that encodes a new member of the IκB family of proteins. *Hum Mol Genet* 3: 793–799

Alkalay I, Yaron A, Hatzubai A, Jung S, Avraham A, Gerlitz O, Pashut-Lavon I, Ben-Neriah Y (1994): In vivo stimulation of IκB-alpha phosphorylation is not sufficient to activate NF-κB. *Mol Cell Biol* 15: 1294–1301

Anderson MT, Staal FJT, Gitler C, Herzenberg LA, Herzenberg LA (1994): Separation of oxidant-initiated and redox-regulated steps in the NF-κB signal transduction pathway. *Proc Natl Acad Sci USA* 91: 11527–11531

Aoki T, Sano Y, Yamamoto T, Inoue JI (1996): The ankyrin repeats but not the PEST-like sequences are required for signal-dependent degradation of IκBα. *Oncogene* 12: 1159–1164

Arenzana-Seisdedos F, Thompson J, Rodriguez MS, Bachelerie F, Thomas D, Hay RT (1995): Inducible nuclear expression of newly synthesized IκBα negatively regulates DNA-binding and transcriptional activities of NF-κB. *Mol Cell Biol* 15: 2589–2696

Arima N, Kuziel WA, Grdina TA, Greene WC (1992): IL-2-induced signal transduction involves the activation of nuclear NF-κB expression. *J Immunol* 149: 83–91

Auphan N, Didonato JA, Rosette C, Helmberg A, Karin M (1995): Immunosuppression by glucocorticoids: Inhibition of NF-κB activity through induction of IκB synthesis. *Science* 270: 286–290

Bachelerie F, Alcami J, Arenzana SF, Virelizier JL (1991): HIV enhancer activity perpetuated by NF-κB induction on infection of monocytes. *Nature* 350: 709–712

Baeuerle PA, Baltimore D (1988a): Activation of DNA-binding activity in an apparently cytoplasmic precursor of the NF-κB transcription factor. *Cell* 53: 211–217

Baeuerle PA, Baltimore D (1988b): IκB: a specific inhibitor of the NF-κB transcription factor. *Science* 242: 540–546

Baeuerle PA, Henkel T (1994): Function and activation of NF-κB in the immune system. *Annu Rev Immunol* 12: 141–179

Baeuerle PA, Lenardo MJ, Pierce JW, Baltimore D (1988): Phorbol-ester induced activation of the NF-κB transcription factor involves dissociation of an apparently cytoplasmic NF-κB/inhibitor complex. *Cold Spring Harbor Symp Quant Biol* LIII: 789–798

Baldi L, Brown K, Franzoso G, Siebenlist U (1996): Critical role for lysines 21 and 22 in signal-induced, ubiquitin-mediated proteolysis of IκB-alpha. *J Biol Chem* 271: 376–379

Ballard DW, Bohnlein E, Lowenthal JW, Wano Y, Franza BR, Greene WC (1988): HTLV-1 tax induces cellular proteins that activate the kappa B element in the IL-2 receptor alpha gene. *Science* 241: 1652–1655

Ballard DW, Walker WH, Doerre S, Sista P, Molitor JA, Dixon EP, Peffer NJ, Hannink M, Greene WC (1990): The v-rel oncogene encodes a kappa B enhancer binding protein that inhibits NF-κB function. *Cell* 63: 803–814

Barroga CF, Stevenson JK, Schwarz EM, Verma IM (1995): Constitutive phosphorylation of IκBα by casein kinase II. *Proc Natl Acad Sci USA* 92: 7637–7641

Beauparlant P, Kwan I, Bitar R, Chou P, Koromilas AE, Sonenberg N, Hiscott J (1994): Disruption of IκBα regulation by antisense RNA expression leads to malignant transformation. *Oncogene* 9: 3189–3197

Beg AA, Ruben SM, Scheinman RI, Haskill S, Rosen CA, Baldwin AS (1992): IκB interacts with the nuclear localization sequences of the subunits of NF-κB. A mechanism for cytoplasmic retention. *Genes Dev* 6: 1899–1913

Beg AA, Finco TS, Nantermet PV, Baldwin AJ (1993): Tumor necrosis factor and interleukin-1 lead to phosphorylation and loss of IκBα: a mechanism for NF-κB activation. *Mol Cell Biol* 13: 3301–3310

Beg AA, Sha WC, Bronson RT, Baltimore D (1995a): Constitutive NF-κB activation, enhanced granulopoiesis and neonatal lethality in IκBα-deficient mice. *Genes Dev* 9: 2736–2746

Beg AA, Sha WC, Bronson RT, Ghosh S, Baltimore D (1995b): Embryonic lethality and liver degeneration in mice lacking the RelA component of NF-κB. *Nature* 376: 167–170

Belka C, Wiegmann K, Adam D, Holland R, Neuloh M, Herrmann F, Kronke M, Brach MA (1995): Tumor necrosis factor (TNF)-alpha activates c-*raf*-1 kinase via the p55 TNF receptor engaging neutral sphingomyelinase. *EMBO J* 14: 1156–1165

Benoist C, Mathis D (1990): Regulation of major histocompatibility complex Class-II genes: X, Y, and other letters of the alphabet. *Annu Rev Immunol* 8: 681–715

Bęraud C, Sun SC, Ganchi P, Ballard DW, Greene WC (1994): Human T-cell leukemia virus type 1 tax associates with and is negatively regulated by the NF-κB2 p100 gene product: Implications for viral latency. *Mol Cell Biol* 14: 1374–1382

Bertrand F, Philippe C, Antoine PJ, Baud L, Groyer A, Capeau J, Cherqui G (1995): Insulin activates nuclear factor nuclear factor kappa B in mammalian cells through a raf-1-mediated pathway. *J Biol Chem* 270: 24435–24441

Betts JC, Cheshire JK, Akira S, Kishimoto T, Woo P (1993): The role of NF-κB and NF-IL6 transactivating factors in the synergistic activation of human serum amyloid A gene expression by interleukin-1 and interleukin-6. *J Biol Chem* 268: 25624–25631

Betts JC, Agranoff AB, Nabel GJ, Shayman JA (1994): Dissociation of endogenous cellular ceramide from NF-κB activation. *J Biol Chem* 269: 8455–8458

Beyaert R, Cuenda A, Vanden Berghe W, Plaisance S, Lee JC, Haegeman G, Cohen P, Fiers W (1996): The p38/RK mitogen-activated protein kinase pathway regulates interleukin-6 synthesis in response to tumour necrosis factor. *EMBO J* 15: 1914–1922

Bjorkoy G, Overvatn A, Diaz-Meco MT, Moscat J, Johansen T (1995): Evidence for a bifurcation of the mitogenic signaling pathway activated by Ras and phosphatidylcholinehydrolizing phospholipase C. *J Biol Chem* 270: 21299–21306

Blair WS, Bogerd HP, Madore SJ, Cullen BR (1994): Mutational analysis of the transcription activation domain of RelA: identiifcation of a highly synergistic minimal acidic activation module. *Mol Cell Biol* 14: 7226–7234

Blank V, Kourilsky P, Israël A (1991): Cytoplasmic retention, DNA binding and processing of the NF-κB p50 precursor are controlled by a small region in its C-terminus. *EMBO J* 10: 4159–4167

Blank V, Kourilsky P, Israël A (1992): NF-κB and related proteins: Rel/dorsal homologies meet ankyrin-like repeats. *Trends Biochem Sci* 17: 135–140

Boehmelt G, Walker A, Kabrun N, Mellitzer G, Beug H, Zenke M, Enrietto PJ (1992): Hormone-regulated v-rel estrogen receptor fusion protein – reversible induction of cell transformation and cellular gene expression. *EMBO J* 11: 4641–4652

Bours V, Franzoso G, Azarenko V, Park S, Kanno T, Brown K, Siebenlist U (1993): The oncoprotein bcl-3 directly transactivates through kappa B motifs via association with DNA-binding p50B homodimers. *Cell* 72: 729–739

Brach MA, de Vos S, Arnold C, Gruss HJ, Mertelsmann R, Herrmann F (1992): Leukotriene B4 transcriptionally activates interleukin-6 expression involving NF-κB and NF-IL-6. *Eur J Immunol* 22: 2705–2711

Brockman JA, Scherer DC, McKinsey TA, Hall SM, Qi X, Lee WY, Ballard DW (1995): Coupling of a signal response domain in IκBα to multiple pathways for NF-κB activation. *Mol Cell Biol* 15: 2809–2812

Brown K, Park S, Kanno T, Franzoso G, Siebenlist U (1993): Mutual regulation of the transcriptional activator NF-κB and its inhibitor, IκBα. *Proc Natl Acad Sci USA* 90: 2532–2536

Brown K, Gerstberger S, Carlson L, Franzoso G, Siebenlist U (1995): Control of IκBα proteolysis by site-specific signal-induced phosphorylation. *Science* 267: 1485–1488

Brownell E, Obrien SJ, Nash WG, Rice NR (1985): Genetic characterisation of human c-rel sequences. *Mol Cell Biol* 5: 2826–2831

Brownell E, Mathieson B, Young HA, Keller J, Ihle JN, Rice NR (1987): Detection of c-rel-related transcripts in mouse hematopoietic tissues, fractionated lymphocyte populations, and cell lines. *Mol Cell Biol* 7: 1304–1309

Bryan RG, Li Y, Lai JH, Van M, Rice NR, Rich RR, Tan TH (1994): Effect of CD28 signal transduction on c-Rel in human peripheral blood T-cells. *Mol Cell Biol* 14: 7933–7942

Bull P, Hunter T, Verma IM (1989): Transcriptional induction of the murine c-rel gene with serum and phorbol-12-myristate-13-acetate in fibroblasts. *Mol Cell Biol* 9: 5239–5243

Bull P, Morley KL, Hoekstra MF, Hunter T, Verma IM (1990): The mouse c-rel protein has an N-terminal regulatory domain and a C-terminal transcriptional transactivation domain. *Mol Cell Biol* 10: 5473–5485

Burkly L, Hession C, Ogata L, Reilly C, Marconi LA, Olson D, Tizard R, Cate R, Lo D (1995): Expression of relB is required for the development of thymic medulla and dendritic cells. *Nature* 373: 531–536

Cano E, Mahadevan LC (1995): Parallel signal processing among mammalian MAPKs. *Trends Biol Sci* 20: 117–122

Cao Z, Henzel WJ, Gao X (1996): IRAK: A kinase associated with the interleukin-1 receptor. *Science* 271: 1128–1131

Carrasco D, Ryseck RP, Bravo R (1993): Expression of relB transcripts during lymphoid organ development-specific expression in dendritic antigen-presenting cells. *Development* 118: 1221–1231

Carter BD, Kaltschmidt C, Kaltschmidt B, Offenhauser N, Bohm-Matthaei R, Baeuerle PA, Barde YA (1996): Selective activation of NF-κB by nerve growth factor through the neurotrophin receptor p75. *Science* 272: 542–545

Chang CC, Zhang JD, Lombardi L, Neri A, Dallafavera R (1994): Mechanism of expression and role in transcriptional control of the proto-oncogene NF-κB 2/LYT-10. *Oncogene* 9: 923–933

Chen ZJ, Hegler J, Palombella VJ, Melandri F, Scherer D, Ballard D, Maniatis T (1995): Signal-induced site-specific phosphorylation targets IκBα to the ubiquitin-proteasome pathway. *Genes Dev* 9: 1586–1597

Chen ZJ, Parent L, Maniatis T (1996): Site-specific phosphorylation of IκBα by a novel ubiquitination-dependent protein kinase activity. *Cell* 84: 853–862

Chirillo P, Falco M, Puri PL, Artini M, Balsano C, Levrero M, Natoli G (1996): Hepatitis B virus pX activates NF-κB-dependent transcription through a Raf-independent pathway. *J Virol* 70: 641–646

Chirmule N, Kalyanaraman VS, Pahwa S (1994): Signals transduced through the CD4 molecule on T lymphocytes activate NF-κB. *Biochem Biophys Res Commun* 203: 498–505

Cogswell PC, Scheinman RI, Baldwin AS (1993): Promoter of the human NF-κB p50/p105 gene – Regulation by NF-κB subunits and by c-REL. *J Immunol* 150: 2794–2804

Collins G, Read MA, Neish AS, Whitley MZ, Thanos D, Maniatis T (1995): Transcriptional regulation of endothelial cell adhesion molecules: NF-κB and cytokine-inducible enhancers. *FASEB J* 9: 899–909

Cordle SR, Donald R, Read MA, Hawiger J (1993): Lipopolysaccharide induces phosphorylation of MAD3 and activation of c-Rel and related NF-κB proteins in human monocytic THP-1 cells. *J Biol Chem* 268: 11803–11810

Crabtree GR (1989): Contingent genetic regulatory events in T lymphocyte activation. *Science* 243: 355–361

Cressman DE, Taub R (1993): IκBα can localize in the nucleus but shows no direct transactivation potential. *Oncogene* 8: 2567–2573

Croston GE, Cao Z, Goeddel DV (1995): NF-κB activation by interleukin-1 (IL-1) requires an IL-1 receptor-associated protein kinase activity. *J Biol Chem* 270: 16514–16517

David-Watine B, Israël A, Kourilsky P (1990): The regulation and expression of MHC Class I genes. *Immunol Today* 11: 286–292

DeMartin R, Vanhove B, Cheng Q, Hofer E, Csizmadia V, Winkler H, Bach FH (1993): Cytokine-inducible expression in endothelial cells of an IκB-alpha-like gene is regulated by NF-κB. *EMBO J* 12: 2773–2779

Deshaies RJ (1995): Make it or break it: the role of ubiquitin-dependent proteolysis in cellular regulation. *Trends Cell Biol* 5: 428–434

Devary Y, Rosette C, Didonato JA, Karin M (1993): NF-κB activation by ultraviolet light not dependent on a nuclear signal. *Science* 261: 1442–1445

Diaz-Meco M, Dominguez I, Sanz L, Dent P, Lozano J, Municio MM, Berra E, Hay RT, Sturgill TW, Moscat J (1994): ζPKC induces phosphorylation and inactivation of IκBα in vitro. *EMBO J* 13: 2842–2848

Didonato J, Mercurio F, Rosette C, Wu-Li J, Suyang H, Ghosh S, Karin M (1996): Mapping of the inducible IκB phosphorylation sites that signal its ubiquitination and degradation. *Mol Cell Biol* 16: 1295–1304

Didonato JA, Mercurio F, Karin M (1994): Phosphorylation of IκBα precedes but is not sufficient for its dissociation from NF-κB. *Mol Cell Biol* 15: 1302–1311

Dobrzanski P, Ryseck RP, Bravo R (1993): Both N- and C-terminal domains of RelB are required for full transactivation: role of the N-terminal leucine zipper-like motif. *Mol Cell Biol* 13: 1572–1582

Dobrzanski P, Ryseck RP, Bravo R (1994): Differential interactions of Rel-NF-κB complexes with IκBα determine pools of constitutive and inducible NF-κB activity. *EMBO J* 13: 4608–4616

Dobrzanski P, Ryseck RP, Bravo R (1995): Specific inhibition of relB/p52 transcriptional activity by the C-terminal domain of p100. *Oncogene* 10: 1003–1107

Dominguez I, Sanz L, Arenzana-Seisdedos F, Diaz-Meco MT, Virelizier J-L, Moscat J (1993): Inhibition of protein kinase C zeta subspecies blocks the activation of an NF-κB-like activity in *Xenopus* laevis oocytes. *Mol Cell Biol* 13: 1290–1295

Donald R, Ballard DW, Hawiger J (1995): Proteolytic processing of NF-κB/IκB in human monocytes. *J Biol* Chem 270: 9–12

Doria M, Klein N, Lucito R, Schneider RJ (1995): The hepatitis B virus HBx protein is a dual specificity cytoplasmic activator of Ras and nuclear activator of transcription factors. *EMBO J* 14: 4747–4757

Du W, Thanos D, Maniatis T (1993): Mechanisms of transcriptional synergism between distinct virus-inducible enhancer elements. *Cell* 74: 887–898

Duckett CS, Perkins ND, Kowalik TF, Schmid RM, Huang ES, Baldwin AS, Nabel GJ (1993): Dimerization of NF-κB2 with RelA(p65) regulates DNA binding, transcriptional activation, and inhibition by an IκB-alpha (MAD-3). *Mol Cell Biol* 13: 1315–1322

Duyao MP, Kessler DJ, Spicer DB, Sonenshein GE (1990): Binding of NF-κB-like factors to regulatory sequences of the c-myc gene. *Curr Topics Microbiol Immunol* 166: 211–220

Eck SL, Perkins ND, Carr DP, Nabel GJ (1993): Inhibition of phorbol ester-induced cellular adhesion by competitive binding of NF-κB in vivo. *Mol Cell Biol* 13: 6530–6536

Emmel EA, Verweij CL, Durand DB, Higgins KM, Lacy E, Crabtree GR (1989): Cyclosporin A specifically inhibits function of nuclear proteins involved in T-cell activation. *Science* 246: 1617–1620

Ernst MK, Dunn LL Rice NR (1995): The PEST-like sequence of IκBα is responsible for inhibition of DNA binding but not for cytoplasmic retention of c-Rel or RelA homodimers. *Mol Cell Biol.* 15: 872–882

Fan CM, Maniatis T (1991): Generation of p50 subunit of NF-κB by processing of p105 through an ATP-dependent pathway. *Nature* 354: 395–398

Finco TS, Baldwin AS (1993): κB site-dependent induction of gene expression by diverse inducers of nuclear factor-κB requires raf-1. *J Biol Chem* 268: 17676–17679

Finco TS, Beg AA, Baldwin AS (1994): Inducible phosphorylation of IκBα is not sufficient for its dissociation from NF-κB and is inhibited by protease inhibitors. *Proc Natl Acad Sci USA* 91: 11884–11888

Folgueira L, McElhinny JA, Bren GD, MacMorran WS, Diaz-Meco MT, Moscat J, Paya CV (1996): Protein kinase C-zeta mediates NF-κB activation in human immunodeficiency virus-infected monocytes. *J Virol* 70: 223–231

Frantz B, Nordby EC, Bren G, Steffan N, Paya CV, Kincaid RL, Tocci MJ, O'Keefe SJ, O'Neill EA (1994): Calcineurin acts in synergy with PMA to inactivate IκB/MAD3, an inhibitor of NF-κB. *EMBO J* 13: 861–870

Franzoso G, Bours V, Park S, Tomita YM, Kelly K, Siebenlist U (1992): The candidate oncoprotein Bcl-3 is an antagonist of p50/NF-κB-mediated inhibition. *Nature* 359: 339–342

Franzoso G, Bours V, Azarenko V, Park S, Tomita-Yamaguchi M, Kanno T, Brown K, Siebenlist U (1993): The oncoprotein bcl-3 can facilitate NF-κB-mediated transactivation by removing inhibiting p50 homodimers from select κB sites. *EMBO J* 12: 3893–3901

Fraser JD, Irving BA, Crabtree GR, Weiss A (1991): Regulation of interleukin-2 gene enhancer activity of the T-cell accessory molecule CD28. *Science* 251: 313–316

Fujimoto K, Yasuda H, Sato Y, Yamamoto K (1995): A role for phosphorylation in the proteolytic processing of the human NF-κB1 precursor. *Gene* 165: 183–189

Fujita T, Nolan GP, Ghosh S, Baltimore D (1992): Independent modes of transcriptional activation by the p50 and p65-subunit of NF-κB. *Genes Dev* 6: 775–787

Fujita T, Nolan GP, Liou HC, Scott ML, Baltimore D (1993): The candidate proto-oncogene bcl-3 encodes a transcriptional coactivator that activates through NF-κB p50 homodimers. *Genes Dev* 7: 1354–1363

Galindo RL, Edwards DN, Gillespie SKH, Wasserman SA (1995): Interaction of the pelle kinase with the membrane-associated protein tube is required for transduction of the dorsoventral signal in *Drosophila* embryos. *Development* 121: 2209–2218

Ganchi PA, Sun SC, Greene WC, Ballard DW (1992): IκB/MAD-3 masks the nuclear localization signal of NF-κB p65 and requires the transactivation domain to inhibit NF-κB p65 DNA binding. *Mol Cell Biol* 3: 1339–1352

Ghosh G, Vanduyne G, Ghosh S, Sigler PB (1995): Structure of NF-κB p50 homodimer bound to a κB site. *Nature* 373: 303–310

Ghosh P, Tan TH, Rice NR, Sica A, Young HA (1993): The interleukin-2 CD28-responsive complex contains at least three members of the NF-κB family – c-Rel, p50, and p65. *Proc Natl Acad Sci USA* 90: 1696–1700

Ghosh S, Baltimore D (1990): Activation in vitro of NF-κB by phosphorylation of its inhibitor IκB. *Nature* 344: 678–682

Ghosh S, Gifford AM, Riviere LR, Tempst P, Nolan GP, Baltimore D (1990): Cloning of the p50 DNA binding subunit of NF-κB: homology to rel and dorsal. *Cell* 62: 1019–1029

Gilmore TD (1995): Regulation of rel transcription complexes. In: *Frontiers in Molecular Biology Eukaryotic Gene Transcription*. Goodbourn S, ed., Oxford University Press, Oxford, England, pp 104–133

Gimble JM, Duh E, Ostrove JM, Gendelman HE, Max EE, Rabson AB (1988): Activation of the human immunodeficiency virus long terminal repeat by herpes simplex virus type 1 is associated with induction of a nuclear factor that binds to the NF-κB/core enhancer sequence. *J Virol* 62: 4104–4112

Green JE (1991): Transactivation of nerve growth factor in transgenic mice containing the human T-cell lymphotropic virus type I tax gene. *Mol Cell Biol* 11: 4635–4641

Grilli M, Chiu JJS, Lenardo M (1993): NF-κB and rel: participants in a multiform transcriptional regulatory system. *Int Rec Cytol* 143: 1–62

Grimm S, Baeuerle PA (1994): Failure of the splicing variant p65Δ of the NF-κB subunit p65 to transform fibroblasts. *Oncogene* 9: 2391–2398

Grosshans J, Bergmann A, Haffter P, Nusslein VC (1994): Activation of the kinase Pelle by Tube in the dorsoventral signal transduction pathway of *Drosophila* embryo. *Nature* 372: 563–566

Grumont RJ, Gerondakis S (1990a): The murine c-rel proto-oncogene encodes two mRNAs the expression of which modulated by lymphoid stimuli. *Oncogene Res* 5: 245–254

Grumont RJ, Gerondakis S (1990b): Murine c-rel transcription is rapidly induced in T-cells and fibroblasts by mitogenic agents and the phorbol ester 12-O-tetradecanoylphorbol-13-acetate. *Cell Growth Differ* 1: 345–350

Grumont RJ, Gerondakis S (1994): Alternative splicing of RNA transcripts encoded by the murine p105 NF-κB gene generates IκBγ isoforms with different inhibitory activities. *Proc Natl Acad Sci USA* 91: 4367–4371

Hambleton J, McMahon M, DeFranco AL (1995): Activation of Raf-1 and mitogen-activated protein kinase in murine macrophages partially mimics lipopolysaccharide-induced signaling events. *J Exp Med* 182: 147–154

Hammarskjold M, Simurda M (1992): Epstein-Barr virus latent membrane protein transactivates the human immunodeficiency virus type 1 long terminal repeat through induction of NF-κB activity. *J Virol* 66: 6496–6501

Hannink M, Temin HM (1990): Structure and autoregulation of the c-rel promoter. *Oncogene* 5: 1843–1850

Hashimoto C, Hudson KL, Anderson KV (1988): The Toll gene of *Drosophila*, required for dorso-ventral embryonic polarity, appears to encode a transmembrane protein. *Cell* 52: 269–279

Haskill S, Beg AA, Tompkins SM, Morris JS, Yurochko AD, Sampson JA, Mondal K, Ralph P, Baldwin AJ (1991): Characterization of an immediate-early gene induced in adherent monocytes that encodes IκB-like activity. *Cell* 1281–1289

Hatada EN, Nieters A, Wulczyn FG, Naumann M, Meyer R, Nucifora G, McKeithan TW, Scheidereit C (1992): The ankyrin repeat domains of the NF-κB precursor p105 and the protooncogene bcl-3 act as specific inhibitors of NF-κB DNA binding. *Proc Natl Acad Sci USA* 89: 2489–2493

Hayashi T, Sekine T, Okamoto T (1993): Identification of a new serine kinase that activates NF kappa B by direct phosphorylation. *J Biol Chem* 268: 26790–26795

Henkel T, Zabel U, van Zee K, Muller JM, Fanning E, Baeuerle PA (1992): Intramolecular masking of the nuclear location signal and dimerization domain in the precursor for the p50 NF-κB subunit. *Cell* 68: 1121–1133

Henkel T, Machleidt T, Alkalay I, Kronke M, Ben Neriah Y, Baeuerle PA (1993): Rapid proteolysis of IκBα is necessary for activation of transcription factor NF-κB. *Nature* 365: 182–185

Herrero JA, Mathew P, Paya CV (1995): LMP-1 activates NF-κB by targeting the inhibitory molecule IκBα. *J Virol* 69: 2168–2174

Higgins KA, Perez JR, Coleman TA, Dorshkind K, McComas WA, Sarmiento UM, Rosen CA, Narayanan R (1993): Antisense inhibition of the p65 subunit of NF-κB blocks tumorigenicity and causes tumor regression. *Proc Natl Acad Sci USA* 90: 9901–9905

Hirai H, Fujisawa J, Suzuki T, Ueda K, Muramatsu M, Tsuboi A, Arai N, Yoshida M (1992): Transcriptional activator tax of HTLV-1 binds to the NF-κB precursor p 105. *Oncogene* 7: 1737–1742

Hirai H, Suzuki T, Fujisawa J, Inoue J, Yoshida M (1994): Tax protein of human T-cell leukemia virus type 1 binds to the ankyrin motifs of inhibitory factor κB and induces nuclear translocation of transcription factor NF-κB proteins for transcriptional activation. *Proc Natl Acad Sci USA* 91: 3584–3588

Hohmann HP, Remy R, Aigner L, Brockhaus M, van Loon A (1992): Protein kinases negatively affect nuclear factor-κB activation by tumor necrosis factor-alpha at two different stages in promyelocytic HL60 cells. *J Biol Chem* 267: 2065–2072

Houldsworth J, Mathew S, Rao PH, Dyomina K, Louie DC, Parsa N, Offit K, Chaganti RSK (1996): REL proto-oncogene is frequently amplified in extranodal diffuse large cell lymphoma. *Blood* 87: 25–29

Hrdlickova R, Nehyba J, Roy A, Humphries EH, Bose HR (1995): The relocalization of v-Rel from the nucleus to the cytoplasm coincides with induction of expression of IκBα and nfkb1 and stabilization of IκBα. *J Virol* 69: 403–413

Hsu H, Xiong J, Goeddel DV (1995): The TNF receptor 1-associated protein TRADD signals cell death and NF-κB activation. *Cell* 81: 495–504

Hsu H, Shu H-B, Pan M-G, Goeddel DV (1996): TRADD-TRAF2 and TRADD-FADD interactions define two distinct TNF receptor 1 signal transduction pathways. *Cell* 84: 266–308

Inoue J, Kerr LD, Kakizuka A, Verma IM (1992): IκBγ, a 70 kd protein identical to the C-terminal half of p110 NF-κB: a new member of the IκB family. *Cell* 68: 1109–1120

Israël A, Yano O, Logeat F, Kieran M, Kourilsky P (1989 a): Two purified factors bind to the same sequence in the enhancer of mouse MHC class I genes: one of them is a positive regulator induced upon differentiation of teratocarcinoma cells. *Nucleic Acids Res* 17: 5245–5257

Israël N, Hazan U, Alcami J, Munier A, Arenzana SF, Bachelerie F, Israël A, Virelizier JL (1989 b): Tumor necrosis factor stimulates transcription of HIV-1 in human T lymphocytes, independently and synergistically with mitogens. *J Immunol* 143: 3956–3960

Israël N, Gougerot Pocidalo M, Aillet F, Virelizier JL (1992): Redox status of cells influences constitutive or induced NF-κB translocation and HIV long terminal repeat activity in human T and monocytic cell lines. *J Immunol* 149: 3386–3393

Jaffray E, Wood KM, Hay RT (1995): Domain organization of IκBα and sites of interaction with NF-κB p65. *Mol Cell Biol* 15: 2166–2172

Jamieson C, Mauxion F, Sen R (1989): Identification of a functional NF-κB binding site in the murine T-cell receptor β2 locus. *J Exp Med* 170: 1737–1743

Jentsch S, Schlenker S (1995): Selective protein degradation: A journey's end within the proteasome. *Cell* 82: 881–884

Joseph CK, Wright SD, Bornmann WG, Randolph JT, Kumar ER, Bittman R, Liu J, Kolesnick RN (1994): Bacterial lipopolysaccharide has structural similarity to ceramide and stimulates ceramide-activated protein kinase in myeloid cells. *J Biol Chem* 269: 17606–17610

Jung M, Zhang Y, Lee S, Dritschilo A (1995): Correction of radiation sensitivity in ataxia telangiectasia cells by a truncated IκBα. *Science* 268: 1619–1621

Kabrun N, Hodgson JW, Doemer M, Mak G, Franza BJ, Enrietto PJ (1991): Interaction of the v-rel protein with an NF-κB DNA binding site. *Proc Natl Acad Sci USA* 88: 1783–1787

Kang SM, Tran AC, Grilli M, Lenardo MJ (1992): NF-κB subunit regulation in nontransformed CD4$^+$ lymphocytes-T. *Science* 256: 1452–1456

Kanno T, Brown K, Siebenlist U (1995): Evidence in support of a role for human T-cell leukemia virus type I tax in activating NF-κB via stimulation of signaling pathways. *J Biol Chem* 270: 11745–11748

Kerr LD (1994): Arrested development: understanding v-*abl*. *Bioessays* 16: 453–456

Kerr LD, Duckett CS, Wamsley P, Zhang Q, Chiao P, Nabel G, McKeithan TW, Baeuerle PA, Verma IM (1992): The proto-oncogene bcl-3 encodes an IκB protein. *Genes Dev* 6: 2352–2363

Kerr LD, Ransone LJ, Wamsley P, Schmitt MJ, Boyer TG, Zhou Q, Berk AJ, Verma IM (1993): Association between proto-oncogene rel and TATA-binding protein mediates transcriptional activation by NF-κB. *Nature* 365: 412–419

Kieran M, Blank V, Logeat F, Vandekerckhove J, Lottspeich F, LeBail O, Urban MB, Kourilsky P, Baeuerle PA, Israël A (1990): The DNA binding subunit of NF-κB is identical to factor KBF1 and homologous to the rel oncogene product. *Cell* 62: 1007–1018

Kitajima I, Shinohara T, Bilakovics J, Brown DA, Xu X, Nerenberg M (1992): Ablation of transplanted HTLV-I Tax-transformed tumors in mice by antisense inhibition of NF-κB. *Science* 258: 1792–1795

Klement JF, Rice NR, Car BD, Abbondanzo SJ, Powers GD, Bhatt H, Chen CH, Rosen CA, Stewart CL (1996): IκBα deficiency results in a sustained NF-κB response and severe widespread dermatitis in mice. *Mol Cell Biol* 16: 2341–2349

Klug CA, Gerety SJ, Shah PC, Chen YY, Rice NR, Rosenberg N, Singh H (1994): The v-abl tyrosine kinase negatively regulates NF-κB/Rel factors and blocks kappa-gene transcription in pre-B-lymphocytes. *Genes Dev* 8: 678–687

Kontgen F, Grumont RJ, Strasser A, Metcalf D, Li R, Tarlinton D, Gerondadis S (1995): Mice lacking the c-rel proto-oncogene exhibit defects in lymphocyte proliferation, humoral immunity, and interleukin-2 expression. *Genes Dev* 9: 1965–1977

Koong AC, Chen EY, Giaccia AJ (1994): Hypoxia causes the activation of nuclear factor κB through the phosphorylation of IκBα on tyrosine residues. *Cancer Res* 54: 1425–1430

Kopp E, Ghosh S (1994): Inhibition of NF-κB by sodium salicylate and aspirin. *Science* 265: 956–959

Kopp EB, Ghosh S (1995): NF-κB and rel proteins in innate immunity. *Adv Immunol* 58: 1–27

Kravchenko VV, Pan Z, Han J, Herbert J-M, Ulevitch RJ, Ye RD (1995): Platelet-activating factor induces NF-κB activation through a G protein-coupled pathway. *J Biol Chem* 270: 14928–14934

Krikos A, Laherty CD, Dixit M (1992): Transcriptional activation of the tumor necrosis factor alpha-inducible zinc finger protein, A20, is mediated by kappa B elements. *J Biol Chem* 267: 17971–17976

Kumar A, Haque J, Lacoste J, Hiscott J, Williams BR (1994): Double-stranded RNA-dependent protein kinase activates transcription factor NF-κB by phosphorylating IκB. *Proc Natl Acad Sci USA* 91: 6288–6292

Kumar S, Rabson AB, Gelinas C (1992): The RxxRxRxxC motif conserved in all rel/κB-proteins is essential for the DNA-binding activity and redox regulation of the v-rel oncoprotein. *Mol Cell Biol* 12: 3094–3106

Kunz J, Hall MN (1993): Cyclosporin A, FK506 and rapamycin: more than just immunosuppression. *Trends Biol Sci* 18: 334–337

Kwak EL, Larochelle DA, Beaumont C, Torti SV, Torti FM (1995): Role for NF-κB in the regulation of ferritin H by tumor necrosis factor-alpha. *J Biol Chem* 270: 152850–15293

Lacoste J, Petropoulos L, Pepin N, Hiscott J (1995): Constitutive phosphorylation and turnover of IκBα in human T-cell leukemia virus type 1-infected and tax-expressing T-cells. *J Virol* 69: 564–569

Laherty C, Hu H, Opipari A, Wang F, Dixit V (1992): The Epstein-Barr virus LMP1 gene product induces A20 zinc finger protein expression by activating nuclear factor κ B. *J Biol Chem* 267: 24157–24160

LeBail O, Schmidt-Ullrich R, Israël A (1993): Promoter analysis of the gene encoding the IκB-alpha/MAD3 inhibitor of NF-κB: positive regulation by members of the rel/NF-κB family. *EMBO J* 12: 5043–5049

LeClair KP, Blanar MA, Sharp PA (1992): The p50 subunit of NF-κB associates with the NF-IL6 transcription factor. *Proc Natl Acad Sci USA* 89: 8145–8149

Legrand-Poels S, Bours V, Piret B, Pflaum M, Epe B, Rentier B, Piette J (1995): Transcription factor NF-κB is activated by photosensitization generating oxidative DNA damage. *J Biol Chem* 270: 6925–6934

Lehmann AR, Carr AM (1995): The ataxia-telangiectasia gene: a link between checkpoint controls, neurodegeneration and cancer. *Trends Genet* 11: 375–377

Lehming N, Thanos D, Brickman JM, Ma J, Maniatis T, Ptashne M (1994): An HMG-like protein that can switch a transcriptional activator to a repressor. *Nature* 371: 175–179

Lenardo M, Pierce JW, Baltimore D (1987): Protein-binding sites in Ig gene enhancers determine transcriptional activity and inducibility. *Science* 236: 1573–1577

Lenardo MJ, Fan CM, Maniatis T, Baltimore D (1989): The involvement of NF-κB in beta-interferon gene regulation reveals its role as widely inducible mediator of signal transduction. *Cell* 56: 287–294

Lernbecher T, Muller U, Wirth T (1993): Distinct NF-κB/Rel transcription factors are responsible for tissue-specific and inducible gene activation. *Nature* 365: 767–770

Lernbecher T, Kistler B, Wirth T (1994): Two distinct mechanisms contribute to the constitutive activation of relB in lymphoid cells. *EMBO J* 13: 4060–4069

Leung K, Nabel GJ (1988): HTLV-1 transactivator induces interleukin-2 receptor expression through an NF-κB-like factor. *Nature* 333: 776–778

Leveillard T, Verma IM (1993): Diverse molecular mechanisms of inhibition of NF-κB/DNA binding complexes by IκB bindings. *Gene Expression* 3: 135–150

Li CCH, Dai R-M, Chen E, Longo DL (1994 a): Phosphorylation of NF-κB-p50 is involved in NF-κB activation and stable DNA binding. *J Biol Chem* 269: 30089–30092

Li CCH, Korner M, Ferris DK, Chen EY, Dai RM, Longo DL (1994 b): NF-κB/Rel family members are physically associated phosphoproteins. *Biochem J* 303: 499–506

Li SF, Sedivy JM (1993): Raf-1 protein kinase activates the NF-κB transcription factor by dissociating the cytoplasmic NF-κB-IκB complex. *Proc Natl Acad Sci USA* 90: 9247–9251

Lin L, Ghosh S (1996): A glycine-rich region in NF-κB p105 functions as a processing signal for the generation of the p50 subunit. *Mol Cell Biol* 16: 2248–2254

Lin YC, Brown K, Siebenlist U (1995): Activation of NF-κB requires proteolysis of the inhibitor IκBα: signal-induced phosphorylation of IκBα alone does not release active NF-κB. *Proc Natl Acad Sci USA* 92: 552–556

Liou HC, Nolan GP, Ghosh S, Fujita T, Baltimore D (1992): The NF-κB p50 precursor, p105, contains an internal IκB-like inhibitor that preferentially inhibits p50. *EMBO J* 11: 3003–3009

Liou HC, Sha WC, Scott ML, Baltimore D (1994): Sequential induction of NF-κB/Rel family proteins during B-cell terminal differentiation. *Mol Cell Biol* 14: 5349–5359

Liptay S, Schmid RM, Nabel EG, Nabel GJ (1994): Transcriptional regulation of NF-κB2: evidence for κB-mediated positive and negative autoregulation. *Mol Cell Biol* 14: 7695–7703

Liu J, Chiles TC, Sen R, Rothstein TL (1991): Inducible nuclear expression of NF-κB in primary B-cells stimulated through the surface Ig receptor. *J Immunol* 146: 1685–1691

Liu J, Perkins ND, Schmid RM, Nabel GJ (1992): Specific NF-κB subunits act in concert with Tat to stimulate human immunodeficiency virus type 1 transcription. *J Virol* 66: 3883–3887

Logeat F, Israël N, Ten R, Blank V, LeBail O, Kourilsky P, Israël A (1991): Inhibition of transcription factors belonging to the rel/NF-κB family by a transdominant negative mutant. *EMBO J* 10: 1827–1832

Los M, Schenk H, Hexel K, Baeuerle PA, Droge W, Schulze-Osthoff K (1995): IL-2 gene expression and NF-κB activation through CD28 requires reactive oxygen production by 5-lipoxygenase. *EMBO J* 14: 3731–3740

Lowenthal JW, Ballard DW, Bohnlein E, Greence WC (1989): Tumor necrosis factor alpha induces proteins that bind specifically to κB-like enhancer elements and regulate interleukin 2 receptor alpha-chain gene expression in primary human T lymphocytes. *Proc Natl Acad Sci USA* 86: 2331–2335

Lozano J, Berra E, Municio MM, Diaz-Meco M, Dominguez I, Sanz L, Moscat J (1994): Protein kinase C zeta isoform is critical for κB-dependent promoter activation by sphingomyelinase. *J Biol Chem* 269: 19200–19202

Lu D, Thompson JD, Gorski GK, Rice NR, Mayer MG, Yunis JJ (1991): Alterations at the rel locus in human lymphoma. *Oncogene* 6: 1235–1241

Machleidt T, Wiegmann K, Henkel T, Schutze S, Baeuerle P, Kronke M (1994): Sphingomyelinase activates proteolytic IκBα degradation in a cell-free system. *J Biol Chem* 269: 13760–13765

MacKichan ML, Logeat F, Israël A (1996): Phosphorylation of p105 PEST sequences via a redox-insensitive pathway up-regulates processing to p50 NF-κB. *J Biol Chem* 271: 6084–6091

Maggirwar SB, Harhaj E, Sun S-C (1995): Activation of NF-κB/rel by tax involves degradation of IκBα and is blocked by a proteasome inhibitor. *Oncogene* 11: 993–998

Maran A, Maitra RK, Kumar A, Dong B, Xiao W, Li G, Williams BR, Torrence PF, Silverman RH (1994): Blockage of NF-κB signaling by selective ablation of an mRNA target by 2-5A antisense chimeras. *Science* 265: 789–792

Mathias S, Younes A, Kan CC, Orlow I, Joseph C, Kolesnick RN (1993): Activation of the sphingomyelin signaling pathway in intact EL4 cells and in a cell-free system by IL-1 beta. *Science* 259: 519–522

Matsusaka T, Fujikawa K, Nishio Y, Mukaida N, Matsushima K, Kishimoto T, Akira S (1993): Transcription factors NF-IL6 and NF-κB synergistically activate transcription of the inflammatory cytokines, interleukin 6 and interleukin 8. *Proc Natl Acad Sci USA* 90: 10193–10197

Mattila PS, Ullman KS, Fiering S, Emmel EA, McCutcheon M, Crabtree GR, Herzenberg LA (1990): The actions of cyclosporin A and FK506 suggest a novel step in the activation of T lymphocytes. *EMBO J* 9: 4425–4433

McElhinny JA, MacMorran WS, Bren GD, Ten RM, Israël A, Paya C (1995): Regulation of IκBα and p105 in monocytes and macrophages persistently infected with human immunodeficiency virus. *J Virol* 69: 1500–1509

McElhinny JA, Trushin SA, Bren GD, Chester N, Paya CV (1996): Casein kinase II phosphorylates IκBα at S-283, S-289, S-293, and T-291 and is required for its degradation. *Mol Cell Biol* 16: 899–906

McKinsey TA, Brockman JA, Scherer DC, Al-Murrani SW, Green PL, Ballard DW (1996): Inactivation of IκBβ by the Tax protein of human T-cell leukemia virus type 1: a potential mechanism for constitutive induction of NF-κB. *Mol Cell Biol* 16: 2083–2090

Mellits KH, Hay RT, Goodbourn S (1993): Proteolytic degradation of MAD3 (IκBα) and enhanced processing of the NF-κB precursor p105 are obligatory steps in the activation of NF-κB. *Nucleic Acids Res* 21: 5059–5066

Menon SD, Guy GR, Tan YH (1995): Involvement of a putative protein-tyrosine phosphatase and IκBα serine phosphorylation in nuclear factor κB activation by tumor necrosis factor. *J Biol Chem* 270: 18881–18887

Mercurio F, Didonato JA, Rosette C, Karin M (1993): p105 and p98 precursor proteins play an active role in NF-κB-mediated signal transduction. *Genes Dev* 7: 705–718

Messina JP, Lawrence DA (1989): Cell cycle progression of glutathione-depleted human peripheral blood mononuclear cells is inhibited at S phase. *J Immunol* 143: 1974–1981

Michaely P, Bennett V (1992): The ANK repeat: a ubiquitous motif involved in macromolecular recognition. *Trends Cell Biol* 2: 127–129

Migliazza A, Lombardi L, Rocchi M, Trecca D, Chang C-C, Antonacci R, Fracchiolla NS, Ciana P, Maiolo AT, Neri A (1994): Heterogeneous chromosomal aberrations generate 3′ truncations of the NFKB2/*lyt*-10 gene in lymphoid malignancies. *Blood* 84: 3850–3860

Miyamoto S, Chiao PJ, Verma IM (1994a): Enhanced IκBα degradation is responsible for constitutive NF-κB activity in mature murine B-cell lines. *Mol Cell Biol* 14: 3276–3282

Miyamoto S, Maki M, Schmitt MJ, Hatanaka M, Verma IM (1994b): Tumor necrosis factor alpha-induced phosphorylation of IκBα is a signal for its degradation but not dissociation from NF-κB. *Proc Natl Acad Sci USA* 91: 12740–12744

Montaner S, Ramos A, Perona R, Esteve P, Carnero A, Lacal JC (1995): Overexpression of PKCζ in NIH3T3 cells does not induce cell transformation nor tumerigenicity and does not alter NFκB activity. *Oncogene* 10: 2213–2220

Mosialos G, Hamer P, Capobianco AJ, Laursen RA, Gilmore TD (1991): A protein kinase-A recognition sequence is structurally linked to transformation by p59$^{v\text{-}rel}$ and cytoplasmic retention of p68$^{c\text{-}rel}$. *Mol Cell Biol* 11: 5867–5877

Mosialos G, Birkenbach M, Yalamanchili R, VanArsdale T, Ware C, Kieff E (1995): The Epstein-Barr virus transforming protein LMP1 engages signaling proteins for the tumor necrosis factor receptor family. *Cell* 80: 389–399

Muller CW, Rey FA, Sodeoka M, Verdine GL, Harrison SC (1995a): Structure of the NF-κB p50 homodimer bound to DNA. *Nature* 373: 311–317

Muller G, Ayoub M, Storz P, Rennecke J, Fabbro D, Pfizenmaier K (1995b): PKC zeta is a molecular switch in signal transduction of TNF-alpha, bifunctionally regulated by ceramide and arachidonic acid. *EMBO J* 14: 1961–1969

Munoz E, Courtois G, Veschambre P, Jalinot P, Israël A (1994): Tax induces nuclear translocation of NF-κB through dissociation of cytoplasmic complexes containing p105 or p100 but does not induce degradation of IκBα/MAD3. *J Virol* 68: 8035–8044

Muroi M, Suzuki T (1993): Role of protein kinase-A in LPS-induced activation of NF-κB proteins of a mouse macrophage-like cell line, J 774. *Cell Signal* 5: 289–298

Murphy TL, Cleveland MG, Kulesza P, Magram J, Murphy KM (1995): Regulation of interleukin 12 p40 expression through an NF-κB half-site. *Mol Cell Biol* 15: 5258–5267

Nabel G, Baltimore D (1987): An inducible transcription factor activates expression of human immunodeficiency virus in T-cells. *Nature* 326: 711–713

Nakanishi H, Brewer KA, Exton JH (1993): Activation of the ζ isozyme of protein kinase C by phosphatidylinositol 3,4,5-trisphosphate. *J Biol Chem* 268: 13–16

Narayanan R, Klement JF, Ruben SM, Higgins KA, Rosen CA (1992): Identification of a naturally occurring transforming variant of the p65 subunit of NF-κB. *Science* 256: 367–370

Narayanan R, Higgins KA, Perez JR, Coleman TA, Rosen CA (1993): Evidence for differential functions of the p50 and p65 subunits of NF-κB with a cell adhesion model. *Mol Cell Biol* 13: 3802–3810

Naumann M, Scheidereit C (1994): Activation of NF-κB in vivo is regulated by multiple phosphorylations. *EMBO J* 13: 4597–4607

Naumann M, Wulczyn FG, Scheidereit C (1993): The NF-κB precursor p105 and the proto-oncogene product Bcl-3 are IκB molecules and control nuclear translocation of NF-κB. *EMBO J* 12: 213–222

Neri A, Chang CC, Lombardi L, Salina M, Corradini P, Maiolo AT, Chaganti RS, Dalla Forera R (1991): B-cell lymphoma-associated chromosomal translocation involves candidate oncogene lyt-10, homologous to NF-κB p50. *Cell* 67: 1075–1087

Neumann M, Tsapos K, Scheppler JA, Ross J, Franza BR (1992): Identification of complex formation between two intracellular tyrosine kinase substrates – human c-rel and the p105 precursor of p50 NF-κB. *Oncogene* 7: 2095–2104

Neumann M, Grieshammer T, Chuvpilo S, Kneitz B, Lohoff M, Schimpl A, Franza BR Jr, Serfling E (1995): RelA/p65 is a molecular target for the immunosuppressive action of protein kinase A. *EMBO J* 14: 1991–2004

Nieman PE, Thomas SJ, Loring G (1991): Induction of apoptosis during normal and neoplastic B-cell development in the bursa of Fabricus. *Proc Natl Acad Sci USA* 88: 5857–5861

Nolan GP, Ghosh S, Liou HC, Tempst P, Baltimore D (1991): DNA binding and IκB inhibition of the cloned p65 subunit of NF-κB, a rel-related polypeptide. *Cell* 64: 961–969

Nolan GP, Fujita T, Bhatia K, Huppi C, Liou HC, Scott ML, Baltimore D (1993): The bcl-3 proto-oncogene encodes a nuclear IκB-like molecule that preferentially interacts with NF-κB p50 and p52 in a phosphorylation-dependent manner. *Mol Cell Biol* 13: 3557–3566

Ohno H, Takimoto G, McKeithan TW (1990): The candidate proto-oncogene bcl3 is related to genes implicated in cell lineage determination and cell cycle control. *Cell* 60: 991–997

Olashaw NE, Kowalik TF, Huang ES, Pledger WJ (1992): Induction of NF-κB-like activity by platelet-derived growth factor in mouse fibroblasts. *Mol Cell Biol* 3: 1131–1139

Orian A, Whiteside S, Israël A, Stancovski I, Schwartz AL, Ciechanover A (1996): Ubiquitin-mediated processing of NF-κB transcriptional activator precursor p105: Reconstitution of a cell free system and identification of the ubiquitin-carrier protein, E2, and a novel ubiquitin-protein ligase, E3, involved in conjugation. *J Biol Chem* 270: 21707–21714

Osborn L, Kunkel S, Nabel GJ (1989): Tumor necrosis factor alpha and interleukin 1 stimulate the human immunodeficiency virus enhancer by activation of the nuclear factor kappa B. *Proc Natl Acad Sci USA* 86: 2336–2340

Pahl HL, Baeuerle PA (1995): A novel signal transduction pathway from the endoplasmic reticulum to the nucleus is mediated by transcription factor NF-κB. *EMBO J* 14: 2580–2588

Paige CJ, Kincade PW, Ralph P (1978): Murine B-cell leukemia line with inducible surface immunoglobulin expression. *J Immunol* 121: 641–647

Paine E, Scheinman RI, Baldwin AS Jr, Raab-Traub N (1995): Expression of LMP1 in epithelial cells leads to the activation of a select subset of NF-κB/rel family protein. *J Virol* 69: 4572–4576

Palombella VJ, Rando OJ, Goldberg AL, Maniatis T (1994): The ubiquitin-proteasome pathway is required for processing the NF-κB1 precursor protein and the activation of NF-κB. *Cell* 78: 773–785

Peng H-B, Libby P, Liao JK (1995): Induction and stabilization of IκBα by nitric oxide mediates inhibition of NF-κB. *J Biol Chem* 270: 14214–14219

Perez JR, Higgins-Sochaski KA, Maltese JY, Narayanan R (1994): Regulation of adhesion and growth of fibrosarcoma celsl by NF-κB/RelA involves transforming growth factor beta. *Mol Cell Biol* 14: 5326–5332

Perez P, Lira SA, Bravo R (1995): Overexpression of relA in transgenic mouse thymocytes: specific increase in levels of the inhibitor protein IκBα. *Mol Cell Biol* 15: 3523–3530

Perkins ND, Schmid RM, Duckett CS, Leung K, Rice NR, Nabel GJ (1992): Distinct combinations of NF-κB subunits determine the specificity of transcriptional activation. *Proc Natl Acad Sci USA* 89: 1529–1533

Perkins ND, Edwards NL, Duckett CS, Agranoff AB, Schmid RM, Nabel GJ (1993): A cooperative interaction between NF-κB and Sp1 is required for HIV-1 enhancer activation. *EMBO J* 12: 3551–3558

Perkins ND, Agranoff AB, Pascal E, Nabel GJ (1994): An interaction between the DNA-binding domains of RelA(p65) and Sp1 mediates human immunodeficiency virus gene activation. *Mol Cell Biol* 14: 6570–6583

Peters J-M (1994): Proteasomes: protein degradation machines of the cell. *Trends Biol Sci* 19: 377–382

Peterson U-M, Bjorklund G, Ip YT, Engstrom Y (1995): The *dorsal*-related immunity factor, Dif, is a sequence-specific *trans*-activator of *Drosophila Cecropin* gene expression. *EMBO J* 14: 3146–3158

Pierce JW, Jamieson CA, Ross JL, Sen R (1995): Activation of IL-2 receptor alpha-chain gene by individual members of the rel oncogene family in association with serum response factors. *J Immunol* 155: 1972–1980

Ray P, Zhang DH, Elias JA, Ray A (1995): Cloning of a differentially expressed IκB-related protein. *J Biol Chem* 270: 10680–10685

Rechsteiner M (1990): PEST sequences are signals for rapid intracellular proteolysis. *Semin Cell Biol* 1: 433–440

Rice NR, Ernst MK (1993): In vivo control of NF-κB activation by IκBα. *EMBO J* 12: 4685–4695

Rice NR, MacKichan ML, Israël A (1992): The precursor of NF-κB p50 has IκB-like functions. *Cell* 71: 243–253

Riviere Y, Blank V, Kourilsky P, Israël A (1991): Processing of the precursor of NF-κB by the HIV-1 protease during acute infection. *Nature* 350: 625–628

Rodriguez MS, Michalopoulos I, Arenzana-Seisdedos F, Hay RT (1995): Inducible degradation of IκBα in vitro and in vivo requires the acidic C-terminal domain of the protein. *Mol Cell Biol* 15: 2413–2419

Rothe M, Sarma V, Dicit VM, Goeddel DV (1995): TRAF2-mediated activation of NF-κB by TNF receptor 2 and CD40. *Science* 269: 1424–1427

Roulston A, Beauparlant P, Rice N, Hiscott J (1993): Chronic human immunodeficiency virus type-1 infections stimulates distinct NF-κB/rel DNA binding activities in myelomonoblastic cells. *J Virol* 67: 5235–5246

Rousset R, Desbois C, Bantignies F, Jalinot P (1996): Effects on NF-κB 1/p105 processing of the interaction between the HTLV-1 transactivator Tax and the proteasome. *Nature* 381: 328–331

Ruben SM, Dillon PJ, Schreck R, Henkel T, Chen CH, Maher M, Baeuerle PA, Rosen CA (1991): Isolation of a rel-related human cDNA that potentially encodes the 65 kD subunit of NF-κB. *Science* 251: 1490–1493

Ruben SM, Klement JF, Coleman TA, Maher M, Chen CH, Rosen CA (1992): I-Rel – A novel rel-related protein that inhibits NF-κB transcriptional activity. *Genes Dev* 6: 745–760

Ryseck RP, Bull P, Takamiya M, Bours V, Siebenlist U, Dobrzanski P, Bravo R (1992): RelB, a new Rel family transcription activator that can interact with p50-NF-κB. *Mol Cell Biol* 12: 674–684

Sachdev S, Rottjakob EM, Diehl JA, Hannink M (1995): IκBα-mediated inhibition of nuclear transport and DNA-binding by rel proteins are separable functions: phosphorylation of C-terminal serine residues of IκBα is specifically required for inhibition of DNA-binding. *Oncogene* 11: 811–823

Sarkar S, Gilmore TD (1993): Transformation by the v-Rel oncoprotein requires sequences carboxy-terminal to the rel homology domain. *Oncogene* 8: 2245–2252

Scheinman RI, Cogswell PC, Lofquist AK, Baldwin AS Jr (1995): Role of transcriptional activation of IκBα in mediation of immunosuppression by glucocorticoids. *Science* 270: 283–286

Schmid RM, Perkins ND, Duckett CS, Andrews PC, Nabel GJ (1991): Cloning of an NF-κB subunit which stimulates HIV transcription in synergy with p65. *Nature* 352: 733–736

Schmitz ML, Silva MAD, Altmann H, Czisch M, Holak TA, Baeuerle PA (1994): Structural and functional analysis of the NF-κB p65 C terminus – an acidic and modular transactivation domain with the potential to adopt an alpha-helical conformation. *J Biol Chem* 269: 25613–25620

Schreck R, Baeuerle P (1991): A role for oxygen radicals as second messengers. *Trends Cell Biol* 1: 39–42

Schreck R, Zorbas H, Winnacker EL, Baeuerle PA (1990): The NF-κB transcription factor induces DNA binding which is modulated by its 65-kD subunit. *Nucleic Acids Res* 18: 6497–6502

Schreck R, Rieber P, Baeuerle PA (1991): Reactive oxygen intermediates as apparently widely used messengers in the activation of the NF-κB transcription factor and HIV-1. *EMBO J* 10: 2247–2258

Schutze S, Potthoff K, Machleidt T, Berkovic D, Wiegmann K, Kronke M (1992): TNF Activates NF-κB by phosphatidylcholine-specific phospholipase-C-induced acidic sphingomyelin breakdown. *Cell* 71: 765–776

Scott ML, Fujita T, Liou HC, Nolan GP, Baltimore D (1993): The p65-subunit of NF-κB regulates IκB by two distinct mechanisms. *Genes Dev* 7: 1266–1276

Sen R, Baltimore D (1986a): Multiple nuclear factors interact with the immunoglobulin enhancer sequences. *Cell* 46: 705–716

Sen R, Baltimore D (1986b): Inducibility of κ immunoglobulin enhancer-binding protein NF-κB by a posttranslational mechanism. *Cell* 47: 921–928

Sha WC, Liou HC, Tuomanen EI, Baltimore D (1995): Targeted disruption of the p50 subunit of NF-κB leads to multifocal defects in immune responses. *Cell* 80: 321–330

Shelton CA, Wasserman SA (1993): Pelle encodes a protein kinase required to establish dorsoventral polarity in the *Drosophila* embryo. *Cell* 72: 515–525

Shiio Y, Sawada J-I, Handa H, Yamamoto T, Inou J-I (1996): Activation of the retinoblastoma gene expression by Bcl-3: implication for muscle cell differentiation. *Oncogene* 12: 1837–1845

Shirakawa F, Mizel SB (1989): In vitro activation and nuclear translocation of NF-κB catalyzed by cyclic AMP-dependent protein kinase and protein kinase C. *Mol Cell Biol* 9: 2424–2430

Siebenlist U, Fransozo G, Brown K (1994): Structure, regulation and function of NF-κB. *Annu Rev Cell Biol* 10: 405–430

Singh S, Aggarwal BB (1995): Protein-tyrosine phosphatase inhibitors block tumor necrosis factor-dependent activation of the nuclear transcription factor NF-κB. *J Biol Chem* 270: 10631–10639

Sokoloski JA, Sartorelli AC, Rosen CA, Narayanan R (1993): Antisense oligonucleotides to the p65 subunit of NF-κB block CD11b expression and alter adhesion properties of differentiated HL-60 granulocytes. *Blood* 82: 625–832

Staal FJ, Roederer M, Herzenberg LA, Herzenberg LA (1990): Intracellular thiols regulate activation of nuclear factor kappa B and transcription of human immunodeficiency virus. *Proc Natl Acad Sci USA* 87: 9943–9947

Stein B, Yang MX (1995): Repression of the interleukin-6 promoter by estrogen receptor is mediated by NF-κB and C/EBPβ. *Mol Cell Biol* 15: 4971–4979

Stein B, Rahmsdorf HJ, Steffer A, Litfin M, Herrlich PH (1989): UV-induced DNA damage is an intermediate step in UV-induced expression of human immunodeficiency virus type 1, collagenase, c-*fos*, and metallothionein. *Mol Cell Biol* 9: 5169–5181

Stein B, Cogswell PC, Baldwin AJ (1993): Functional and physical associations between NF-κB and C/EBP family members: a Rel domain-bZIP interaction. *Mol Cell Biol* 13: 3964–3974

Stephens RM, Rice NR, Hiebisch RR, Bose HR Jr, Gilden RV (1983): Nucleotide sequence of v-*rel*: The oncogene of reticuloendotheliosis virus. *Proc Natl Acad Sci USA* 80: 6229–6233

Sun SC, Ganchi PA, Ballard DW, Greene WC (1993): NF-κB controls expression of inhibitor IκBα – Evidence for an inducible autoregulatory pathway. *Science* 259: 1912–1915

Sun SC, Elwood J, Beraud C, Greene WC (1994a): Human T-cell leukemia virus type I Tax activation of NF-κB/Rel involves phosphorylation and degradation of IκBα and RelA (p65)-mediated induction of the c-rel gene. *Mol Cell Biol* 14: 7377–7384

Sun SC, Ganchi PA, Beraud C, Ballard DW, Greene WC (1994b): Autoregulation of the NF-κB transactivator relA (p65) by multiple cytoplasmic inhibitors containing ankyrin motifs. *Proc Natl Acad Sci USA* 91: 1346–1350

Sun SC, Maggirwar SB, Harhaj E (1995): Activation of NF-κB py phosphatase inhibitors involves the phosphorylation of IκBα at phosphatase 2A-sensitive sites. *J Biol Chem* 270: 18347–18351

Suzuki YJ, Mizuno M, Packer L (1994): Signal transduction for nuclear factor-κ B activation – proposed location of antioxidant-inhibitable step. *J Immunol* 153: 5008–5015

Ten RM, Paya CV, Israël N, LeBail O, Mattei MG, Virelizier JL, Kourilsky P, Israël A (1992): The characterization of the promoter of the gene encoding the p50 subunit of NF-κB indicates that it participates in its own regulation. *EMBO J* 11: 195–203

Tewari M, Dixit VM (1996): Recent advances in tumor necrosis factor and CD40 signalling. *Curr Opin Genet Dev* 6: 39–44

Thevenin C, Kim SJ, Rieckmann P, Fujiki H, Norcross MA, Sporn MB, Fauci AS, Kehrl JH (1990): Induction of nuclear factor-kappa B and the human immunodeficiency virus long terminal repeat by okadaic acid, a specific inhibitor of phosphatases 1 and 2A. *New Biol* 2: 793–800

Thompson JE, Philips RJ, Erdjument-Bromage H, Tempst P, Ghosh S (1995): IκBβ regulates the persistent response in a biphasic activation of NF-κB. *Cell* 80: 573–582

Tobin D, Nilsson M, Toftgard R (1996): Ras-independent activation of rel-family transcription factors by UVB and TPA in cultured keratinocytes. *Oncogene* 12: 785–793

Toledano MB, Leonard WJ (1991): Modulation of transcription factor NF-κB binding activity by oxidation-reduction in vitro. *Proc Natl Acad Sci USA* 88: 4328–4332

Tong-Starksen SE, Luciw PA, Peterlin BM (1989): Signaling through T lymphocyte surface proteins, TCR/CD3 and CD28, activates the HIV-1 long terminal repeat. *J Immunol* 142: 702–707

Tong-Starksen SE, Welsh TM, Peterlin BM (1990): Differences in transcriptional enhancers of HIV-1 and HIV-2. *J Immunol* 145: 4348–4354

Torcia M, Bracci-Laudiero L, Lucibello M, Nencioni L, Labardi D, Rubartelli A, Cozzolino F, Aloe L, Garaci E (1996): Nerve growth factor is an autocrine survival factor for memory B lymphocytes. *Cell* 85: 345–356

Traenckner EBM, Wilk S, Baeuerle PA (1994): A proteasome inhibitor prevents activation of NF-κB and stabilizes a newly phosphorylated form of IκBα that is still bound to NF-κB. *EMBO J* 13: 5433–5441

Traenckner EBM, Pahl HL, Henkel T, Schmidt KN, Wilk S, Baeuerle PA (1995): Phosphorylation of human IκBα on serines 32 and 36 controls IκBα proteolysis and NF-κB activation in response to diverse stimuli. *EMBO J* 14: 2876–2883

Ullman KS, Northrop JP, Verweij CL, Crabtree GR (1990): Transmission of signals from the T lymphocyte antigen receptor to the genes responsible for cell proliferation and immune function: The missing link. *Annu Rev Immunol* 8: 421–452

Vassilli P (1992): The pathophysiology of tumor necrosis factors. *Annu Rev Immunol* 10: 411–452

Verma IM, Stevenson JK, Schwarz EM, Van Antwerp D, Miyamoto S (1995): Rel/NF-κB/IκB family: intimate tales of association and dissociaton. *Genes Dev* 9: 2723–2735

Wang P, Wu P, Siegel MI, Egan RW, Billah MM (1995): Interleukin (IL)-10 inhibits nuclear factor κB (NF-κB) activation in human monocytes. *J Biol Chem* 270: 9558–9563

Watanabe M, Muramatsu MA, Hirai H, Suzuki T, Fujisawa Y, Yoshida M, Arai KI, Arai N (1993): HTLV-I encoded tax in association with NF-κB precursor p105 enhances nuclear localization of NF-κB p50 and p65 in transfected cells. *Oncogene* 8: 2949–2958

Watanabe M, Muramatsu MA, Tsuboi A, Arai K (1994): Differential response of NF-κB1 p105 and NF-κB2 p100 to HTLV-I encoded tax. *FEBS Lett* 342: 115–118

Weih F, Carrasco D, Bravo R (1994): Constitutive and inducible Rel/NF-κB activities in mouse thymus and spleen. *Oncogene* 9: 3289–3297

Weih F, Carrasco D, Durham SK, Barton DS, Rizzo CA, Ryseck RP, Lira SA, Bravo R (1995): Multiorgan inflammation and hematopoietic abnormalities in mice with a targeted disruption of RelB, a member of the NF-κB/Rel family. *Cell* 80: 331–340

Westendorp MO, Shatrov VA, Schulze-Osthoff K, Frank R, Kraft M, Los M, Krammer PH, Droge W, Lehmann V (1995): HIV-1 Tat potentiates TNF-induced NF-κB activation and cytotoxicity by altering the cellular redox state. *EMBO J* 14: 546–554

Westwick JK, Bielawska AE, Dbaibo G, Hannun A, Brenner DA (1995): Ceramide activates the stress-activated protein kinases. *J Biol Chem* 270: 22689–22692

Whalen AM, Steward R (1993): Dissociation of the dorsal-cactus complex and phosphorylation of the dorsal protein correlate with the nuclear localization of dorsal. *J Cell Biol* 123: 523–534

White DW, Gilmore TD (1993): Temperature-sensitive transforming mutants of the v-rel oncogene. *J Virol* 67: 6876–6881

White DW, Roy A, Gilmore TD (1995): The v-rel oncoprotein blocks apoptosis and proteolysis of IκBα in transformed chicken spleen cells. *Oncogene* 10: 857–868

Whiteside ST, Ernst MK, LeBail O, Laurent-Winter C, Rice N, Israël A (1995): N- and C-terminal sequences control degradation of MAD3/IκBα in response to inducers of NF-κB activity. *Mol Cell Biol* 15: 5339–5345

Whitley MZ, Thanos D, Read MA, Maniatis T, Collins T (1994): A striking similarity in the organization of the E-selectin and beta interferon gene promoters. *Mol Cell Biol* 14: 6464–6475

Wiegmann K, Schutze S, Kampen E, Himmler A, Machleidt T, Kronke M (1992): Human 55-kDa receptor for tumor necrosis factor coupled to signal transduction cascades. *J Biol Chem* 267: 17997–18001

Wilhelmsen KC, Eggleton K, Temin HM (1984): Nucleic acid sequences of the oncogene v-*rel* in reticuloendotheliosis virus strain T and its cellular homolog, the proto-oncogene c-*rel*. *J Virol* 52: 172–182

Wulczyn FG, Naumann M, Scheidereit C (1992): Candidate proto-oncogene bcl-3 encodes a subunit-specific inhibitor of transcription factor NF-κB. *Nature* 358: 597–599

Xie QW, Kashiwabara Y, Nathan C (1994): Role of transcription factor NF-κB/Rel in induction of nitric oxide synthase. *J Biol Chem* 269: 4705–4708

Yang YL, Reis LFL, Pavlovic J, Aguzzi A, Schafer R, Kumar A, Williams BRG, Aguet M, Weissmann C (1995): Deficient signaling in mice devoid of double-stranded RNA-dependent protein kinase. *EMBO J* 14: 6095–6106

Yang Z, Costanzo M, Golde DW, Kolesnick RN (1993): Tumor necrosis factor activation of the sphingomyelin pathway signals nuclear factor κB translocation in intact HL-60 cells. *J Biol Chem* 268: 20520–20523

Yodoi J, Uchiyama T (1992): Diseases associated with HTLV-I: virus, IL-2 receptor dysregulation and redox regulaton. *Immunol Today* 13: 405–411

Zabel U, Baeuerle PA (1990): Purified human IκB rapidly dissociate the complex of the NF-κB transcription factor with its cognate DNA. *Cell* 61: 255–265

Zabel U, Henkel T, Silva MD, Baeuerle PA (1993): Nuclear uptake control of NF-κB by MAD-3, an IκB protein present in the nucleus. *EMBO J* 12: 201–211

Zhu L, Jones PP (1990): Transcriptional control of the invariant chain gene involves promoter and enhancer elements common to and distinct from major histocompatibility complex Class II genes. *Mol Cell Biol* 10: 3906–3916

Zimmermann K, Dobrovnik M, Ballaun C, Bevec D, Hauber J, Bohnlein E (1991): Transactivation of the HIV-1 LTR by the HIV-1 Tat and HTLV-I Tax proteins is mediated by different cis-acting sequences. *Virology* 182: 874–888

6

Oncogenes as Transcriptional Regulators
Vol. 1: Retroviral Oncogenes
ed. by M. Yaniv and J. Ghysdael

Structure/Function and Oncogenic Conversion of Fos and Jun

ANDREW J. BANNISTER AND TONY KOUZARIDES

Introduction

The extracellular environment is constantly changing and consequently eukaryotic cells have to respond to these changes. The intracellular signals generated by these stimuli invariably require a cellular response at the gene transcriptional level, which involves the functional mobilization of numerous transcription factors. The AP-1 transcription factor is a key player in mediating this environmental response.

AP-1 consists of a collection of distinct, but related, dimeric protein complexes which have binding sites in a wide variety of promoters. The prototypes for this family of factors are the c-Fos and c-Jun proteins, which heterodimerize and regulate transcription from promoters containing AP-1 DNA binding sites (TGAC/GTCA; also referred to as *T*PA-*R*esponsive *E*lements, TREs; Curran and Franza, 1988).

The c-*fos* and *c-jun* genes were originally identified because of their oncogenic viral counterparts. Transduction of the *c-fos* gene has occurred in two murine osteosarcoma viruses, FBJ.MuSV and FBR.MuSV (Finkel et al, 1966; Finkel et al, 1975). v-Jun was identified as the oncogene in ASV17, a virus isolated from a spontaneous chicken sarcoma (Maki et al, 1987). Using the viral genes as probes, both c-Fos and c-Jun were isolated by screening DNA libraries. Subsequent work revealed that c-Fos and c-Jun are members of two related families; c-Fos is closely related to Fra1, Fra2, and FosB whereas c-Jun is related to JunB and JunD (Kouzarides and Ziff, 1989a; Angel and Karin, 1991). It is beyond the scope of this chapter to comprehensively cover all members of both families. Therefore, this review will focus on the prototypes of the AP-1 family, namely c-Fos and c-Jun.

The *c-fos* and *c-jun* genes are immediate-early genes, which respond rapidly to many different extracellular stimuli without the need for *de novo* protein synthesis (reviewed in Angel and Karin, 1991; Angel and Herrlich, 1994). The mRNA, and the encoded proteins, have very short half-lives leading to a transient appearance of c-Fos and c-Jun proteins. Posttranslational modification then occurs which stimulates the activity of the c-Fos/c-Jun complex. AP-1 DNA binding sites are found in the promoters of a variety of genes, and confers to them the ability to respond to a wide spectrum of extracellular agents (Angel and Karin, 1991; Franklin et al, 1992; Dixit et al, 1989; Radler-Pohl et al, 1993). AP-1 activity has been implicated in many different cellular processes. Broadly speaking, AP-1 function is involved in both proliferative and differentiation pathways. Thus, AP-1 regulated genes include differentiation associated genes, such as the plasminogen activator and the keratin 18 genes (Rickles et al, 1989; Oshima et al, 1990), and proliferation associated genes, such as TGFα and TGFβ (Ryseck and Bravo, 1991; Kim et al, 1990).

In this chapter, we will describe how the structure of the c-Fos and c-Jun proteins relates to their function and relate this information to the mechanism by which the viral Fos and Jun proteins may mediate oncogenic transformation.

Structure and Function of c-Fos and c-Jun

DNA-binding domain

c-Fos and c-Jun belong to the bZIP super-family of transcription factors, which dimerize and bind DNA via their basic/leucine zipper (bZIP) region (Halazonetis et al, 1988; Kouzarides and Ziff, 1988, 1989b; Nakabeppu et al, 1988; Sassone-Corsi et al, 1988; Gentz et al, 1989; Schuermann et al, 1989; Turner and Tjian, 1989).

Dimerization

The leucine zipper dimerization interface of this family of proteins provides a surface for multiple interactions between family members. The c-Jun protein can form either homodimers or heterodimers with c-Fos but, c-Fos itself is unable to form homodimers. These simple dimerization rules also hold true for other family members. Thus, all Fos members can heterodimerize with any member of the c-Jun family, but they can not form homodimers. All c-Jun members can homodimerize and can form heterodimers with any c-Fos family member (for review, see Kouzarides and Ziff, 1989a). In addition, c-Jun has the ability to dimerize with bZIP families other than c-Fos. For example, it can complex with the ATF or CREB families (Benbrook and Jones, 1990; Chatton et al, 1994).

The ability of any of these factors to dimerize is determined by the leucine zipper, a helical region, composed of a heptad repeat of leucines (Landschultz et al, 1988; Kouzarides and Ziff, 1988). An additional repeat of hydrophobic residues at the fourth position, following the first leucine, makes the zipper analogous to coiled-coils of intermediate filaments which contain a similar repeat of hydrophobic amino acids called the 4-3 repeat. The leucine zipper is not promiscuous in its interactions, but has determinants which dictate dimerization specificity (Kouzarides and Ziff, 1989b; Sellers and Struhl, 1989). As shown by the crystal structure, discrimination between different partners is provided by a number of charged residues within the zipper, (Glover and Harrison, 1995). Compatable electrostatic interactions with another zipper allow interaction to proceed. However, c-Fos is unable to form homodimers because when two c-Fos monomers come together, there is an electrostatic repulsion between non-hydrophobic residues in their leucine zippers. Thus, c-Fos is an obligate heterodimeric protein.

DNA-binding

In the bZIP family of proteins, a region rich in basic amino acids occurs immediately N-terminal to the leucine zipper. This region of the bZIP domain is responsible for sequence-specific DNA binding (Nakabeppu and Nathans, 1989; Kouzarides and Ziff, 1989b). Following dimerization via the leucine zipper, a bZIP dimeric protein binds DNA because residues within the two basic regions make base-specific contacts with the DNA (Glover and Harrison, 1995). These contacts are within the major groove of DNA and it is the partners within a given dimer which dictate the sequence specificity of the DNA binding site. For example, dimers consisting solely of c-Jun family members bind, with varying degrees of affinity, to either TRE sites or CRE sites. However, c-Jun/ATF2 or c-Jun/CREB dimers have a preference for CRE sites, whereas c-Fos family/c-Jun family heterodimers have a higher affinity for TRE sites (Hai and Curran, 1991; Hadman et al, 1993; Halazonetis et al, 1988; Ryseck and Bravo, 1991).

Transcriptional activation domains

Both c-Fos and c-Jun are potent transcriptional activator proteins. Many studies have tried to establish the regions within each protein which function as transcriptional activation domains. These studies show that both c-Fos and c-Jun contain a number of activation domains and these domains are modular in structure.

c-Fos activation domains

Transcriptional activation domains have been mapped both N- and C-terminal to the c-Fos bZIP DNA binding domain (Figure 6.1 (A)) (Abate

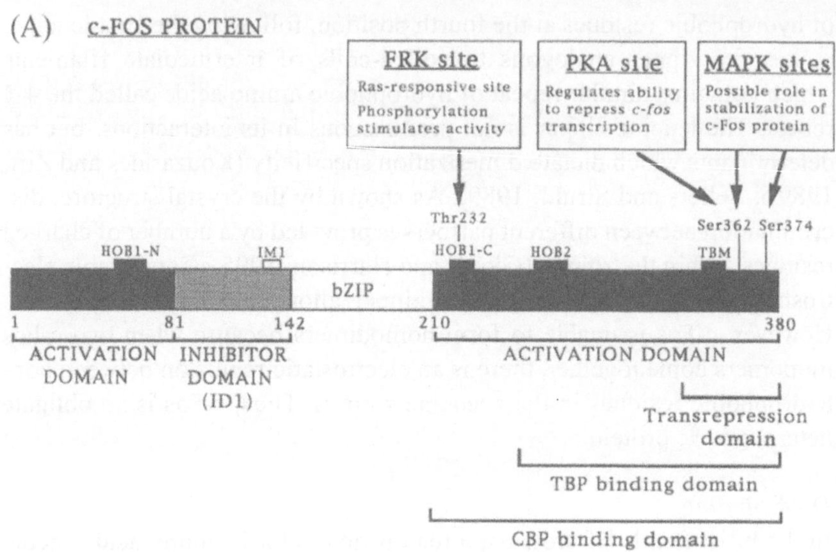

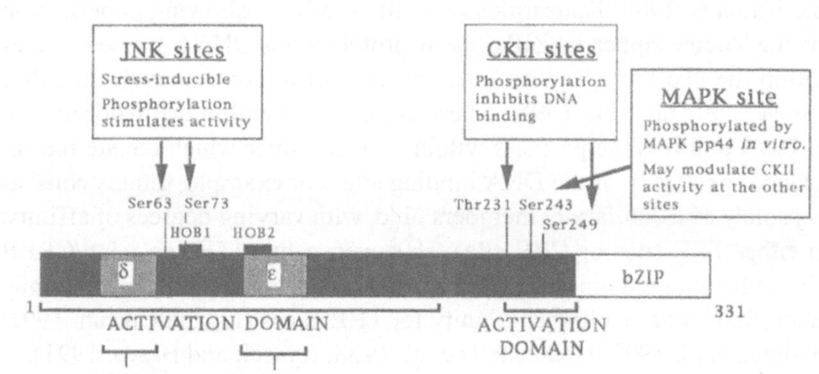

Figure 6.1 Diagrammatic representation of the c-Fos and c-Jun proteins. (A) Structural features of c-Fos. The positions of motifs, HOB1-N, HOB1-C, HOB2, TBM and IM1 are indicated. Domains implicated in activation and inhibition of activation are shown. Functional phosphorylation sites and the relevant kinase are indicated. Regions in the C-terminus, required for *c-fos* transrepression, TBP, and CBP binding, are shown. bZIP; basic/zipper DNA binding domain. (B) Structural features of c-Jun. The positions of the HOB1 and HOB2 motifs are indicated. Activation domains and inhibitor binding domains are indicated. Functional phosphorylation sites, and the relevant kinase, are indicated. Regions in the N-terminus, required for JNK and CBP binding, are shown. bZIP; basic/zipper DNA binding domain.

et al, 1991; Sutherland et al, 1992; Jooss et al, 1994; Brown et al, 1995). Detailed analysis of the C-terminal activation functions revealed the presence of three 'Fos activation modules' (FAMs) within amino acids 210–380 (Sutherland et al, 1992). Each 'module' has very little independent activity as a Gal4-fusion but can activate transcription synergistically when combined with a second module. Two of these modules contain motifs (HOB1 and HOB2) which are also present in c-Jun. The third activation module contains a motif, TBM, which mediates binding to the TATA-binding protein, TBP, both in vitro and in vivo (Metz et al, 1994). Interestingly, TBP binding to the TBM is significantly increased by the HOB1 and HOB2 motifs and this may explain, at least in part, how the c-Fos activation modules cooperate to increase transcription. Indeed, the HOB2 motif may represent a second TBP binding site since a HOB2 motif, present within C/EBP, has been shown to contact TBP (Nerlov and Ziff, 1995).

Cooperativity between activation modules may also reflect the cooperation in binding distinct proteins required for activation. For example, the C-terminal domains of c-Fos bind the CBP co-activator protein, and the TBP protein, using distinct sequences within cooperating activation modules (Bannister and Kouzarides, 1995).

The c-Fos N-terminus also contains activation domains (Abate et al, 1991; Jooss et al, 1994; Brown et al, 1995). A region at the very N-terminus appears to be composed of two cooperating modules present within the first 81 amino acids (Brown et al, 1995). One of these modules contains a HOB1 motif (HOB1-N) (Figure 6.1 (A)).

c-Jun activation domains

The c-Jun protein, contains at least two activation domains (Figure 6.1 (B)), the most potent being at the very N-terminus (Baichwal and Tjian, 1990; Bohmann and Tjian, 1989; Abate et al, 1991; Angel et al, 1989). This N-terminal activation domain was originally shown to be composed of three subdomains, I, II, and III (Angel et al, 1989). Subsequent work identified that regions II and III were homologous to the HOB1 and HOB2 motifs present in the c-Fos C-terminal activation domain (Sutherland et al, 1992). Importantly, the HOB regions of c-Jun are interchangeable with those of c-Fos in terms of transcriptional cooperativity, indicating functional similarity.

As with c-Fos, CBP is able to function as a coactivator for c-Jun activity in vivo (Arias et al, 1994; Bannister et al, 1995). This requires the N-terminal activation domain of c-Jun which interacts with the CREB binding domain of CBP (Bannister et al, 1995). Two phosphorylation sites within the c-Jun N-terminal activation domain (Ser63 and Ser73) are required for efficient binding to CBP, although the need for phosphorylation of these sites still has to be determined.

Inhibitor domains

c-Fos

An inhibitor domain has been identified in the N-terminus of c-Fos (ID1, Figure 6.1 (A)) (Brown et al, 1995). This inhibitor domain specifically inhibits activation domains containing the HOB1 motif. A motif (IM1) present within this domain, and conserved in the c-Fos related protein, FosB, is essential for inhibitor activity.

c-Jun

The N-terminal activation domain of c-Jun has a repressor region which is termed the δ region (amino acids 32–59, Figure 6.1 (B)) (Baichwal and Tjian, 1990; Bohmann and Tjian, 1989). This region has been shown to bind a cell-type specific factor which represses c-Jun function. The repressing effect of the δ region on c-Jun activity is augmented by a region C-terminal to the δ domain, termed the ε region (amino acids 100–127) (Baichwal et al, 1992).

Regulation of c-Fos and c-Jun Activity

The potency of c-Fos/c-Jun activity is under tight control in the cell. Post-translational events are used to regulate functions such as DNA binding, activation, localization and mRNA/protein turn-over.

Control of DNA binding and transactivation by phosphorylation

c-Jun

Following the stimulation of cells with various agents, an increase in the phosphorylation status of the c-Jun N-terminal activation domain is observed, which is concomitant with an increase in c-Jun's ability to activate transcription (Figure 6.1 (B)) (Binetruy et al, 1991). The main phosphorylation sites required for this potentiation of activity are Ser 63 and Ser 73 (the latter Ser forming part of the HOB1 motif) (Pulverer et al, 1991, Smeal et al, 1991; Sutherland et al, 1992). Both of these residues are immediately followed by a proline residue, so it was originally suggested that they might be substrates for the proline-directed mitogen activated protein kinase (MAPK) family. This proved to be the case following the identification of new members of the MAPK family, namely the JNKs (c-*J*un *N*-terminal *K*inases) (Derijard et al, 1994; Kallunki et al, 1994). It was demonstrated that JNKs bind with high affinity to the δ region of c-Jun and efficiently phosphorylate Ser63 and Ser73 both in vitro and in vivo. These kinases are activated by a wide array of cellular stresses and their increased activity correlates with increased AP-1 activity in vivo. Direct evidence that phosphorylation of Ser63/73 is required for transcriptional activation was shown with an altered kinase specificity mutant of c-Jun (Smeal et al, 1994). In these experiments, amino acids preceeding Ser63 and Ser73 were mutated such that PKA, instead of JNK, could phosphorylate Ser63 and Ser73. The activation capacity of this c-Jun mutant is efficiently potentiated in vivo by PKA, demonstrating the importance of phosphorylation of these Ser residues.

The DNA binding capacity of c-Jun is also under phosphorylation control. In resting cells, c-Jun is phosphorylated at three sites (Ser243, Ser249 and Thr239 or Thr231) near its DNA binding domain (Figure 6.1 (B)) (Boyle et al, 1991; Lin et al, 1992). Upon stimulation of the cells with tumour promoters or mitogens, dephosphorylation of these sites occurs. Originally, a predominantly cytoplasmic kinase, glycogen synthase kinase III (GSK III), was identified which could phosphorylate these sites in vitro. Phosphorylation of these sites inhibits c-Jun homodimer DNA binding activity but does not affect c-Fos/c-Jun heterodimers (Boyle et al, 1991). Subsequent work revealed that nuclear casein kinase II (CKII) phosphorylates two of these sites (Thr231 and Ser249) in vivo, and this was sufficient to block DNA binding (Lin et al, 1992). The role of the third site (Ser243) still needs to be fully established but it is possible that phosphorylation of this site by a second kinase could promote phosphorylation of the other two sites by CKII. In this regard, it should be noted that Ser243 has been reported to be a phospho-acceptor site for MAPK p44 (Figure 6.1 (B)) (Baker et al, 1992). It is noteworthy that the DNA-dependent protein kinase phosphorylates Ser249, at least in vitro (Bannister et al, 1993), although the physiological consequence of this reaction remains uncertain.

Interestingly, phosphorylation of the N-terminal activation domain of c-Jun leads to a decrease in the phosphorylation of the C-terminal sites (Papavassiliou et al, 1995). How this is achieved is still unclear, but it may reflect an altered conformation of c-Jun following phosphorylation of Ser63 and Ser73, leading to enhanced phosphatase activity at the C-terminal sites (alternatively, it could lead to decreased CKII accessibility/activity).

c-Fos

The transactivation potential of c-Fos is also regulated by phosphorylation although no effect on DNA binding has yet been identified. The HOB1-C motif in c-Fos (Figure 6.1 (A)) is phosphorylated by a mitogen-responsive proline-directed protein kinase (FRK), which may be distinct from JNK and MAPK (Deng and Karin, 1994). FRK was shown to phosphorylate Thr232 within the c-Fos HOB1-C motif. Moreover, this amino acid was shown to be important for the efficient stimulation of c-Fos transcriptional activity by mitogens or activated Ras, suggesting that phosphorylation at this site mediates the effect of these agents. Direct evidence for this came from analysis of an altered kinase specificity mutant of c-Fos (Bannister et al, 1994). Changing the context of c-Fos Thr232 to that of a PKA site generates a HOB1-C motif whose activity is stimulated by PKA, indicating that phosphorylation of Thr232 is sufficient to stimulate activity.

c-Fos is able to repress transcription from its own promoter, in a DNA-binding-independent manner (Ofir et al, 1990). Three clusters of serine residues in the c-Fos C-terminus are involved in this process, and it has been shown that PKA phosphorylates Ser362 in one of these clusters (Tratner et al, 1992). Mutation of this serine, to a non-phosphorylatable alanine, results in loss of the repression activity of c-Fos on its own promoter. c-Fos is also phosphorylated in this region by the MOS/MEK/ERK pathway, presumably by ERK1 or ERK2 (Chen et al, 1993; Okazaki and Sagata, 1995). Thus, there may exist cross-over between signal transduction pathways regulating c-Fos's ability to repress its own promoter.

Regulation of c-Fos/c-Jun DNA binding by redox potential

As discussed above, phosphorylation of the c-Jun DNA binding domain prevents c-Jun from binding DNA but does not inhibit c-Fos/c-Jun heterodimers. Another mechanism also exists in the cell to regulate c-Fos/c-Jun heterodimer (and c-Jun/c-Jun homodimer) binding to DNA. Redox changes affect the in vitro DNA binding ability of c-Fos and c-Jun (Abate et al, 1990; Bannister et al, 1991). This sensitivity to redox state is due to the presence of a conserved cysteine residue within the c-Fos (Cys154) and c-Jun (Cys269) DNA binding domains. Under reduced conditions these cysteines have sulphydral side-groups, allowing c-Fos/c-Jun or c-Jun/c-Jun dimers to bind to DNA. However, if the redox equilibrium shifts towards the oxidative state then a disulphide bridge forms between the cysteine residues of each basic motif within a dimer, thereby preventing DNA binding (Bannister et al, 1991). In the absence of chemical reducing agents, a ubiquitous nuclear redox factor (Ref-1) promotes c-Fos/c-Jun and c-Jun/c-Jun DNA binding (Xanthoudakis et al, 1992). Interestingly, Ref-1 also possesses an independent activity, that of an apurinic/apyrimidinic (AP) endonuclease DNA repair activity.

AP-1 mRNA and protein turn-over

During the mitotic cell-cycle, the levels of AP-1 proteins vary greatly. Two key stages at which the intracellular concentration of proteins can be influenced are (i) mRNA turnover and (ii) protein turnover.

The mRNA transcript from the *c-fos* gene is highly labile with a half-life of approximately 10 min. One AU-rich element (ARE) in the *c-fos* 3′ non-coding region is sufficient to mediate degradation. At least one degradation route is via a polysome-associated mechanism which is coupled to ongoing translation (Winstall et al, 1995). In addition to the 3′ UTR, sequences within the *c-fos* protein coding region are also able to mediate degradation via a mechanism coupled to translation (Schiavi et al, 1994).

The c-Fos protein contains, within the last 21 amino acids, a PEST sequence which is a short stretch of amino acids involved in the degradation of numerous, rapidly degraded proteins (Rogers et al, 1986). This sequence makes the c-Fos protein highly labile and deletion of the c-Fos PEST sequence significantly lengthens the protein half-life (Tsurumi et al, 1995). A mechanism for the effect of the PEST sequence on stability has recently been revealed. The PEST sequence targets c-Fos for degradation by the 26S proteasome in a ubiquitin-dependent manner (Tsurumi et al, 1995). Thus, degradation is regulated by phosphorylating Ser362 and Ser374 within the PEST sequence This phosphorylation occurs via the Mos/Map kinase pathway, leading to stabilization of the c-Fos protein (Okazaki and Sagata, 1995).

The c-Jun protein can also affect the stability of c-Fos. Hetero-dimerization of c-Fos with unphosphorylated c-Jun, has little or no effect on c-Fos stability, but the addition of three kinases, MAPK, CKII and CDC2 kinase leads to a c-Jun-mediated, destabilisation of c-Fos via the proteasome-ubiquitin system (Tsurumi et al, 1995). Interestingly, activated c-Jun, purified from phorbol-ester stimulated cells, is not able to target c-Fos for degradation (Papavassiliou et al, 1992). This suggests that the c-Jun DNA binding domain may be involved since phorbol esters induce the dephosphorylation of the CKII sites near the DNA binding domain (Boyle et al, 1991).

The c-Jun protein also has a relatively short half-life- Degradation of c-Jun in vivo is ubiquitin-dependent and requires the c-Jun δ region (Treier et al, 1994). Indeed, this domain is sufficient, when tethered to a heterologous protein, to target that protein for ubiquitin-dependent degradation. It is known that a cell-type specific inhibitor of c-Jun's transcriptional activity also binds to the δ region (Baichwal and Tjian, 1990). Clearly, an inhibitor of this type and the δ-binding protein(s) of the degradation machinery could be one and the same. However, this possibility is excluded by a c-Jun mutant lacking the ε domain (amino acids 100–127) which does not bind the inhibitor protein (Bohmann and Tjian, 1989) but it does become ubiquitinated with a wild-type efficiency (Treier et al, 1994).

Regulation of c-Fos/c-Jun intracellular localization

The transport of c-Fos and c-Jun proteins from the cytoplasm, where they are synthesised, to the nucleus is not a passive event. In serum starved cells, c-Fos accumulates in the cytoplasm and no protein can be detected in the nucleus (Vriz et al, 1992). However, upon serum stimulation of the cells, c-Fos translocates to the nucleus. In a separate study, c-Fos was again found to be retained in the nucleus in quiescent cells but it was stimulated to enter the nucleus following activation of PKA (Roux et al, 1990). With this in mind, it is interesting to note that a predominantly cytoplasmic protein, IP-1, interacts with c-Fos (and/or c-Jun), an interaction which is reversible upon phosphorylation of IP-1 by PKA (Auwerx and Sassone-Corsi, 1991). Whether this PKA regulated interaction is involved in the PKA-dependent, cellular localization of c-Fos (or c-Jun), still remains to be determined.

The cellular localization of c-Jun seems to be somewhat less regulated than that of c-Fos. In fact, the nuclear translocation of c-Jun seems to occur throughout the cell cycle at a relatively constant rate (Chida and Vogt, 1992).

Mechanisms of c-Fos and c-Jun Mediated Transcriptional Activity

Activation via protein-protein interaction

The AP-1 transcription factor complex can increase the rate of RNA Pol II mRNA synthesis by mechanisms involving direct contact with the basal transcriptional machinery. Such interactions may stabilize the Pol II holoenzyme onto the promoter. In addition, there is evidence that c-Fos and c-Jun activity is mediated indirectly by the binding of coactivators which may contact the basal machinery or may catalyse the accessibility of a chromatin template to the transcriptional apparatus.

One of the protein-protein interactions which is essential for efficient c-Fos induced transcriptional activation, is the interaction with TBP, a protein present within the basal transcription factor complex, TFIID (Metz et al, 1994). This interaction is mediated via the TBM motif in c-Fos (Figure 6.1 (A)), and is crucial for activation since deletion of TBM inhibits c-Fos induced activation of a TATA-containing promoter. However, c-Fos missing TBM sequences can still activate a promoter lacking a TATA-box and driven only by an initiator element (Metz et al, 1994). Thus, it appears c-Fos is able to activate transcription by at least two distinct mechanisms; one involves direct contact with TBP via the TBM sequences, and the other probably involves an interaction with a distinct 'initiator' specific protein which would be mediated by one of the other c-Fos activation domains.

c-Fos and c-Jun also stimulate transcription, at least partly, through contacting the CBP coactivator protein (Bannister and Kouzarides, 1995; Bannister et al, 1995). Recently, CBP was shown to possess an intrinsic histone acetyltransferase (HAT) activity (Bannister and Kouzarides, 1996; Ogryzko et al, 1996). Therefore, by targeting CBP to AP-1 dependent promoters, c-Fos and c-Jun may induce a conformational change in the surrounding nucleosomal structure and thereby stimulate gene transcription. In addition, CBP may be bringing into the AP-1 complex additional activation domains, which it is known to possess (Kwok et al, 1994), or it may allow additional contacts with the basal machinery via TFIIB contact. The c-Fos:CBP interaction may not rely on prior phosphorylation of c-Fos. There is some evidence that the c-Jun:CBP interaction depends on prior phosphorylation of c-Jun (Arias et al, 1994) but this issue is still to be resolved (Bannister et al, 1995).

What is clear from the analysis of both c-Fos and c-Jun is that the activity of a number of their activation domains is stimulated by phosphorylation. Most characterised is the phosphorylation of the HOB1 motifs. In c-Jun, this phosphorylation is mediated by a kinase, JNK, which directly binds the activation domain. Thus, JNK could be considered a 'co-factor'. The phosphorylation imposed by JNK on the HOB1 motif may act as a catalyst for the binding of a second 'co-factor', such as CBP, or one of the other less characterised factors which bind the c-Jun activation domain (Oehler et al, 1992).

The positive effect of co-activators may be counterbalanced with the negative effect of proteins which bind c-Fos/c-Jun and inhibit their activity. Evidence for such inhibitor proteins exists for c-Jun (the δ-region binding cell-type inhibitor; Baichwal and Tjian, 1989) and c-Fos (Brown et al,

1995). How these proteins may act is unclear but a reasonable speculation is that they may act on regulating the binding of co-activators at the HOB1 motif of c-Fos and c-Jun. Certainly, for c-Fos, the inhibitor factor binds a domain which specifically represses HOB1 containing activation domains (Brown et al, 1995).

Cooperation with adjacent protein binding sites

Many genes contain multiple enhancer elements within their promoters. From promoter characterization studies, it has become evident that two or more unrelated enhancer elements are often found together, suggesting that cooperative interactions may occur between different classes of transcription factors. Exemplified below are examples of cooperative interactions involving the AP-1 transcription factor.

Ets/Ap-1 sites

The Ets family of transcription factors contain at least 10 members which have a related DNA binding domain. Numerous, functionally important enhancer elements contain adjacent binding sites for Ets and AP-1 transcription factors which are able to activate transcription cooperatively when both factors are present (reviewed in Bassuk and Leiden, 1995). This was first demonstrated for the Ets/AP-1 sites in the PEA3 element of the polyoma virus enhancer (Wasylyk et al, 1990). When c-ets1 is expressed in the presence of c-Fos and c-Jun, the factors cooperate for transcriptional activation. This cooperation is due, at least in part, to a direct interaction between the DNA binding domain of c-Ets1 and the basic region of c-Jun (Bassuk and Leiden, 1995). Other Ets transcription factors, including Elf1, PU.1 and Fli1, also make direct contacts with c-Jun (Bassuk and Leiden, 1995).

Cooperation still occurs if the Ets/AP-1 DNA binding sites are interdigitated, rather than adjacent to each other. This is the case in the B lymphocyte-specific immunoglobulin heavy chain (IgH) 3′ enhancer which binds a novel complex, nuclear factor of activated B-cells (NFAB; Grant et al, 1995). NFAB is composed of the Ets factor Elf1 and the AP-1 factors c-Fos and JunB, and requires both an intact Ets DNA site and an intact AP-1 site for efficient transcriptional activation.

The previous example of Ets/AP-1 cooperation culminates in tissue-specific gene expression. Cellular differentiation is also regulated by Ets/Ap-1 interactions. In *Drosophila*, Jun interacts with the *Drosophila* ets domain protein pointed, thereby inducing neuronal differentiation in the eye (Treier et al, 1995). Again this interaction is cooperative in nature but it can be antagonized by another ets domain protein YAN, which inhibits eye development.

NF-AT/AP-1 sites

Nuclear factor of activated T-cells (NFAT) is a transcription factor family, each composed of two components; one responds to PKC whilst the other responds to a rise in the levels of intracellular Ca^{2+} (for review see Rao, 1994). The component responding to PKC contains c-Fos and c-Jun family members, whereas the Ca^{2+} sensitive component is a member of a lymphoid-specific transcription factor family (including NFATp and NFATc) (Jain et al, 1992; Rao, 1994). The mechanism of NFAT regulated gene expression is commonly analysed using the interleukin2 (IL-2) promoter as a model system. This promoter contains an NFAT binding site which is important for the response of the gene to T-cell activation. In vitro, and in the absence of NFAT(p/c), a c-Fos/c-Jun complex binds very weakly to this site. Likewise, in the absence of c-Fos/c-Jun, NFAT(p/c) shows very weak binding to the site (Yaseen et al, 1994). However, when NFAT(p/c) and c-Fos/c-Jun are both present, cooperative DNA binding is observed which is probably a consequence of a direct physical interaction between NFAT(p/c) and c-Fos/c-Jun. Indeed, a region of c-Fos, amino acids 118–138, immediately N-terminal to the basic DNA binding domain, has been identified as being required to form a NFAT(p/c):c-Fos/c-Jun complex (Yaseen et al, 1994). This region of c-Fos is not required for c-Fos/c-Jun binding to an AP-1 site.

Cross-talk between c-Fos/c-Jun and different transcription factors via a single DNA binding site

To generate prolonged genetic responses a eukaryotic cell has to be able to integrate diverse extracellular signals. For example, exposure of a cell to steroid hormones can lead to differentiation, whereas exposure to mitogens or tumour promoters can lead to growth/proliferation. Steroid hormones often inhibit the effect of exposure to these proliferative signals. Conversely, growth factors such as EGF can inhibit differentiation induced by steroid hormones such as cortisol. This suggests that cross-talk between transcription factors is occurring which is regulated by different intracellular signal transduction pathways.

How these antagonistic regulatory mechanisms are achieved has been the subject of intense research. Some light has been shed by work involving the regulation of the collagenase gene which is stimulated by AP-1 but repressed by glucocorticoids. The DNA sequence required for both these effects is the AP-1 binding site in the gene promoter (Jonat et al, 1990). The hormone activated glucocorticoid receptor (GR) can not bind directly to this site (Konig et al, 1992). In vitro, DNA binding by c-Fos/c-Jun is inhibited by GR and appears to be mediated, in the main, by an interaction between GR and c-Fos (Kerppola et al, 1993). However, in vivo the ability of AP-1 to bind to the collagenase AP-1 site appears not to be affected by the presence or absence of GR (Konig et al, 1992). In either case, it is clear that the GR does not directly bind to the AP-1 site, and that GR either directly or indirectly contacts AP-1. It is noteworty that the reciprocal regulation also occurs; that is to say, AP-1 can repress, in a DNA binding independent manner, glucocorticoid activated gene transcription via a GR DNA binding site (Jonat et al, 1990; Lucibello et al, 1990).

The above picture is somewhat complicated by the fact that, in certain circumstances, glucocorticoids can actually enhance AP-1 activity. This appears to depend upon (i) the AP-1 species binding to the promoter and (ii) the promoter context of the regulatory site(s) (Shemshedini et al, 1991). Using the proliferin gene promoter as a reporter in F9 cells, c-Jun/c-Jun homodimer activity is enhanced by the presence of GR, whereas c-Fos/c-Jun heterodimer activity is repressed (Diamond et al, 1990). In the latter situation, the basic region of c-Fos is required for repression (Kerppola et al, 1993).

c-Fos and c-Jun are capable of stimulating the DNA binding activity (and therefore activation potential) of certain transcription factors. This is exemplified by the interaction of c-Fos/c-Jun with NF-κB. Under many conditions NF-κB is activated in parallel with c-Fos and c-Jun, but different intracellular signal transcription pathways can be involved. Over-expressing c-Fos or c-Jun in cells leads to an increased transcription of NF-κB regulated gene promoters. This effect is mediated via the NF-κB DNA binding site and is due to enhanced DNA binding by NF-κB (Stein et al, 1993). It appears that c-Fos/c-Jun interact with the rel homology domain of NF-κB, via their leucine zippers, thereby stimulating NF-κB DNA binding (Stein et al, 1993). As with c-Fos/c-Jun:GR regulated gene expression there is a reciprocity to this effect as over-expression of NF-κB in cells leads to an increase in the transcription of AP-1 regulated genes. The net effect is an increase in transcription of the relevant gene which is dependent on incoming signals to both NF-κB and c-Fos/c-Jun.

Extracellular signals can cause certain cells to initiate a genetic response leading to cellular differentiation. Via an interaction with the muscle specific transcription factor MyoD, c-Jun can inhibit myogenic differentiation (Li et al, 1992; Bengal et al, 1992). The bZIP region and the N-terminal activation domain of c-Jun both contact the helix-loop-helix of MyoD leading to repression of MyoD activity. Once again, this effect is reciprocal in that MyoD can repress c-Jun mediated activity from AP-1 responsive promoters.

Involvement of c-Fos and c-Jun in Oncogenesis

Both c-Fos and c-Jun have oncogenic viral counterparts which were identified because they induce tumours in mice and chickens respectively. The viral proteins contain numerous differences with respect to their cellular counterparts, and most of these changes appear to stimulate the transforming ability of the viral protein. However, when over-expressed in the appropriate cells even c-Fos and c-Jun are capable of promoting transformation.

c-Fos induced transformation

Transgenic mice have been invaluable in determining some of the in vivo roles of c-Fos. c-Fos is crucial for bone remodeling as it is involved in determination of cell differentiation along the osteoclast/macrophage lineage (Grigoriadis et al, 1995). The flip-side to differentiation is tumourogenesis, and when c-Fos is over-expressed in mice it leads to osteosarcomas or chondrosarcomas (reviewed in Grigoriadis et al, 1995). In addition to the over-expressing animal model, over-expression of c-Fos (or v-Fos) in cultured fibroblasts (and cartilage, bone and muscle cells) leads to their transformation (Jenuwein et al, 1985). The bZIP region of c-Fos is essential for transformation although it is not sufficient (Jenuwein and Muller, 1987). This suggests that AP-1 dependent transcription is involved in the transformation process and that activation domains are required.

A region in the N-terminal 110 amino acids is important for c-Fos induced transformation (Jooss et al; 1994). This region contains an autonomous activation domain (Jooss et al, 1994; Brown et al, 1995). Mutagenesis of the NTM motif (amino acids 60–84), which corresponds to the HOB1-N region (Figure 6.1 (A)), affects both activation and transformation by c-Fos. Thus, there is a correlation between the requirement for activation, and the requirement for transformation, within the N-terminus of c-Fos.

The C-terminus of c-Fos also contains a potent transcriptional activation domain. Like the c-Fos N-terminal region, there is a strong correlation in this region between transcriptional activity and the ability to transform rodent fibroblast cell lines (Wisdom and Verma, 1993). The most potent activation domain in the C-terminus contains the HOB1 and HOB2 motifs (Sutherland et al, 1992). However, the FBR-v-Fos protein has several deletions in the HOB1/2 region but still retains a C-terminal activation domain which induces transformation (Wisdom and Verma 1993). Whether this viral protein has other changes which compensate for the loss of HOB1/2 or whether other regions are involved still remains to be determined.

c-Jun induced transformation

In contrast to c-Fos, c-Jun is a relatively weak inducer to transformation. Indeed, mice overexpressing v-Jun do not develop spontaneous tumours. However, these mice do develop tumours during the process of wound healing suggesting that the process of tissue repair produces a factor which can cooperate with v-Jun to induce transformation (Schuh et al, 1990). In vitro, only chicken embryo fibroblasts (CEF) can be transformed by v-Jun. c-Jun also transforms CEF cells in vitro but to a lesser extent than its viral counterpart (Bos et al, 1990).

Mammalian cells are resistant to transformation by either c- or v-Jun. However, if Ha-Ras is coexpressed, then mammalian cells are efficiently transformed (Schutte et al, 1989; Alani et al, 1991). The bZIP domain is essential for this process, again suggesting that a functional AP-1 DNA binding activity is required. Contradictory evidence exists concerning the contribution of the activity of c-Jun's activation domain towards transformation. In order to cotransform rat embryo cells with Ha-Ras an active c-Jun N-terminal activation domain is required (Alani et al, 1991). However, in CEF cells there is an inverse correlation between transformation and the activity of the c-Jun N-terminal activation domain (Havarstein et al, 1992). The apparent difference in these results may be explained by cell-type specific requirements for transformation. For instance, transformation of rat embryo cells may require AP-1 to induce transformation-mediating genes, whereas in CEF cells transformation may be due to the failure of AP-1 to stimulate the transcription of growth attenuating genes.

Oncogenic conversion of c-Fos and c-Jun

The viral counterparts of c-Fos and c-Jun are very efficient at promoting transformation. Each viral protein contains numerous changes with respect to the cellular homologue, allowing them to induce transformation more readily.

Now that a lot more is known about the function of the c-Fos/c-Jun proteins, and how function relates to structure, we can begin to piece together a picture of why the viral forms are more oncogenic. A number of point mutations and deletions have been introduced during the process of viral transduction of c-Fos and c-Jun (Figure 6.2). In this section we attempt to decipher how these changes lead to a protein with increased oncogenic potential, given the evidence for function provided in the previous sections.

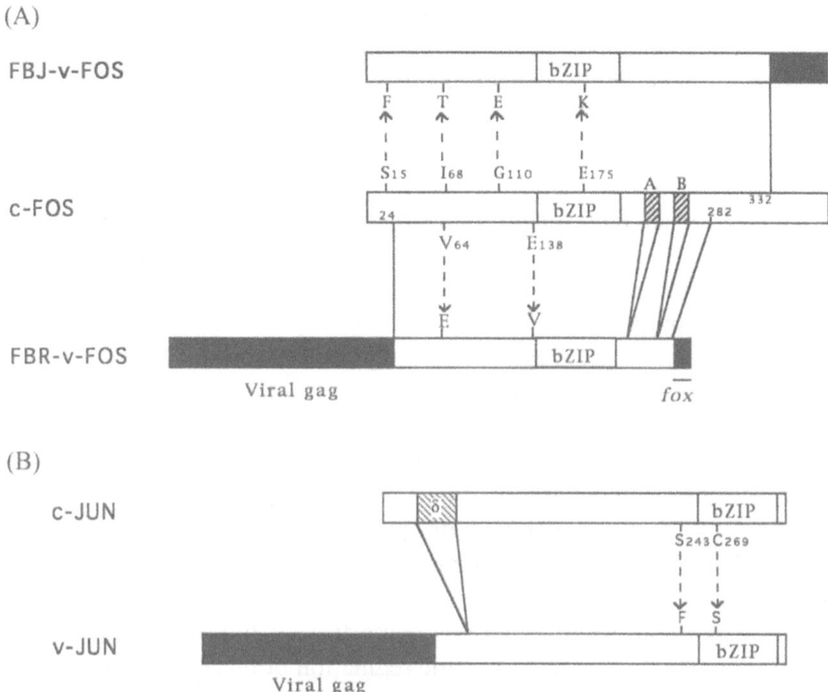

Figure 6.2 Comparison of c-Fos and c-Jun with their viral derivatives. (A) Relationship of c-Fos to FBJ-v-Fos and FBR-v-Fos. The major differences between c-Fos and the v-Fos's are indicated. FBJ-v-Fos contains a mutation which leads to a loss of the last 48 amino acids and replacement with 49 new ones (black box). The positons of point mutations throughout the protein are indicated. In FBR-v-Fos, 310 amino acids of the viral gag protein replace the first 24 amino acids of the c-Fos protein (N-terminal black box). In addition, eight amino acids from the cellular locus, *fox*, substitute for the last 98 amino acids (C-terminal black box). Regions A and B in c-Fos, represent amino acids 228–240 and 259–267, respectively, which are deleted from FBR-v-Fos. (B) Relationship of c-Jun to v-Jun. The major differences between c-Jun and v-Jun are indicated. The v-Jun coding sequence is fused, in-frame, at the C-terminal end of the 220 amino acid viral gag coding sequence (black box). The δ region in c-Jun (amino acids 32–58; hatched box) is deleted from v-Jun. Functionally characterised point mutations are indicated.

Elevation of intracellular concentrations of Fos and Jun

The c-Fos and c-Jun proteins, as well as their respective mRNAs, have very short half-lives. This is to ensure that the mitogenic signals mediated by these proteins are transient. The virus life cycle on the other hand, demands a prolonged and sustained mitogenic signal. Thus, retroviruses that have transduced *c-fos* and *c-jun* have removed some of the cues which signal the destruction of the proteins and their mRNA. For example, both *v-jun* and *v-fos* mRNAs are lacking the cellular 3′ UTR sequences which dictate mRNA instability (Bos et al, 1990). In v-Jun, one of the major differences is the absence of the δ region which is required for ubiquitin-mediated degradation of the c-Jun protein (Treier et al, 1994). In v-Fos, a deletion of C-terminal sequences (found in both viral forms) has resulted in the removal of PEST sequences which are found associated with short-lived proteins. In addition, this 'deletion' removes residues essential for the ability of c-Fos to repress its own promoter (Ofir et al, 1990).

All these changes result in the increase of Fos and Jun protein in the nucleus. Furthermore, by removing regulatory signals from the protein products (rather than activating the *c-fos* and *c-jun* promoter) the virus has ensured that the increase in protein levels is likely to be prolonged, because the cell has no means of keeping it under control.

Deregulation of Fos and Jun activity

Having more of the Fos and Jun proteins in the cell is not productive if this protein is going to be inactivated by a normally operating cellular regulatory system. The viral Fos and Jun proteins therefore possess changes which allow them to escape cellular controls which would otherwise lower their potency.

Two very good examples of this come from v-Jun, where changes in v-Jun allow the protein to escape the regulation of its DNA binding functions. The first is the change of Cys269 (present within the DNA binding domain) to Ser. This change generates a v-Jun protein which is no longer under the influence of changing redox potential (Oehler et al, 1993). However, this conservative change still allows v-Jun to bind DNA as efficiently as c-Jun. The second example is the Ser243 to Phe change adjacent to the c-Jun DNA binding domain. This mutation prevents CKII from phosphorylating the nearby Thr231 and Ser249 residues, whose phosphorylation normally inhibits DNA binding. Accordingly, v-Jun is not responsive to CKII-mediated inhibition of DNA binding, and consequently it has constitutive DNA binding ability.

The v-Fos protein has also sustained a change which results in its altered regulation. Under serum starved conditions, when cells are in a quiescent state, the c-Fos protein is retained in the cytoplasm, moving into the nucleus following a mitogenic cue. In contrast, the v-Fos protein is nuclear regardless of the state of the cell (Roux et al, 1990).

Selectivity of activation domains

Paradoxically, some of the major deletions found in v-Fos and v-Jun, result in the elimination of what have been characterised as potent activation domains. For example, the δ region of c-Jun, the HOB1/2 region of c-Fos ((A) and (B), Figure 6.2), and the very C-terminus of c-Fos have activation capacity, yet they are deleted from the viral proteins. Nevertheless, the overall transcriptional activation capacity of the viral proteins is as potent, if not more so, than the cellular counterparts.

The explanation for this paradox may lie in an observation made by Oliviero et al (1992), who showed that linking the Fos/Jun DNA binding domain to a heterologous activation domain from GCN4 does not generate a transforming protein. In other words, a specific activating function present in Fos and Jun is necessary for transformation. Since we know from a number of studies (Angel et al, 1989; Abate et al, 1991; Sutherland et al, 1992; Jooss et al, 1994; Brown et al, 1995) that the c-Fos and c-Jun proteins contain several independently acting activation domains, it would not be surprising if only a subset of these are necessary for transformation. The N-terminal activation domain of c-Fos (surrounding HOB1-N), which is known to be essential for transformation, has not been deleted by the viruses. This activation domain may be required to stimulate the activity of genes necessary to mediate uncontrolled proliferation. Other activation domains, which may drive genes required for differentiating pathways, may not be necessary for transformation. Indeed, it may be necessary to delete these differentiation-specific domains to avoid a conflict of signals within a cell which has high levels of unregulatable v-Fos protein.

The viruses that have transduced Fos and Jun have had a very long time to mold the proteins to their needs. Although we are beginning to understand how they have achieved their goal, there are probably many more details, which at the moment, we can not appreciate. The only way to do so is to thoroughly understand the multifaceted functions of the cellular products. This should keep us busy for a while longer!

References

Abate C, Patel L, Rauscher FJ, Curran T (1990): Redox regulation of Fos and Jun DNA-binding activity in vitro. *Science* 249: 1157–1161

Abate C, Luk D, Curran T (1991): Transcriptional regulation by Fos and Jun in vitro: interaction among multiple activator and regulatory domains. *Mol Cell Biol* 11: 3624–3632

Alani R, Brown P, Binetry B, Dosaka H, Rosenberg RK, Angel P, Karin M, Birrer MJ (1991): The transactivating domain of the c-Jun proto-oncogene is required for cotransformation of rat embryo cells. *Mol Cell Biol* 11: 6286–6295

Angel P, Herrlich P (eds) (1994): *The Fos and Jun Families of Transcription Factors*, CRC Press, Boca Raton

Angel P, Karin M (1991): The role of Jun, Fos and the AP-1 complex in cell-proliferation and transformation. *Biochim Biophys Acta* 1072: 129–257

Angel P, Smeal T, Meek J, Karin M (1989): Jun and v-Jun contain multiple regions that participate in transcriptional activation in an interdependent manner. *New Biol* 1: 35–43

Arias J, Alberts AS, Brindle P, Claret FX, Smeal T, Karin M, Feramisco J, Montminy M (1994): Activation of cAMP and mitogen responsive genes relies on a common nuclear factor. *Nature* 370: 226–229

Auwerx J, Sassone-Corsi P (1991): IP-1: A dominant inhibitor of Fos/Jun whose activity is modulated by phosphorylation. *Cell* 64: 983–993

Baichwal VR, Tjian R (1990): Control of c-Jun activity by interaction of a cell-specific inhibitor with regulatory domain δ: Differences between v- and c-Jun. *Cell* 63: 815–825

Baichwal VR, Park A, Tjian R (1992): The cell-type-specific activator region of c-Jun juxtaposes constitutive and negatively regulated domains. *Genes Dev* 6: 1493–1502

Baker SJ, Kerppola TK, Luk D, Vandenberg MT, Marshak DR, Curran T, Abate C (1992): Jun is phosphorylated by several protein kinases at the same sites that are modified in serum-stimulated fibroblasts. *Mol Cell Biol* 12: 4694–4705

Bannister AJ, Kouzarides T (1995): CBP-induced stimulation of c-Fos activity is abrogated by E1A. *EMBO J* 14: 4758–4762

Bannister AJ, Kouzarides T (1996): The CBP co-activator is a histone acetyltransferase. *Nature* 384: 641–643

Bannister AJ, Cook A, Kouzarides T (1991): In vitro DNA binding activity of Fos/Jun and BZLF1 but not C/EBP is affected by redox changes. *Oncogene* 6: 1243–1250

Bannister AJ, Gottlieb TM, Kouzarides T, Jackson SP (1993): c-Jun is phosphorylated by the DNA-dependent protein kinase in vitro; definition of the minimal kinase recognition motif. *Nucleic Acids Res* 21: 1289–1295

Bannister AJ, Brown HJ, Sutherland JA, Kouzarides T (1994): Phosphorylation of the c-Fos and c-Jun HOB1 motif stimulates its activation capacity. *Nucleic Acids Res* 22: 5173–5176

Bannister AJ, Oehler T, Wilhelm D, Angel P, Kouzarides T (1995): Stimulation of c-Jun activity by CBP: c-Jun residues Ser63/73 are required for CBP induced stimulation in vivo and CBP binding in vitro. *Oncogene* 11: 2509–2514

Bassuk AG, Leiden JM (1995): A direct physical association between ETS and AP-1 transcription factors in normal human T-cell. *Immunity* 3: 223–237

Benbrook DM, Jones NC (1990): Heterodimer formation between CREB and JUN protein. *Oncogene* 5: 295–302

Bengal E, Ransone L, Scharfmann R, Dwarki VJ, Tapscot SJ, Weintraub H, Verma IM (1992): Functional antagonism between c-Jun and MyoD proteins: a direct physical association. *Cell* 68: 507–519

Binetruy B, Smeal T, Karin M (1991): Ha-Ras augments c-Jun activity and stimulates phosphorylation of its activation domain. *Nature* 351: 122–127

Bohmann D, Tjian R (1989): Biochemical analysis of transcriptional activation by Jun: Differential activity of c- and v-Jun. *Cell* 59: 709–717

Bos TJ, Monteclaro FS, Mitsunobu F, Ball AR, Chang CHW, Nishimura T, Vogt PK (1990): Efficient transformation of chicken embryo fibroblasts by c-Jun requires structural modification in coding and noncoding sequences. *Genes Dev* 4: 1677–1687

Boyle WJ, Smeal T, Defize LHK, Angel P, Woodgett JR, Karin M, Hunter T (1991): Activation of protein kinase C decreases phosphorylation of c-Jun at sites that negatively regulates its DNA-binding activity. *Cell* 64: 5730–584

Brown HJ, Sutheriand JA, Cook A, Bannister AJ, Kouzarides T (1995): An inhibitor domain in c-Fos regulates activation domains containing the HOB1 motif. *EMBO J* 14: 124–131

Chatton B, Bocco JL, Goetz J, Gaire M, Lutz Y, Kedinger C (1994): Jun and Fos heterodimerize with ATFa, a member of the ATF/CREB family and modulate its transcriptional activity. *Oncogene* 9: 375–385

Chen R-H, Abate C, Curran T (1993): Phosphorylation of the c-Fos transrepression domain by mitogen-activated protein kinase and 90-kDa ribosomal S6 kinase. *Proc Natl Acad Sci* 90: 10952–10956

Chida K, Vogt PK (1992): Nuclear translocation of viral Jun but not of cellular Jun is cell cycle dependent. *Proc Natl Acad Sci* 89: 4290–4294

Curran T, Franza BR (1988): Fos and Jun: the AP-1 connection. *Cell* 55: 395–397

Deng T, Karin M (1994): c-Fos transcriptional activity stimulated by Ha-Ras activated protein kinase distinct from JNK and ERK. *Nature* 371: 171–175

Derijard B, Hibi M, Wu IH, Barrett T, Su B, Deng T, Karin M, Davis RJ (1994): JNK1: A protein kinase stimulated by UV light and Ha-Ras that binds and phosphorylates the c-Jun activation domain. *Cell* 76: 1025–1037

Diamond MI, Miner JN, Yoshinaga SK, Yamamoto KR (1990): c-Jun and c-Fos levels specificy positive or negative glucocorticoid regulation from a composite GRE. *Science* 249: 1266–1272

Dixit VM, Marks RM, Sarma V, Prochownik EV (1989): The antimitogenic action of tumor necrosis factor is associated with increased AP-1/c-jun proto-oncogene transcription. *J Biol Chem* 264: 16905–16909

Finkel MP, Biskis BO, Jinkins PB (1966): Virus induction of osteosarcomas in mice. *Science* 151: 698–701

Finkel MP, Reilly CA, Biskis BO (1975): Viral etiology of bone cancer. *Front Radiat Theor Oncol* 10: 28–39

Franklin CC, Sanchez V, Wagner F, Woodgett JR, Kraft AS (1992): Phorbol ester-induced amino-terminal phosphorylation of human Jun but not JunB regulates transcriptional activation. *Proc Natl Acad Sci* 89: 7247–7251

Gentz R, Rauscher FJ, Abate C, Curran T (1989): Parallel association of Fos and Jun leucine zippers juxtaposes DNA binding domains. *Science* 243: 1695–1699

Glover JNM, Harrison SC (1995): Crystal structure of the heterodimeric bZIP transcription factor c-Fos/c-Jun bound to DNA. *Nature* 373: 257–261

Grant PA, Thompson CB, Pettersson S (1995): IgM receptor-mediated transactivation of the IgH 3′ enhancer couples a novel Elf1-AP-1 protein complex to the developmental control of enhancer function. *EMBO J* 14: 4501–4513

Grigoriadis AE, Wang ZQ, Cecchini MG, Hofstetter W, Felix R, Fleisch HA, Wagner EF (1994): c-Fos: A key regulator of osteoclast-macrophage lineage determination and bone remodelling. *Science* 266: 443–448

Grigoriadis AE, Wang ZQ, Wagner EF (1995): Fos and bone cell development: lessons from a nuclear oncogene. *Trends Genet* 11: 436–441

Hadman M, Loo M, Bos TJ (1993): In vivo viral and cellular Jun complexes exhibit differential interaction with a number of in vitro generated AP-1 and CREB-like target sequences. *Oncogene* 8: 1895–1903

Hai T, Curran T (1991): Cross-family dimerization of transcription factors Fos/Jun and ATF/CREB alters DNA binding specificity. *Proc Natl Acad Sci* 88: 3720–3724

Halazonetis TD, Georgopoulos K, Greenberg ME, Leder P (1988): c-Jun dimerizes with itself and c-Fos, forming complexes of different DNA binding affinities. *Cell* 55: 917–924

Havarstein LS, Morgan IM, Wong W-Y, Vogt PK (1992): Mutations in the Jun delta region suggest an inverse correlation between transformation and transcriptional activation. *Proc Natl Acad Sci* 89: 618–622

Jain J, McCafrey PG, Valge-Archer V, Rao A (1992): Nuclear factor of activated T-cells contains Fos and Jun. *Nature* 356: 801–804

Jenuwein T, Muller R (1987): Structure-function analysis of fos protein: A single amino acid change activates the immortalizing potential of v-fos. *Cell* 48: 647–657

Jenuwein T, Muller D, Curran T, Muller R (1985): Extended life span and tumorigenicity of nonestabished mouse connective tissue cells transformed by the fos oncogene of FBR-MuSV. *Cell* 41: 629–637

Jonat C, Rahmsdorf HJ, Park KK, Cato ACB, Gebel S, Ponta H, Herrlich P (1990): Anti-tumour promotion and antiinflammation: down modulation of AP-1 (Fos/Jun) activity by glucocorticoid hormone. *Cell* 62: 1189–1204

Jooss KU, Funk M, Muller R (1994): An autonomous N-terminal domain in Fos plays a crucial role in transformation. *EMBO J* 13: 1467–1475

Kallunki T, Su B, Tsigelny I, Sluss HK, Derijard B, Moore G, Davis R, Karin M (1994): JNK2 contains a specificity-determining region responsible for efficient c-Jun binding and phosphorylation. *Genes Dev* 8: 2996–3007

Kerppola TK, Luk D, Curran T (1993): Fos is a preferential target of glucocorticoid receptor inhibition of AP-1 activity in vitro. *Mol Cell Biol* 13: 3782–3791

Kim SY, Angel P, Lafyatis R, Hattori K, Kim KY, Sporn MB, Karin M, Roberts AB (1990): Autoinduction of transforming growth factor beta 1 is mediated by the AP-1 complex. *Mol Cell Biol* 10: 1492–1497

Konig H, Ponta H, Rahmsdorf HJ, Herrlich P (1992): Interference between pathway specific transcription factors: Glucocorticoids antagonize phorbol ester-induced AP-1 activity without altering AP-1 site occupatio in vivo. *EMBO J* 11: 2241–2246

Kouzarides T, Ziff E (1988): The role of the leuzine zipper in the fos-jun interaction. *Nature* 336: 646–651

Kouzarides T, Ziff E (1989a): Behind the Fos and Jun leucine zipper. *Cancer cells* 1: 71–76

Kouzarides T, Ziff E (1989b): Leucine zippers of fos, jun and GCN4 dictate dimerization specificity and thereby control DNA binding. *Nature* 340: 568–571

Kovary K, Bravo R (1992): Existence of different Fos/Jun complexes during the G0-to-G1 transition and during exponential growth in mouse fibroblasts: Differential role of Fos protein. *Mol Cell Biol* 12: 5015–5023

Kwok RPS, Lundblad JR, Chrivia JC, Richards JP, Bachinger HP, Brennan RG, Roberts SGE, Green MR, Goodman RH (1994): Nuclear protein CBP is a coactivator for the transcription factor CREB. *Nature* 370: 223–226

Landschultz WH, Johnson PF, McKnight SL (1988): The leucine zipper: a hypothetical structure common to a new class of DNA-binding proteins. *Science* 240: 1759–1764

Li L, Chambard J-C, Karin M, Olson EN (1992): Fos and Jun repress transcriptional activation by myogenin and MyoD: the amino terminus of Jun can mediate repression. *Genes Dev* 6: 676–689

Lin A, Frost J, Deng T, Smeal T, Al-Alawi N, Kikkawa U, Hunter T, Brenner D, Karin M (1992): Casein kinase II is a negative regulator of c-Jun DNA binding and AP-1 activity. *Cell*.70: 777–789

Lucibello FC, Slater EP, Jooss KU, Beato M, Muler R (1990): Mutual transrepression of Fos and the glucocorticoidreceptor: involvement of a functional domain in Fos which is absent in FosB. *EMBO J* 9: 2827–2834

Maki Y, Bos TJ, Davis C, Starbuck M, Vogt PK (1987): Avian sarcoma virus 17 carries the jun oncogene. *Proc Natl Acad Sci* 84: 2848–2852

Metz R, Bannister AJ, Sutherland JA, Hagemeier C, O'Rourke EC, Cook A, Bravo R, Kouzarides T (1994): c-Fos-induced activation of a TATA-box-containing promoter involves direct contact with TATA-box-binding protein. *Mol Cell Biol* 14: 6021–6029

Nakabeppu Y, Ryder K, Nathans D (1988): DNA binding activities of three murine Jun proteins: Stimulation by Fos. *Cell* 55: 907–915

Nakabeppu Y, Nathans D (1989): The basic region of Fos mediates specific DNA binding. *EMBO J* 8: 3833–3841

Nerlov C, Ziff EB (1995): CCAAT/enhancer binding protein-a amino acid motifs with dual TBP and TFIIB binding ability co-operate to activate transcription in both yeast and mammalian cells. *EMBO J* 14: 4318–4328

Oehler T, Angel P (1992): A common intermediary factor (p52/54) recognizing "acidic blob"-type domains is required for transcriptional activation by the Jun proteins. *Mol Cell Biol* 12: 5508–5515

Oehler T, Pintzas A, Stumm S, Darling A, Gillespie D, Angel P (1993): Mutation of a phosphorylation site in the DNA-binding domain is required for redox-independent transactivation of AP1-dependent genes by v-Jun. *Oncogene* 8: 1141–1147

Ofir R, Dwarki VJ, Rashid D, Verma IM (1990): Phosphorylation of the C-terminus of Fos protein is required for transcriptional transrepression of the c-fos promoter. *Nature* 348: 80–82

Ogryzko VV, Schiltz RL, Russanova V, Howard BH, Nakatani Y (1996): The transcriptional Coactivators p300 and CBP are histone acetyltransferases. *Cell* 87: 953–959

Okazaki K, Sagata N (1995): The Mos/MAP kinase pathway stabilizes c-Fos by phosphorylation and augments its transforming activity in NIH 3T3 cells. *EMBO J* 14: 5048–5059

Oshima RB, Abrams L, Kulesh D (1990): Activation of an intron enhancer within the keratin 18 gene by expression of c-fos and c-jun in undifferentiated F9 embryonal carcinoma cells. *Genes Dev* 4: 835–848

Papavassiliou AG, Treier M, Chavrier C, Bohmann D (1992): Targeted degradation of c-Fos, but not v-Fos, by a phosphorylation-dependent signal on c-Jun. *Science* 258: 1941–1944

Papavassiliou AG, Treier M, Bohmann D (1995): Intramolecular signal transduction in c-Jun. *EMBO J* 14: 2014–2019

Pulverer BJ, Kyriakis JM, Avruch J, Nikolakaki E, Woodgett JR (1991): Phosphorylation of c-jun mediated by MAP kinases. *Nature* 353: 670–674

Radler-Pohl A, Sachsenmaier C, Gebel S, Auer HP, Bruder JT, Rapp U, Angel P, Rahmsdorf HJ, Herrlich P (1993): UV-induced activation of AP-1 involves obligatory extranuclear steps including RAF-1 kinase. *EMBO J* 12: 1005–1012

Rao A (1994): NF-ATp: a transcription factor required for the co-ordinate induction of several cytokine genes. *Immunol Today* 15: 274–281

Rickles RJ, Darrow AL, Strickland S (1989): Differentiation-responsive elements in the 5′ region of the mouse tissue plasminogen activator gene confer two-stage regulation by retinoic acid and cAMP in teratocarcinoma cells. *Mol Cell Biol* 9: 1691–1704

Rogers S, Wells R, Rechsteiner (1986): Amino acid sequences common to rapidly degraded proteins: the PEST hypothesis. *Science* 234: 364–368

Roux P, Blanchard J-M, Fernandez A, Lamb N, Jeanteur P, Piechaczyk M (1990): Nuclear localization of c-Fos, but not v-Fos proteins, is controlled by extracellular signals. *Cell* 63: 341–351

Ryseck RP, Bravo R (1991): Integrity of FosB leucine zipper is essential for its interaction with Jun proteins. *Oncogene* 6: 533–542

Sassone-Corsi P, Ransone L, Lamph WW, Verma I (1988): Direct interaction between Fos and Jun nuclear oncoproteins: Role of the "leucine zipper domain". *Nature* 336: 692–695

Schuermann M, Neuberg M, Hunter JB, Jenuwein T, Ryseck R-P, Bravo R, Muller R (1989): The leucine repeat motiv in Fos protein mediates complex formation with Jun-AP1 and is required for transformation. *Cell* 56: 507–516

Schiavi SC, Wellington CL, Shyu AB, Chen CY, Greenberg ME, Belasco JG (1994): Multiple elements in the c-fos protein coding region facilitate mRNA deadenylation and decay by a mechanism coupled to translation. *J Biol Chem* 269: 3441–3448

Schuh AC, Keating SJ, Monteclaro FS, Vogt PK, Breitman ML (1990): Obligatory wounding requirement for tumorigenesis in v-jun transgenic mice. *Nature* 346: 756–757

Schutte J, Minna JD, Birrer MJ (1989): Deregulated expression of human c-jun transforms primary rat embryo cells in cooperation with an activated c-Ha-ras gene and transforms Rat-1a cells as a single gene. *Proc Natl Acad Sci* 86: 2257–2261

Sellers JW, Struhl K (1989): Changing Fos oncoprotein to a Jun-dependent DNA-binding protein with GCN4 dimerisation specificity by swapping "leucine zippers". *Nature* 341: 74–76

Shemshedini L, Knauthe R, Sassone-Corsi P, Pornon A, Gronenmeyer H (1991): Cell specific inhibitory and stimulatory effects of Fos and Jun on transcription activation by nuclear receptors. *EMBO J* 10: 3839–3849

Smeal T, Binetruy B, Mercola DA, Birrer M, Karin M (1991): Oncogenic and transcriptional cooperation with Ha-Ras requires phosphorylation of c-Jun on serines 63 and 73. *Nature* 354: 494–496

Smeal T, Hibi M, Karin M (1994): Altering the specificity of signal transduction cascades: positive regulation of c-Jun transcriptional activity by protein kinase A. *EMBO J* 13: 6006–6010

Stein B, Baldwin AS, Ballard DW, Green WC, Angel P, Herrlich P (1993): Cross-coupling of the NF-kBp65 and Fos/Jun transcription factors produces potentiated biological functiion. *EMBO J* 12: 3879–3891

Sutherland JA, Cook A, Bannister AJ, Kouzarides T (1992): Conserved motifs in Fos and Jun define a new class of activation domain. *Genes Dev* 6: 1810–1819

Tratner I, Ofir R, Verma IM (1992): Alteration of a cAMP-dependent protein kinase phosphorylation site in the c-Fos protein augments its transforming potential. *Mol Cell Biol* 12: 998–1006

Treier M, Stszewski LM, Bohmann D (1994): Ubiquitin-dependent c-Jun degradation in vivo is mediated by the δ domain. *Cell* 78: 787–798

Treier M, Bohmann D, Mlodzik M (1995): JUN cooperates with the ETS domain protein pointed to induce photoreceptor R7 fate in the *Drosophila* eye. *Cell* 83: 753–760

Tsurumi C, Ishida N, Tamura T, Kakizuka A, Tanaka K (1995): Degradation of c-Fos by the 26S proteasome is accelerated by c-Jun and multiple protein kinases. *Mol Cell Biol* 15: 5682–5687

Turner R, Tjian R (1989): Leucine repeats and an adjacent DNA binding domain mediate the formation of functional c-Fos-c-Jun heterodimers. *Science* 243: 1689–1694

Vriz S, Lemaitre J-M, Leibovici M, Thierry N, Mechali M (1992): Comparative analysis of the intracellular localization of c-Myc, c-Fos, and replicative proteins during cell cycle progression. *Mol Cell Biol* 12: 3548–3555

Wasylyk B, Wasylyk C, Flores P, Begue A, Leprince D, Stehelin D (1990): The c-ets proto-oncogenes encode transcription factors that cooperate with c-Fos and c-Jun for transcriptional activation. *Nature* 346: 191–193

Winstall E, Gamache M, Raymond V (1995): Rapid mRNA degradation mediated by the c-fos 3′ AU-rich element and that mediated by the granulocyte-macrophage colony-stimulating factor 3′ AU-rich element occur through similar polysome-associated mechanisms. *Mol Cell Biol* 15: 3796–3804

Wisdom R, Verma IM (1993): Transformation by Fos proteins requires a C-terminal transactivation domain. *Mol Cell Biol* 13: 7429–7438

Xanthoudakis S, Miao G, Wang F, Pan Y-CE, Curran T (1992): Redox activation of Fos-Jun DNA binding activity is mediated by a DNA repair enzyme. *EMBO J* 11: 3323–3335

Yasen NR, Park J, Kerppola T, Curran T, Sharma S (1994): A central role for Fos in human B- and T-cell NFAT: an acidic region is required for in vitro assembly. *Mol Cell Biol* 14: 6886–6895

Index

ankyrin repeat, 172
AP-1, 138
apoptosis, 1
Avian Erythroblastosis Virus (AEV), 119

basic/helix-loop-helix/"leucine zipper" (b/HLH/Z), 3
Bcl-3, 178

cancer, 1
CBP, 134
cell cycle, 1, 92
cell, erythrocytic, 120
c-erbA, 122
c-rel, 171
childhood pre B-ALL, 69
coactivator, 98
cross-talk, 237

differentiation, 149
dimerization, 226
DNA-binding, 227

E box, 5
ERG-2, 33
erythroleukemia, 64, 119
ETS domain, 29
ewing sarcoma, 64
EWS gene, 64

fibroblast, 120
FLI-1, 33
fluctuation, conformational, 96

Friend virus, 62

helix-turn-helix (HTH), 95
HRE, 130

IϰB, 174
inhibitor, 100
inhibitor domain 230

leukemia, chronic myelomonocytic, 67

Max, 5
MB1, 13
MB2, 16
myogenic cell, 121

neuroepithelioma, 64
neutronal cell, 121
NF-ϰB, 167
nuclear receptor, 122

oligomerization, 68
oncogenesis, 239

phosphorylation, 36, 101, 193, 230
proteasome, 193
protein-protein interaction, 234
protein turn-over, 233
proteolysis, stimulus-induced, 192
PU-1/spi-1, 29, 51

RAR, *see* retinoic acid receptor
Ras signaling pathway, 36
redox potential, 203, 232
rel, 167
repression, 1
retinoic acid receptor(RAR), 123
RXR, 123

serum response element (SRE), 54
sevenless signaling, 35
signaling pathway, 199

target gene, 104
Tax, 190
TEL, 67
TEL-ABL, 67
TEL-AML-1, 68
TEL-PDGFR☏, 67
TFIIB, 134
thyroid hormone receptor(TR), 123
transactivation, 1
transcriptional activation domain, 227
transformation, 9, 239

ubiquitination, 194

v-erbB, 119
v-ets oncogene, 29
v-rel, 171